DATE DUE

4-6-07	

MATHEMATICAL PEOPLE
Profiles and Interviews

MATHEMATICAL PEOPLE

Profiles and Interviews

Donald J. Albers and G. L. Alexanderson, editors

Menlo College   University of Santa Clara

Introduction by Philip J. Davis

CONTEMPORARY
BOOKS, INC.
CHICAGO ▪ NEW YORK

Mathematical People is published in collaboration with the Mathematical Association of America.

Library of Congress Cataloging-in-Publication Data

Mathematical People.

   "Published in cooperation with the Mathematical Association of America"—T.p. verso.
   Includes index.
   1. Mathematicians—Biography.   I. Albers, Donald J., 1941–   II. Alexanderson, Gerald L.   III. Mathematical Association of America.
QA28.M37   1984   510′.92′2 [B]   84-21602
ISBN 0-8092-4976-6

CIP—Kurtitelaufnahme der Deutschen Bibliothek

Mathematical people: profiles and interviews/ed. by Donald J. Albers and G. L. Alexanderson. With an introd. by Philip J. Davis. Publ. in collab. with the Mathemat. Assoc. of America.—Basel; Boston; Stuttgart; Birkhäuser, 1985.
   ISBN 0-8092-4976-6

NE: Albers, Donald J. [Hrsg.]

Published by Contemporary Books, Inc.
180 North Michigan Avenue, Chicago, Illinois 60601
Manufactured in the United States of America
Library of Congress Catalog Card Number: 84-21602
International Standard Book Number: 0-8092-4976-6
Published simultaneously in Canada by Beaverbooks, Ltd.
195 Allstate Parkway, Valleywood Business Park
Markham, Ontario L3R 4T8 Canada

This edition published by special arrangement with Burkhäuser Boston, Inc.

PHOTO AND ILLUSTRATION CREDITS

p. 5 (left), The University of Chicago Office of News and Information; p. 20, Daniel Wheeler; pp. 23, 28, John Blaustein; p. 37, James H. White; p. 42, Simon J. Fraser; pp. 43, 44, 46, 47, 48, 49, Richard Guy; pp. 57, 59, 61, 62, Dave Logothetti; pp. 66, 72, 78, News and Publications Service, Stanford University; p. 82, Burton Halpern Public Relations; p. 85, Steven Swanson; pp. 87, 91, University Communications, University of Santa Clara; p. 97, Scott Morris/OMNI; pp. 100, 104, Peter Renz; p. 106, Scott Morris/OMNI; pp. 110, 112, 114, 115, 116, Ken Regan–CAMERA 5; pp. 152, 155, 157, 162, Adrian N. Bouchard; p. 168, New York University; pp. 182, 185, 201, Tom Black; pp. 216, 219, 220, Benoit B. Mandelbrot; p. 223, Richard Voss; p. 248, Stella Pólya; p. 250, Bernadette Boyd; p. 253, Rose Mandelbaum; p. 256, City University of New York; pp. 265, 267, Mina Rees; pp. 273, 275, 277 (2), Springer-Verlag, Inc.; p. 300, Bill Ray; pp. 304, 306, Blanche Smullyan; p. 310, University of London; pp. 344, 349, Princeton University; p. 354, Steve Larson; p. 356, Los Alamos Photo Laboratory; p. 358, HERBLOCK.

(All other photographs were supplied by interview and profile subjects.)

MATHEMATICAL PEOPLE

Additional Acknowledgements

Several of the interviews and profiles in this volume first appeared in the *College Mathematics Journal*: Birkhoff (March 1983), Chern (November 1983), Conway (November 1982), Coxeter (January 1980), Erdös (September 1981), Gardner (September 1979), Halmos (September 1982), Kemeny (January 1983), Kline (June and September 1979), Knuth (January and March 1982), Pollak (June 1984), Pólya (January 1979), Reid (September 1980), Robbins (January 1984), Tucker (June 1983), Ulam (June 1981). The profile of Graham first was published in *Discover* (September 1982), and the profile of Smullyan in *The Smithsonian* (June 1982).

TABLE OF CONTENTS

PREFACE

From the time of Plato's Academy with its motto "Let no one ignorant of mathematics enter here" to the Age of Enlightenment, educated people were expected to be interested in the latest developments in mathematics as well as those in the sciences, literature, philosophy, music, and fine arts. The year Leibniz' first work on the differential calculus appeared in the *Acta Eruditorum* (1684), the journal also carried articles on theology, archeology, linguistics, philosophy, and the anatomy of snakes. With the increase in abstraction and complexity in the nineteenth and twentieth centuries mathematics became more and more difficult, even for the otherwise well-educated and informed. Today mathematicians can rarely communicate to other scholars or scientists, or even to other mathematicians, the ideas in mathematics that are exciting to them. Along with this intellectual isolation has come a feeling on the part of many outside mathematics that mathematicians are remote, unapproachable, aloof, and maybe even a bit strange.

We hope through the interviews and profiles included in this volume to dispel, at least in part, this notion. We claim that mathematicians are no stranger than certain philosophers we have known, or certain musicians, or certain artists, or . . . well, we need not go on. We hope to demonstrate that mathematicians are a remarkably diverse group with a wide range of intellectual interests and a full spectrum of personalities. It is true that mathematics does require great concentration and intellectual power of a rather special type, so the creative mathematician may display a single-mindedness beyond that observed in a wider population, but those who persist through the present volume will come to see, we believe, that mathematicians are still an interesting bunch. They may share more traits with artists than with experimental scientists. We leave it to the reader to judge.

The genesis of this volume coincided with the 90th birthday of George Pólya, when the first editor of the volume, the then newly appointed Editor of the *Two-Year College Mathematics Journal*, asked the second editor to do an interview of Pólya for the *Journal*. This turned out to be of interest to the readers and led to a series of interviews carried out by a number of different mathematicians. These interviews have been published in the *Journal* over the past five years. The present volume includes all of those interviews and contains a number of new interviews done just for the present volume.

Clearly the editors would not claim that those interviewed are the best, most productive, or even the most interesting of twentieth century mathematicians. They are certainly among the best, most productive, and more interesting. That's why they were chosen. But there are many others who would qualify for inclusion but who were not willing to be interviewed, or were not available. Or the editors did not know someone who could approach them. Or they were just overlooked. We are grateful to those who agreed to be interviewed and we apologize to those who should be included but are not. Perhaps, at another time, for another volume or another journal, these omissions can be rectified.

The reader will note that not all of the entries are interviews. A few profiles and one autobiographical sketch have been used instead, largely because they were already available.

The object has been rather different from what one would compile in an oral history project. There has been no attempt to cover in detail the subject's mathematical achievements. Instead, while trying to give some notion of what sort of mathematics the person has done, we try to concentrate on the human side. We have tried to discover why the person chose mathematics, why the person is excited by mathematics, and why the person thinks that what he or she does is important.

Naturally, such a project has involved far more people than the two editors who put the final collection together. We wish to thank all those who agreed to be interviewed, all those who did the interviewing or wrote profiles, and the many who have helped in the production effort: Geri Albers,

Lisa Albers, Ken Allen, Joan Bailey, Margery Douglas, Elena Giulini, Judith Grabiner, Raoul Hailpern, Dave Jackson, Yvonne Pasos, Alan Ringold, Gerald D. Silverberg, Rosemary Uribe, and Judith Goodstein and Mary Terrall of the Caltech Archives.

And of course, we wish to thank Klaus Peters, President of Birkhäuser, Boston, and the officers and committee heads of the Mathematical Association of America who made the whole project possible: Richard D. Anderson, the late Edwin F. Beckenbach, Leonard Gillman, Ivan Niven, David P. Roselle, Alan Tucker, and Alfred B. Willcox.

Both of us were inspired by *Men of Mathematics*. We here offer special thanks to its author, the late E. T. Bell.

D. J. Albers, Menlo College
G. L. Alexanderson, University of Santa Clara
January, 1985

FOREWORD

Mathematicians focus their primary attention, and rightly so, on mathematical ideas, their origin, their continuing creation, and their dissemination. This history of mathematics plays a somewhat peripheral role, and biographical studies are rare, although they do get a wide audience, as witness *Men of Mathematics* by E. T. Bell and the more recent biographies by Constance Reid. However, these were not about living mathematicians. This present volume, by contrast, offers us insights into the lives of some leading contemporary scholars, obtained for the most part through direct questioning. Why did they pursue mathematics, not some other field? How did they get into their particular speciality? All the questions addressed are related directly or indirectly to mathematics.

The outcome is a collection of lively sketches that illuminate the mathematical scene, especially in the United States and Canada, in the last several decades. The book is a noteworthy example of living history gathered directly from several of the luminaries who have played key roles in the mathematical life of this continent. Would that similar vignettes were available of such leaders of an earlier time as George David Birkhoff, Eliakim Hastings Moore, and Oswald Veblen. This collaborative effort between the *Mathematical Association of America* and the publisher, Birkhäuser Boston, is a welcome addition to the literature on the human side of mathematics. Our special thanks go to the writers for their skillful and sensitive treatment of the subject matter, and to the mathematicians who cooperated by "sitting for their portraits."

IVAN NIVEN, President
Mathematical Association of America

INTRODUCTION
Reflections on Writing
the History of Mathematics

Philip J. Davis

The essays, interviews, and reminiscences that are contained in this book present profiles of some of the finest contemporary mathematical minds. They link men and women of extraordinary achievement with both their material and their environment. Although the writing is relaxed, informal, and delightfully easy to read, we come away exhilarated, feeling that we too have been present at moments when great mathematics was created and have shared the feelings of the creators.

At the level of shop talk we will learn which mathematician has written more than nine hundred papers and how he views this accomplishment. We will learn what mathematician is skilled at parlor tricks and how this relates to his professional work. We will read of the mathematician who says, quite flatly, that the computer is important, but not to mathematics; and we will read the exact opposite view expressed. We will know, if we read on, which mathematician, much more than most, explained in his papers how he got his results.

Shop talk aside, it is inevitable that having turned the last leaf of the last profile one finds oneself in a historical mood. One begins to think of generations of people and of the relationship between mental worlds and the historical events against which our lives are played out.

Since the sixteenth century, the world has become mathematized, scientized, technologized at an increasing clip. Our comforts, our customs, our mental states, our imagery, our sense of what is and what is not, our fate, even, have all become increasingly tied to the latest developments in technology. Despite this, books of general history do not deal adequately with science and technology. History has traditionally been the chronicle of the struggles of dynasties or of power blocs. One can read in history books about social arrangements, about economic arrangements; one can learn how powerful are the forces of the ideas of freedom, of glory, of nationalism, of religious faith; but one finds very little indeed of a precise nature as to how mathematics and technology grew, and what the interaction was between them and the general culture. Taking the popular *History of the Modern World* by Palmer and Colton (Knopf) as an example, one finds in it one chapter, 20 pages out of 900, like a plum in a pudding, devoted to the growth of science. Our college history departments are only now beginning, tentatively and timorously, to hire scholars who have more than a beginner's knowledge of scientific matters.

Reasons for this neglect are not hard to find. In the first place, the ideas of science are not easily comprehended, and the history of those ideas forms a web that is far more tangled than the marriages of the Hapsburgs. For one and the same person to understand both science and "conventional" history requires a great deal. Secondly, even when a writer has a grasp of both history and of the history of science, it appears extraordinarily difficult to link the two. There is a school of thought that denies the possibility of doing so in a meaningful way. The ideas of science and of mathematics, it asserts, form an independent dimension of human intelligence and experience. These ideas appear gratuitously on the stage of history; and though they may create subsequent history, they are random events, unexplainable by the history to that point in time. Granted that conventional history is, in part, an artistic construction; that, according to Jacob Burckhardt, famous philosopher of history, "the outlines of a culture and its mentality may present a different picture to every beholder", then it would appear that one merely adds another layer of artistic conjecture with every attempt to paint scientific lines into the canvas of general history.

Before explaining what my own feelings (and hopes) are in this matter, I will back up a bit and say a few words about how the history of mathematics has been written. At a primitive level we can compile chronological lists of people who had mathematical thoughts: Thales, Pythagoras, Parmenides, Zeno . . . , putting in as many names as we find worthy and determining their dates as accurately as we can. This might be compared to early historical writings, which were often nothing more than lists of kings that ancient scribes collated. A list of mathematicians, carried forward to modern times, might surprise quite a few people, for the average person hardly thinks that mathematics has a history; rather, that all of it was revealed in a flash to some ancient mathematical Moses or reclaimed from a handbag at the left luggage room in Waterloo Station.

We want more than lists. And more is, in fact, available. In dealing with what is available, the historian of mathematics has several tasks. The first is that of interpretation and description, where the material is to be considered almost in the sense of static isolation. The historian of mathematics must identify material as mathematics and tell us what that mathematics purports to be; he must build a coherent and continuous historical structure of mathematical ideas. For ancient material this may be a job of the first magnitude.

The second task is to try to discover the genesis of mathematical ideas and to describe their interrelation one with the other and with the outside world.

The third task is for the historian to communicate his discoveries and insights to his readers. This obvious point is not so much that there are scholars who study and study and fail to publish, but that each writer of history must imagine for himself a readership of a certain kind and try to make sense to that readership.

Let us discuss the last task first. Begin by asking: what has the readership been for mathematical history? It is read by some few mathematicians and scientists, some professional historians of science, some philosophers, and some people who are interested in the history of ideas. Then, there is the raggle-taggle remainder: younger students who very occasionally sign up for a course on the History of Mathematics and people who prefer the *Encyclopedia Britannica* to Agatha Christie for a bed-book.

The intelligent layman, although his life is increasingly mathematized, doesn't know that mathematics has a history and wouldn't care to read about it if he did know. The general historian doesn't much care about it. And now comes the low blow: the average scientist, the average teacher of mathematics at any level, and the average research mathematician do not read the history of mathematics. It is not thought necessary to their professional careers. Researchers go forward from recent material, occasionally referring, as needed, to a paper written twenty-five or fifty years ago.

Ask the next question: Who writes mathematical history?* Lumping biographies of mathematicians together with the history of mathematical ideas (perhaps one shouldn't do this), we find that it is written by all kinds of people from poets to professional biographers to scientific Grub Streeters, to math buffs, to mathematicians of great accomplishment.

I have enjoyed Muriel Rukeyser's biography of Thomas Hariot (1560–1621). Rukeyser, a poet, writes with poetic mistiness about her attempt to reconstruct this swashbuckling Elizabethan mathematician from the dusty manuscripts in the library of the Duke of Northumberland. If, here and there, a statement is made about mathematics that betrays her ignorance, it can serve to remind the professionals that the bulk of humankind who live in great ignorance of mathematics still have reactions to its very existence.

I have enjoyed equally well (but in a different sense) papers of Clifford Truesdell straightening me out on some point in Euler.

It is pleasant to be able to report that the number of highly professional mathematicians who have attempted history has been on the increase. For the general history of post-Renaissance mathematics, we should list among such authors the names of E. T. Bell, Morris Kline, Clifford Truesdell, Howard Eves, Oystein Ore, C. B. Boyer, Herman Goldstine, André Weil, D. J. Struik, R. L. Wilder, Salomon Bochner, L. C. Young, Harold Edwards, and numerous others, including the short historical notices prepared by Nikolaus Bourbaki.

Grossly, one may divide authors into two categories: those who understand the mathematics they write about and those who do not. History can be written from the outside and from the inside. Surveying what is available in its wide variety, and given my own background, I find it not too difficult to reach the conclusion that the most significant mathematical history is that written by insiders for the inside. Let me designate it as Inside History.

How is such material to be characterized? First of all, it assumes that the reader has a fair amount

---

*I have been informed that there is no comprehensive history of the histories of mathematics. I should think that this might be a worthwhile scholarly project.

of understanding of the mathematics under discussion. Now this would be a reasonable assumption for inside readers and for material up to, say, 1900. After that time, there has developed such a plethora of mathematical material and such a degree of specialization that the insider knows only the "taste" of the next field but lacks deeper knowledge. The historian cannot simultaneously write a history and a mathematical text book, and so even the insider skips what he doesn't understand and strains out the list of names and the surface linkages. Thus we are in a dilemma: one cannot really understand Inside History unless one is on the inside, and not even insiders are completely inside. The necessity for popularization therefore exists at all levels.

Inside History comes forward in time. Having distinguished a number of major lines of development, it follows these developments through. At the same time, it tries to give snapshot pictures of how the whole mathematical scene at a particular era shaped up. It assumes there is progress in mathematics; it points to the resolution of old problems and to the birth of new ideas. In carrying the action forward to the present, it assumes that the past was aimed, like an arrow, at the problem. It interprets the past in terms of the present, rather than on its own terms. It assumes that the present is the justified completion of the past.

Inside History tends to strip away from its narrative all considerations that are extra-mathematical. Inside History thinks that mathematics is culture-free; that it has a universality which renders it everywhere valid and interpretable;* that the little man on Mars—or wherever he now has taken up residence—will inevitably come up with the sequence of digits 3.14159 . . . and think it quite as interesting as we do. Just as it has been said of the novels of Jane Austen that one could not tell from them that the Napoleonic Wars had been fought, so in our mathematical histories, with the possible exception of one aspect, one can hardly tell that the mathematics was created by people living in historic times or by people who had an intellectual life that extended beyond the mathematical. The one exception is that connections are drawn—not as often as might be, but sufficiently often—to the concerns of science. Connections are made to astronomy, to geography, to navigation, to surveying, to physics, and to engineering. But excluded from such narratives, whether by ignorance on the part of the author or by self-censorship, are any intimations that mathematics has any relationship to or derives any inspiration from the irrational, the mystical, the metaphysical, the theological, the ritual, or from the general activities of mankind, including all the things that the current generation of writers happen to think are foolish or reprehensible.

Inside History moves from great name to great name and tends to draw the major lines of development in such a way that they pass through the great names.

Inside History purports to give the ultimate and unique explanation of how mathematics came to be what it obviously seems to be.

In virtue of the way Inside History treats the relationship between the present and the past, I fall in with the suggestion of Professor Joan Richards that it be called Whig History. The term was introduced by historian Herbert Butterfield in 1931 and alludes to the tendency to "emphasize certain principles of progress in the past and to produce a story which is the ratification, if not the glorification, of the present."

Whig history of mathematics is now the finest kind we have and the kind that most of us are comfortable with. It has great strengths and certain inadequacies. It distorts the past by not describing the past in its own complex nature. It ascribes to the past a teleological direction. It omits from its description a great deal of connecting tissue that is not formalized, written-down mathematics. It does not treat (or treats in a slapdash way) the relationship between mathematics and business, war, religion and theology, metaphysics, cosmology, ethics. These omissions are absolutely scandalous in that they lead to an inadequate account of the sources of mathematical inspiration.

In the case of general history, Whig history is not the only way in which history has been written. Among alternative ways are history as a guide to conduct: the lessons of history. There is history as it really was: the total recovery of the past. There is history as "Heilsgeschichte": history as the road to salvation and the linking of events with the universal sacred dramas. There is history as science: history has laws that can be discovered and that have the same validity as scientific laws.

Now the history of mathematics may be written, with some obvious modifications, along all these lines. When Raymond Wilder analyzes and categorizes the varieties of pressure that lead to progress in mathematics, he is, in part, attempting a science of the history of mathematics. When André Weil advocates the study of mathematical history so we can learn from the thought of the masters, he is

---

*Ulf Grenader in "Mathematical Experiments on the Computer," (Academic Press, 1982) writes, "Sometimes it may be hard to understand one's own (APL) code six months after it has been written."

setting up the Old Masters as guides to future mathematical conduct and creativity. When the Scottish mathematician Colin Maclaurin (1698–1746) wrote that the purpose of doing mathematics was to reveal and reflect God's glory, then the link to the sacred and the hermetic has been made.

* * *

In his essay "Great Men and Their Environment," the American psychologist and philosopher William James wrote, "The community stagnates without the impulse of the individual; the impulse dies away without the sympathy of the community." I should like now to raise the question: Is it possible to write history of mathematics along the lines suggested by this quotation? I should like to think so, but I am not at all certain it can be done. Call it social history or social-psychological history. I often call it Jamesian history.

If Whig history is, in part, a fiction, then Jamesian history may be an impossibility. Many scholars have asserted as much, and they deprecate, if they do not actually hold in contempt, what little has been done along these lines.

Having despaired of the possibility of such history, I will attempt to set down some of its features. Is this paradoxical? Mathematics, of course, often proceeds in just this way. The requirements for $\sqrt{-1}$ imply the impossibility of its existence; mathematicians ignore this and proceed anyway. I undertake this exercise to liven my personal hope that someone might hit upon a mode within which the impossibility may be dispelled.

A first and perhaps foremost requirement is that *we achieve an understanding of the willingness of the community to mathematize and to be mathematized.* I emphasize particularly the second part of the statement, for—make no mistake about it—mathematics, whether it is thought to have been put into social practice by reason or by whim, contains ideas, patterns, and procedures which are potent vehicles of social organization and change. The heart-cry "Do not fold, staple, or punch me" can be taken as the battle cry of one party in the struggle between the "two cultures."

Consider, for example, the tendency currently sweeping the western world to use credit-card money instead of cash money. Now this favors mathematics (at the primitive level of bookkeeping and data processing). It is not the result of a conspiracy of computer manufacturers in cahoots with mathematicians. It was slow in finding acceptance, having apparently been started some twenty-five years ago by four restaurateurs in New York City. Here is a conflict; but the "Zeitgeist" apparently favors mathematics.

Why did it take until the 1500s for the community to allow itself to be stochasticized? In the Roman Empire, apparently, certain statistics were collected pertaining to accidents. But no one thought to divide through by the base population to arrive at stable rates. The mathematics of elementary combinatorial probability is simple enough. Why were not the odds for dicing laid out in the third century B.C.?

Is there a relation, as some have claimed, between the non-Euclidean geometry of the 1800s and the doctrine of ethical relativism which subsequently became popular? A case can be and has been made for such a connection.

Why do we, today, allow our military strategies to be so mathematized and computerized when the difference of one bit in a program may send all down the road to oblivion? What is the basis of our confidence or our cynicism?

Why did mathematics dry up under the influence of Roman civilization?

Why did not ancient Oriental mathematics discover proof?

Here we have touched on the two most famous questions of this type of history. The list may be extended indefinitely.

It is difficult enough to write within mathematics itself how certain ideas grow from seed. Descartes tells us that the idea of the total mathematization of science came to him in a vision couched in secular symbols and evoking a religious reaction. Poincaré, centuries later, tells us that the solution to a certain problem came to him as he stepped aboard a streetcar. (And if visions and streetcars are now scarce, shall we throw up our hands?) How much more difficult it is then to link a specific creation to the community. Yet we must try. In doing so, we will write many foolish things. I do not go along with the Wittgensteinian ukase: "Wovon man nicht sprechen kann, darüber muss man schweigen." ("If you can't talk about it, shut up.") This condemns us to the minimal communication of the monk's cell, to vows of silence that hardly enrich our intellectual life.

A second requirement brings us to individuals. *We must achieve a proper understanding of the relationship between the great men and those of lesser rank.* When Newton remarked modestly that if he saw further than most men it was because he stood on the shoulders of giants, he was expressing a truth which is all too often ignored. The giants stand on the shoulders of lesser giants, and the whole pyramid would collapse without the firm support of the community.

A third requirement is the *avoidance of parochialism*. Too often, particularly when the author is a professional mathematician well versed in some specialty, he writes history as though it were the manifest destiny of the past to lead with trumpets and flags precisely to his specialty. For example, I have felt that most histories of the theory of functions of a complex variable are inadequate insofar as they underplay (or are totally unaware of) the contributions and the driving force that came from dynamical astronomy. Avoidance of parochialism requires *a proper understanding of dead motives*. Why were certain things done? The reasons we ascribe are too often our reasons and not the reasons of the time. To write mathematical history "wie es echt gewesen" ("the way it really happened") may be impossible, but the attempt rewards us when we discover what different stuff was packed into the minds of the past.

* * *

A mathematician sits somewhere in space and time and creates mathematics. He may be working on paper or on papyrus. He may be drawing figures on sand or on a computer scope. He may be carving symbols on stone. He may be in a scientific laboratory. He may be at a blackboard before a group or he may be by himself. He may be in jail or in a lighthouse. He may be blind. He may have access to books or he may not. Whatever he creates, if it is disseminated and proves to be intelligible to other men, it has de facto integrity.

What is his relationship to his material? From whence does it derive? What does he think he is doing? Toward what end is it directed? Why does he do it? What are the consequences of his scribblings? What is the nature of his conceptualization? What are his criteria for success? If it is thought that his symbols have some applicability to the greater world, how does this come about?

One man works with numbers, another man works with space. A third man works with logic, another with patterns. Someone else works with differential equations and yet another with computer languages. Is there a common element to these inquiries, a common attitude or psychology? What distinguishes them? What is important? What is not? What is alive? What is now irrelevant? What is beautiful? What is not? What is simple? What is complex? What makes it so? What does the experience of being a mathematician amount to?

From material such as now follows, the historians of the future will draw their conclusions.

Bibliography

BERNAL, J. D. *The Social Function of Science*. New York: Macmillan, 1939.

BLOOR, DAVID. *Knowledge and Social Imagery*. Boston: Rutledge and Keegan Paul, 1976.

BOCHNER, S. *The Role of Mathematics in the Rise of Science*. Princeton: Princeton University Press, 1966. (In particular, Chapter 2.)

CLARK, G. N. *Science and Social Welfare in the Age of Newton*. Oxford: Oxford University Press, 1949.

DASTON, L. J. "Mathematics and Moral Sciences: The Rise and Fall of the Probability of Judgments, 1785–1840." In Jahnke and Otte, 1981.

DAUBEN, J. W. "Mathematics in Germany and France in the Early Nineteenth Century: Transmission and Transformations." In Jahnke and Otte, 1981.

DAVIS, P. J., and REUBEN HERSH. *The Mathematical Experience*. Cambridge: Birkhäuser Boston, 1981. (Paperback edition, Boston: Houghton Mifflin, 1982.)

GOLDSTINE, HERMAN H. *The Computer from Pascal to von Neumann*. Princeton: Princeton University Press, 1972.

———. *A History of Numerical Analysis from the Sixteenth Century through the Nineteenth Century*. New York: Springer, 1977.

GRABINER, J. V. "Changing Attitudes Toward Mathematical Rigor: Lagrange and Analysis in the Eighteenth and Nineteenth Centuries." In Jahnke and Otte, 1981.

GRATTAN-GUINNESS, I. "Mathematical Physics in France, 1800–1835." In Jahnke and Otte, 1981.

HACKING, IAN. *The Emergence of Probability*. New York: Cambridge University Press, 1975.

HESSEN, B. *The Social and Economic Roots of Newton's Principia*. New York: Howard Fertig, 1971 (reprint edition).

JAHNKE, H. N., and M. OTTE, eds. *Epistemological and Social Problems of the Sciences in the Early Nineteenth Century*. Dordrecht: D. Reidel, 1981. (In particular, Section III.)

JAMES, WILLIAM. "Great Men and Their Environment." In *Selected Papers on Philosophy*, p. 165–97. London: J. M. Dent and Sons, 1917.

KLINE, M. *Mathematical Thought from Ancient to Modern Times*. Oxford: Oxford University Press, 1972.

LÜTZEN, JESPER. *The Prehistory of the Theory of Distributions*. New York: Springer, 1982.

MEHRTENS, H. "Mathematicians in Germany Circa 1800." In Jahnke and Otte, 1981.

NEEDHAM, JOSEPH. *Science and Civilization in China, Vol. III*. Cambridge: Cambridge University Press, 1959.

PURCELL, EDWARD A. *The Crisis of Democratic Theory*. Lexington, Kentucky: University of Kentucky Press, 1973.

RICHARDS, JOAN L. "The Reception of a Mathematical Theory: Non-Euclidean Geometry in England, 1868–1883." In *Natural Order: Historical Studies of Scientific Culture*, edited by B. Barnes and S. Shapin. Beverly Hills, CA: Sage Publications, 1979.

RUKEYSER, MURIEL. *The Traces of Thomas Hariot*. New York: Random House, 1970.

SCHARLAU, W. "The Origins of Pure Mathematics." In Jahnke and Otte, 1981.

SHAPIRO, BARBARA J. *Probability and Certainty in Seventeenth-Century England*. Princeton: Princeton University Press, 1983.

STRUIK, D. J. *A Concise History of Mathematics*. New York: Dover, 1967.

SWETZ, F. J. "Whatever Happened to the History of Mathematics?" *Am. Math. Monthly* 89 (1982): 695–96.

WEIL, ANDRÉ. *History of Mathematics: Why and How, Collected Papers, Vol. III*. New York: Springer, 1980 (pp. 434–42).

WILDER, R. L. *The Evolution of Mathematical Concepts*. New York: J. Wiley & Sons, 1968.

———. *Mathematics as a Cultural System*. New York: Pergamon, 1981.

YOUNG, L. C. *Mathematicians and Their Times*, Math. Studies No. 48, North Holland, 1981.

Journals

Historia Mathematica

Historical Studies in the Physical Sciences

Isis

Scripta Mathematica

Social Studies of Science

MATHEMATICAL PEOPLE
Profiles and Interviews

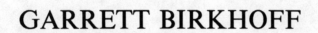

GARRETT BIRKHOFF

GARRETT BIRKHOFF

Interviewed by G. L. Alexanderson and Carroll Wilde

In 1976, as part of the recognition of the American bicentennial, the annual meeting of the Mathematical Association of America in San Antonio was dedicated to the history of American mathematics. One of the principal speakers was Professor Garrett Birkhoff. He has unique qualifications to provide a first hand account of much of the 20th century history, since not only has he had a most distinguished mathematical career himself, but he grew up with additional insights gained from his father, George David Birkhoff, who was perhaps the first American to be recognized as one of the world's leading mathematicians.

Garrett Birkhoff's San Antonio paper, "Some Leaders in American Mathematics: 1891–1941," can be found in *The Bicentennial Tribute to American Mathematics 1776–1976* (D. Tarwater, ed.; MAA). A professor of mathematics at Harvard for 45 years, he has made important contributions in an amazingly broad range of mathematical fields and has written, sometimes with co-authors, a number of pioneering books: *Lattice Theory*, 1940, 3rd edition, 1967; *Survey of Modern Algebra* (with Saunders Mac Lane), 1941, 4th edition, 1977; *Hydrodynamics*, 1950, revised edition, 1960; *Jets, Wakes & Cavities* (with E. Zarantello), 1957; *Ordinary Differential Equations* (with G.-C. Rota), 1962, 3rd edition, 1978; *Algebra* (with Saunders Mac Lane), 1967, revised edition, 1979; *Modern Applied Algebra* (with Thomas Bartee), 1970; and *Source Book in Classical Analysis*, 1973.

His father, George D. Birkhoff, was also for many years professor of mathematics at Harvard and in 1938, to mark the semicentennial of the American Mathematical Society (founded in 1888 as the New York Mathematical Society), he wrote an earlier version of what his son was to do in 1976. It was a paper titled "Fifty Years of American Mathematics" and appeared in the AMS Semicentennial Publications. It is interesting to note that even in 1938 he could already cite his son's significant contributions, especially to lattice theory.

George D. Birkhoff (1884–1944) was also known for the breadth of his interests. He published approximately 190 papers in classical analysis, difference equations, dynamical systems, physical theories, the four color problem, aesthetic measure and even, in one case, number theory. His books were *Relativity and Modern Physics*, 1923; *The Origin, Nature and Influence of Relativity*, 1925; *Dynamical Systems*, 1927; *Aesthetic Measure*, 1933; and *Basic Geometry* (with R. Beatley), 1940. Though he seemed to be concerned mainly with problems close to celestial mechanics and physics, he nevertheless wrote in his 1938 essay that "it is well not to forget that many of the most astonishing mathematical developments began as a pure *jeu d'esprit*."

The following conversation took place in January, 1982, at the Naval Postgraduate School, Monterey, California.

MP: *What strikes me about your career is the incredible range of interests, from lattice theory through mathematical physics. How did you happen to shift over the years or was it really a shift? Were you always interested in all these aspects of mathematics?*

Birkhoff: I was always interested in everything. In fact, my father was concerned that as a boy, I couldn't decide what I wanted to do; he thought I should make a choice. Finally, when I entered college, he told me in words of one syllable that I would have to earn my living when I graduated, and I had better make use of my four years to prepare myself for a profession. It was at that point that I decided to become a mathematician. I liked mathematics, and my father's being a mathematician was no reason I should not become one too.

Actually, advised by my father, I prepared in college for a career as a mathematical physicist, taking two courses in mathematics and one in physics each of four years. (In those days, mechanics was considered a mathematics course.) In my junior year, I took a course from George W. Pierce in

"Electric Oscillations and Electric Waves"* (Maxwell's electromagnetic theory), and in my senior year one from E. C. Kemble in the then new quantum mechanics. It was through these courses, and a half-course in potential theory with Kellogg, that I came to understand partial differential equations. I gave my first talk in a physics seminar, as a senior.

But in connection with my undergraduate thesis, I also read a lot of modern abstract mathematics, such as Lebesgue theory, point-set topology, and so on. As a senior, I also secretly discovered finite groups in the library, and fell in love with them.

When I went to Cambridge University on a Henry Fellowship after graduating, I planned to work on mathematical physics, specializing in quantum mechanics. However, my course with Kemble at Harvard had not prepared me for Dirac's lectures. Kemble had not suggested that there might be undiscovered elementary particles; his course had left me believing that quantum mechanics was exclusively concerned with solving the Schrödinger equation. Not until after I heard a lecture on positrons by Carl Anderson in the spring of 1933, did it dawn on me that Dirac's lectures were aimed at describing particles that hadn't yet been observed! By then I had switched from mathematical physics to abstract algebra. My main sources were van der Waerden's *Moderne Algebra* and Speiser's *Gruppentheorie*, both of which I had learned about from Carathéodory in the summer of 1932.

MP: *Where was Carathéodory at that time?*

Birkhoff: He was in Munich. I had spent July of 1933 in Munich working on groups by myself, and I called on him. I had read and greatly admired his paper "Über das lineare Mass von Punktmengen", in connection with my thesis. He gave me tea with his son and daughter, and then showed me his imposing library. He said, "If you're interested in group theory, you should study Speiser's book. And if you want to know more about algebra, read van der Waerden." I followed his good advice, greatly helped by Philip Hall after I arrived at Cambridge University.

Lattice Theory

By January, 1933, I had begun thinking about what I called lattices. I changed my "research supervisor" from Fowler, a mature mathematical physicist, to Philip Hall, a young but already notable group theorist. Hall was a very stimulating and generous person who encouraged me in abstract algebra. I decided that I could achieve more researchwise in that area than in mathematical physics at that time. Van der Waerden's book made modern algebra seem like a field just opening up and blossoming and its approach dominated my research for the next seven or eight years, with special emphasis on lattice theory.

My ideas about lattices developed gradually. Philip Hall did not know of the important work of Dedekind on "Dualgruppen", although he did call my attention to Fritz Klein's related papers on "Verbände". It was my father who, when he told Ore at Yale about what I was doing some time in 1933, found out from Ore that my lattices coincided with Dedekind's Dualgruppen. Ore had edited Dedekind's collected works. I was lucky to have gone beyond Dedekind before I discovered his work. It would have been quite discouraging if I had discovered all my results anticipated by Dedekind.

I could talk on, but H. J. Mehrten's excellent *Die Entstehung der Verbandstheorie* records the history of lattice theory to 1940 very well. Anyone interested can find the story there; Mehrtens wrote to every active participant who was still alive in 1975, and carefully analyzed and coordinated their recollections.

MP: *The Colloquium Series publication (Lattice Theory) came out in* 1940.

Collaboration with Mac Lane

Birkhoff: Yes, and Birkhoff and Mac Lane in 1941. Morse had told me that nobody under 30 should write a book. So I thought it over and wrote two! I think the first thing to be said about Birkhoff-Mac Lane is that it made algebra interesting to many students who were not in mathematics at all. It was very popular and created a real revolution; I considered it most unfortunate that one studied algebra and geometry exclusively in high school, followed by three years of calculus in college, with little visible connection between them.

---

*The title of Pierce's book (McGraw-Hill, 1920) and his course. Pierce had collected $2,000,000 in a patent suit against A. T. & T.

MP: *The picture has remained bad in many places practically until today. Some still see only calculus, differential equations and advanced calculus in the first years and conclude that this is mathematics. Some years ago we put an abstract algebra course in the sophomore year for our students, and I must say that it isn't only Birkhoff–Mac Lane that students read: your Lattice Theory is one of the most often checked out books in the mathematics collection in our library.*

When I went to school there were essentially two books we read in algebra–van der Waerden (Moderne Algebra) and Birkhoff–Mac Lane (Survey of Modern Algebra). I think many who developed mathematically at that time still reach for Birkhoff–Mac Lane when an algebra question arises. How did you and Mac Lane get together to write that book?

Mac Lane   **Birkhoff**

Birkhoff: My father recognized Mac Lane's exceptional qualities, and got him invited as a Benjamin Peirce Instructor in 1934–1936. He returned to Harvard in 1938, the year after I had given a course in modern algebra on the undergraduate level for the first time. Although my course was well attended, I was much more research-oriented than teaching-oriented. Mac Lane had had much more teaching experience than I, and I think the popularity of our book owes more to him than to me. His problems and his organization of linear algebra were especially timely. My conservative inclusion of material from the then traditional "college algebra" and "theory of equations" courses (Bôcher, Dickson, Fine) may have helped with its initial success, as did my recognition that Galois theory used vector spaces.

Our collaboration involved some compromises. When I taught "modern algebra" in "Math. 6" the first time, in 1937–38, I began with sets and ended with groups. The next year Mac Lane put group theory first, and set theory (Boolean algebra) last! That was characteristic of his freshness, his initiative, and his lack of respect for conformity; but it came as a slight shock to me at the time. After teaching the course again the next year, I suggested that we co-author a book, usable by our colleagues, so that we wouldn't have to alternate teaching it forever, and he agreed. One of us would draft a chapter and the other would revise it. The longer chapters are his; the shorter ones mine.

MP: *Tell us about your other collaborators.*

Birkhoff: I have been very lucky in meeting and knowing so many outstanding mathematicians of all ages. I wrote joint papers with Philip Hall and John von Neumann in the 1930's, and, of course, Birkhoff–Mac Lane is another example. I enjoy people and try to learn from them. I would really have to go over a complete list of my papers to be able to name all of my collaborators. I feel that I owe a great deal to other people. Appreciating contributions of others is an important aspect of mathematics.

MP: *I would like to ask you which of your many achievements has given you the greatest sense of satisfaction. What are you most proud of having accomplished?*

Birkhoff: These are for me very different questions. I am proud of having contributed to the defeat of Hitler through my work on shaped charges, and to the understanding of modern technology through my work on scientific computing. But my greatest satisfaction has come from my part in establishing lattice theory as a recognized branch of mathematics, along with universal algebra, in the 1930's. I am looking forward to writing a fourth edition of my book *Lattice Theory* during the 1980's!

Undergraduate Education

MP: *We would like to back up in time, if we may, to explore the foundations of your remarkable career, beginning with your undergraduate education.*

Birkhoff: The story begins with a tribute to my father: not only did he ask stimulating questions, mathematical or otherwise, but also in college I had four of his Ph.D.'s as instructors: Morse, Whitney, Walsh, and Brinkmann. This was surely a unique experience, and I was totally unaware of it at the time. Another teacher was Kellogg who, as it happens, was our next door neighbor. So my Harvard undergraduate mathematical training was almost an inside job.

My undergraduate mathematics also had an unusual beginning. After a year at boarding school in Lake Placid, my sister and I joined our parents in Europe. At the end of the first week, my father asked me "What are you going to do this summer?" I said, "I suppose that I'll improve my French, go to some museums and absorb European culture." He said, "You're going to learn the calculus!" So he bought me a second-hand, dactylographed French calculus book which I studied through the summer, more or less. In the fall I was exposed to Morse and Whitney in second-year calculus; they brought out vividly the contrast between its intuitive plausibility and its theoretical complexity. For example, they digressed to construct a classic pathological function whose mixed partial derivatives $\partial^2 u/\partial x\,\partial y$ and $\partial^2 u/\partial y\,\partial x$ are *not* equal. They also noted that a smooth function having m local maxima ("peaks") and p local minima ("pits") in the interior of a disk, and vanishing identically on the boundary, ordinarily had $m + p - 1$ saddle-points ("passes"). Such digressions probably seemed like irrelevant distractions to many students, but I found them fascinating.

MP: *Were Whitney and Morse using a text in your second-year calculus course?*

Birkhoff: Yes, Osgood's *Introduction to the Calculus*; I only learned recently that our corrector was Harry Blackmun, now a Supreme Court Justice! As part of freshman tutorial, I also read Osgood and Graustein's *Analytic Geometry*, from which I learned linear algebra. The following year I had advanced calculus with Brinkmann, using Osgood's splendid *Advanced Calculus*. At the same time, I took a graduate course in complex function theory with Walsh. I realized that this was the course that was intended to make or break prospective mathematicians. I managed to get an A +, after which I felt like a professional mathematician.

Harvard had a tutorial system—it was new in those days—and concentrators writing senior honors theses had the illusion that they were doing research. I tried to publish my undergraduate thesis in the *Transactions*, but Tamarkin rejected it. He wrote me a very nice letter and said that efforts like mine should be judged from two points of view. Judged by professional standards, it fell short. Most of my results about what I called a "counted point-set", and which has recently been called a "multiset", followed easily from known properties of integer-valued functions. But judged as a first research effort by an aspiring young mathematician, it was very promising. I did get a small piece of it published a year later.*

---

* *Bull. Amer. Math. Soc. 39* (1933), 601–7.

MP: *Was that your first publication?*

Birkhoff: Yes, my first publication.

MP: *Who were your most stimulating teachers?*

Birkhoff: I found most of my undergraduate courses at Harvard inspiring, and felt that I had a galaxy of stimulating teachers in all subjects. Among my mathematics teachers, especially stimulating were Morse, Whitney, Walsh and Kellogg. The theoretical explanations of the calculus by Morse and Whitney made it possible for me to appreciate the beauties of complex function theory as taught by Walsh, a year later.

Kellogg, after introducing me to analytical mechanics in my freshman year, tried to teach me potential theory as a sophomore. Though I was not ready for it then, his classic *Potential Theory*, which came out later that same year, has contributed greatly to shaping my ideas about elliptic boundary value problems.

Many of my professors took a personal interest in helping their students to learn. I have mentioned G. W. Pierce's role in teaching me the basic mathematics of electricity and magnetism. He was a clear and interesting teacher as well as a notable inventor, but his teaching assistant used to fill the board with formulas without giving motivation. Assuming that all the material was in Pierce's book I just cut the class for the last month. Then as the final exam approached I called on Pierce and said: "I suppose the final exam will be based on material in your book, and that if I know what's in it I should be all right." He said, "By no means. I have totally revised the book and have a big set of lecture notes." He lent me his notes, thereby salvaging me. Then on the final exam one problem was to determine a certain number. I claimed that the number was not unique. Since I got an A + in the course, and Pierce asked me to speak later in a graduate seminar, I assume I was right.

At Cambridge University in 1932–33, Hardy was by far my most stimulating teacher.

MP: *Did many Harvard professors write their dissertations with earlier Harvard professors?*

Birkhoff: I don't think so. Before 1914, most American mathematicians wrote Ph.D. theses in Europe. And after 1930, our department tried consciously to avoid inbreeding. Although Loomis wrote his thesis with Walsh, and Mackey's thesis adviser was Stone, Stone had left Harvard before Mackey joined the Harvard faculty. In my own graduate years, I was closer to von Neumann than to anyone on the Harvard faculty, including my father. Zariski was the first Harvard mathematician after my father to have Ph.D. students stay on as members of the faculty. We always felt that inbreeding was a weakness, and that the Chicago mathematics department was suffering from it in the late 1930's and early 1940's. Each of Moore, Dickson, and Bliss was succeeded by his leading Ph.D.

MP: *Albert was a student of Dickson?*

Birkhoff: Yes, and Graves was a student of Bliss, who in turn had been a student of Bolza. Other, less distinguished Chicago mathematics Ph.D.'s had also stayed on as members of the department. At Harvard the policy has generally been to avoid that sort of succession, because it tends to get very political. I didn't regard myself as anybody's student.

Childhood and Early Education

MP: *May we back up once again, this time to your boyhood and pre-college experience? There is no question that you had a very famous father, and children of famous parents often have a difficult time of it —they're under pressures and expectations are high. Did you feel this when you were growing up?*

Birkhoff: That's a very good question. I sometimes think I was trained to be precocious. I was educated at home till I was eight and I graduated from grammar school at eleven. By that time my family was aware enough of Wiener's unhappy youth to take me out of school for a year before I entered high school. I spent the year getting physically stronger riding a bicycle around and playing baseball and skating at the Cambridge rink. It's a miracle that I didn't get killed on my bicycle. I remember a truck screeching its brakes as I came whizzing around a corner. I did get a scar on my forehead from another bicycle accident. I spent six years after graduating from grammar school before entering Harvard at the age of 17. I am glad that I became physically strong enough to participate with my classmates in sports and so on. So I seemed like all the others in college. There was a period in my early life when I definitely was very advanced, especially in things like mathematics. We began algebra in 8th grade and the teacher had me correct the papers to keep me out of mischief.

MP: *Did your mother have a career?*

Birkhoff: Mother went to the University of Illinois and was trained as a librarian. She practiced her profession until she married and then gave it up; this was common in those days.

MP: *Did she have a wide range of intellectual interests?*

Birkhoff: Though she read a lot, I think she was more social than intellectual, really. She and my father essentially divided responsibilities. Having friends, social friends, and caring for the children were primarily her responsibility. However, my father was a stimulating person at all ages. He took my sister and me on excursions, particularly when we were very young, and made life interesting for us somewhat later by telling us exciting stories, having opinions on controversial questions, and so on.

MP: *You mentioned a six-year gap between your elementary school education and the time you entered Harvard, but you didn't mention where you went to high school.*

Birkhoff: I spent five years at Browne and Nichols. I lost a year because I didn't have any foreign language and some of the things you get in private school. My sister went to private grammar school and public high school—partly for principle, partly for thrift. I went to public grammar school and private high school. I had a good mathematics teacher there: Harry Gaylord, who had written a trigonometry book with Bôcher. He was lucid and forceful. In algebra, he used the Dalton Plan under which you could progress as fast and as far as you wanted. So I finished high-school algebra my second year there. I started geometry in my third year, but we went to Europe in February, so my Euclidean geometry was a little sloppy. A year later my father wanted to go around the world on his "Aesthetic Measure" project. So my parents suggested I take my college board examinations a year early. After passing them, I had a wonderful fifth year at Lake Placid where I did much skiing. It was a splendid year physically and for general maturing. I was there while my family went around the world.

George D. Birkhoff's Education

MP: *I would like to ask about your father a little bit, particularly his education.*

Birkhoff: My father also went to a very superior high school, the Lewis Institute in Chicago, later part of the Illinois Institute of Technology, and he told me his opinion that my high school education was not nearly as stimulating as his. The Lewis Institute was semi-private—there were adjustable fees depending on the financial status of the students. The teachers were, I think, very enthusiastic and that's where he fell in love with mathematics.

MP: *Both your father and Veblen took their doctorates with E. H. Moore at Chicago, though they had done their undergraduate work at Harvard. Why was that?*

Birkhoff: My father's uncle, Garrett Droppers, went to Harvard in the class of 1892. He was an economist who was for a time president of the University of Nebraska, and was made Minister to Greece by Woodrow Wilson. He had gone to Harvard and married a Cambridge girl. It may have been his example that made my father go to Harvard after two years as a Chicago undergraduate. Perhaps it was because he had mastered so many of the courses at Chicago by that time and wanted new stimuli.

At Harvard he was primarily under the influence of Bôcher. He and Bôcher were very congenial, and I think his liking for French mathematics was due partly to Bôcher and partly to his admiration for Poincaré, whose geometric theorem* he proved. This was mentioned by Constance Reid in her book; Courant described it to me.

MP: *Courant said when he was in Göttingen that it was published in an obscure American journal. Which journal was that?*

Birkhoff: It's in the *Transactions*, 1912. You can find it in my father's collected works.

MP: *Scarcely an obscure journal today!*

---

*G. D. Birkhoff states the problem thus, "Let us suppose that a continuous one-to-one transformation T takes the ring R, formed by concentric circles C_a and C_b of radii a and b respectively ($a > b > 0$), into itself in such a way as to advance the points of C_a in a positive sense, and the points of C_b in the negative sense, and at the same time to preserve areas. Then there are at least two invariant points."

Garrett Birkhoff as a Harvard freshman in 1928

Birkhoff: He was a lifelong friend of Oswald Veblen. His friendship with Veblen and R. G. D. Richardson, and his cordial relations with Bliss and Dickson at Chicago dating back to his undergraduate years, were all a real part of the vitality and strength of American mathematics during those years.

MP: *I am surprised and impressed that the two greatest American mathematicians in that period— Birkhoff and Veblen—took their Ph.D.'s at Chicago, which had been established in 1892, only a few years before they went there, whereas Harvard dates from 1636.*

Birkhoff: Chicago had a remarkable faculty. E. H. Moore was given a free hand by William Rainey Harper, and he enticed to Chicago two of Klein's students who were high school teachers in Germany —Bolza and Maschke—both superb teachers. There was also Michelson: my father took a course in electricity and magnetism from him. The University of Chicago was just an exciting place to be at that time.

MP: *But it became an exciting place practically overnight. The story of the University of Chicago is incredible.*

Birkhoff: Yes, it is. It shows you what forty million dollars could do.

MP: *At the time.*

Birkhoff: It was a marvelous thing that happened there and I guess Harper should get lots of credit.

MP: *I would like to go back to the reason your father returned to Chicago rather than going on to Göttingen—or to Paris since his tastes were perhaps more French.*

Birkhoff: Language may have had something to do with it. Also, he was already doing research when he entered college. He was trying to solve the Fermat problem with Vandiver at about that time, a problem that Vandiver worked on all his life. I should have saved the letter Vandiver gave me—he gave me a bag of letters from my father ten years ago. In one of them, my father wrote: "We will solve Fermat's problem and then some!"

MP: *He did have a taste for hard problems. I recall that he also worked on the four color problem for years.*

Birkhoff: He always regarded himself as competing with the greatest mathematicians of all time. It was an aspiration of his to be a great mathematician. He was also fascinated by mathematics and he was not afraid of difficult problems. So he took on the most notable ones. Throughout his four college years, two at Chicago and two at Harvard, where he got his M.A. at 21, he was doing research. If you look at his published works you can see that he was trying to solve problems and publish their solutions. After two years at Chicago, he probably wanted new exposures, so he went to Harvard where his uncle had gone.

He and Osgood were never very congenial. He felt that Osgood was more or less a martinet, who had the German teacher-dominating-student approach. Bôcher was much more flexible and informal, and was generally considered a more inspiring, if less systematic, teacher. Bôcher also liked differential equations. My father claimed, and others I have talked to have shared this opinion, that Osgood never mastered the Lebesgue integral. It is my private opinion that Hilbert also never mastered the Lebesgue integral. When I was an undergraduate I inherited my father's prejudices: I never took a course with Osgood and I learned the Lebesgue integral.

MP: *I do not recall that you mentioned Lebesgue's name in the lecture* yesterday.*

Birkhoff: I thought I did, if briefly. Baire's thesis, Borel measure, and the Lebesgue integral, which is essential for the Riesz-Fischer theorem, are all part of the French pre-history of functional analysis.

Interest in Applied Algebra

MP: *Let us move back to your writing for a few minutes. Birkhoff-Mac Lane was a pioneering text and more accessible to a larger number of students than van der Waerden was. But Birkhoff-Bartee was also a pioneering work. We have seen a number of applied algebra books since, but that [one] was, to my knowledge, the earliest.*

Birkhoff: The book was Bartee's idea, and he selected the applications. We wrote it in the late 1960's. During the preceding decade, Louis Solomon, Neal Zierler, and other Harvard Ph.D.'s had worked on coding theory. My contribution came in correlating the material with the basic principles of the abstract algebra.

MP: *That was a very exciting book when it came out because it opened up vistas for a lot of people.*

Birkhoff: Possibly I helped to stimulate interest in those areas by an article in the *Monthly*[†], and by a symposium on more advanced topics. A number of Harvard Ph.D.'s were involved in algebraic coding theory.

MP: *For many people, Birkhoff-Bartee represented their first encounter with many of those ideas. When did it come out?*

Birkhoff: It came out in 1970. We had taught the material for four years before that. I had always felt that Bourbaki missed many ideas in algebra, and this gave me something of a crusading spirit. Bourbaki emphasized linear and multilinear algebra, to the virtual exclusion of finite groups, Boolean algebra, lattices and combinatorics. The *Monthly* article on current trends in algebra gave a picture of algebra as I saw it developing, and I think that subsequent events have more or less substantiated that.

---

* A lecture on the evolution of functional analysis at the Naval Postgraduate School in Monterey. The substance of this talk will be published in a paper with Erwin Kreyszig. The first on "The Development of Functional Analysis: 1903–1933" will appear in *Historia Mathematica* .

[†] "Current Trends in Algebra", *Amer. Math. Monthly 80*, 760–82.

MP: *I recall a meeting of the MAA in San Antonio in 1970 where a panel made up of Mac Lane, Jacobson and Herstein discussed the appropriate content and the aims of an undergraduate course in abstract algebra. The views ranged from Herstein's rather modest expectations to Jacobson's description of what seemed to many to be a two-year graduate sequence. Where do you stand in this spectrum?*

Birkhoff: Where I stand in this spectrum is this. First of all, I think we should mention Mac Lane-Birkhoff (*Algebra*). Both Mac Lane and I decided that to tamper with the *Survey* would be a mistake. So his idea was that the *Algebra* would be an updated *Survey*. He was able to teach that to freshmen very successfully because he is a marvelous teacher.

MP: *And the University of Chicago probably has marvelous freshmen too!*

Birkhoff: Yes. But I have never felt that this material is what one should teach freshmen. I feel strongly that it is very dangerous for mathematics to detach itself from the rest of the world; to be part of the world around one is much healthier. I observed a great decline in enrollment in the undergraduate algebra course at Harvard since we adopted Herstein instead of Birkhoff-Mac Lane; Herstein's appeal is primarily for professional pure mathematicians. And we all know that opportunities are much more numerous in computing and statistics than in pure mathematics. Perhaps the greatest contribution of Birkhoff-Mac Lane, as contrasted with van der Waerden, is that it relates algebra to useful mathematical techniques, such as matrices and linear algebra, but without sacrificing rigor. Actually, the swing back to pure algebra in most later texts represents a retrogression to van der Waerden: groups, rings, and fields. This retrogression was exactly what my article on "Current Trends in Algebra" and Birkhoff-Bartee were aimed at correcting: the idea that van der Waerden's *Moderne Algebra* was what algebra was for all time. As I get older, I am more and more convinced that van der Waerden, stemming from Emmy Noether and Hilbert, really tries to glorify number theory at the expense of real and complex algebra, which are actually far more substantial subjects.

Interest in Mechanics

MP: *I am still interested in the range of your interests in mathematics, from pure to applied mathematics.*

Birkhoff: I was always interested in practical applications. As a Harvard freshman taking "analytical mechanics" with Kellogg, I was intrigued by his explanation of the "sting" felt by baseball batters in hitting a ball, when they failed to hold the bat at the "center of percussion" relative to the point of impact (and the bat's center of gravity). This stimulated me to write an essay on the bounce of a spinning tennis ball hitting the court at an arbitrary angle. (Kellogg was our next-door neighbor, and we had played tennis together occasionally.)

But my initiation into serious applied mathematics concerned the analysis of two important weapons of World War II. The first of these was the then secret "proximity fuze", a device which determined the *distance* to a target by timing the reflection of radio waves. If the target was 10 meters away, a radar echo would come back after about 33 nanoseconds. I first worked on these proximity fuzes with a secret committee consisting of Morse, von Neumann, and one or two others, and set up by Warren Weaver. We studied their effectiveness for anti-aircraft use, and that involved me in probabilistic work. Our task was to estimate the factor by which the 'probability of kill' of an enemy airplane by an anti-aircraft shell would be increased if the usual 'time of flight' fuzes were replaced by proximity fuzes.

Later, at the Ballistic Research Laboratory of the Aberdeen Proving Ground, I had the privilege of having a desk in the office of R. H. Kent, who was our leading expert in Army Ordnance. In his company I saw early X-ray shadowgraphs of exploding shaped charges (the bazooka), and within 24 hours developed a simple explanation of why they were so effective in the "bazooka", our main infantry-operated anti-tank weapon. This was my most substantial contribution to the war effort; G. I. Taylor had proposed the same explanation a few months earlier in England.

Still later, through the Navy, I became involved in trying to develop theoretical models of the entry of torpedoes into water and "skip-bombing". But most important of all, I realized that the computer was going to influence profoundly the nature of applied mathematics. Of course, von Neumann was a friend of mine and we discussed informally many problems, especially compressible flows and shock waves around projectiles.

MP: *World War II was then very influential.*

Garrett Birkhoff

Birkhoff: Yes. Mathematical physics and engineering are very different, and I have been more attracted to engineering mathematics, which is concerned with practical problems of immediate importance, than to mathematical physics. Most of my "applied" interests have been really the result of circumstances. During the war I found that my ability to diagnose fluid mechanics, even with a rather limited knowledge of it, but knowing Kellogg's *Potential Theory*, was very useful to both the Army and the Navy. I decided that if opportunity permitted after the war, I would try to see what could be done scientifically, instead of sort of on a crash basis, to treat some of the questions I had become involved in. That's what motivated my book *Jets, Wakes, and Cavities*. My *Hydrodynamics* resulted from an invitation to give lectures at the University of Cincinnati. I talked about interesting questions in fluid mechanics, and pointed out that hydrodynamics is not what Lamb thought it was. You don't really get any idea about the realities of hydrodynamics from Lamb (Sir Horace Lamb: *Hydrodynamics*). But my book on *Jets, Wakes, and Cavities*, which I wrote with generous support from the Navy through the 50's, was a direct outgrowth of my wartime activities.

Another interest of mine is *numerical* linear algebra. I became interested because this subject is crucial in solving large systems of linear equations. This interest did not become active until some years after I had begun my first hydrodynamics phase. My activity in scientific computing really began in 1948, when I gave to David Young as a thesis topic solving the Dirichlet problem on a computer, thus automating Southwell's relaxation methods. Varga succeeded Young, working for me as a research assistant. When he went to Westinghouse, I was asked to be a consultant there; this was 1954.

MP: *That's a long way from lattices.*

Birkhoff: It is, but at Westinghouse, Varga and I related vector lattices to nuclear reactors. In treating elliptic boundary value problems numerically, the main cost comes in solving the approximating linear difference equations. If you have a thousand equations in a thousand unknowns, you know there exists a solution, but how do you compute it? Von Neumann was never able to solve that problem satisfactorily. We are having a meeting here (at the Naval Postgraduate School) in January, 1983, which really continues the development that began with Young's thesis. In turn, this was written only a few years after von Neumann first foresaw the great influence that digital electronic computers would have on the solution of scientific, technical and engineering problems.

Computing is a fascinating field, but debugging a program is a terrible bore. It is very much more efficient to have canned programs that have been debugged and apply to a wide variety of problems, and then to modify or supplement them for special purposes, than to start from scratch. One outstanding package of subroutines is NASTRAN, developed to put a man on the moon, for structures that we used for space vehicles. ELLPACK is another, designed to solve elliptic problems; it is better from an educational point of view, partly because it is not nearly so massive. NASTRAN has nearly 350 thousand lines of programming and you really have to search through the package forever. It would be like going to the Library of Congress to learn how to blow a whistle. ELLPACK is an intermediate size package, 35,000 lines long. With it you can do for $5 what would have cost $10,000 30 or 40 years ago.

George D. Birkhoff's Work on the Four Color Problem

MP: *Today, one seems to need computers even to solve the four color problem!*

Birkhoff: Yes, to show that each of nearly 2,000 ring-shaped maps is "reducible". The concept of a "reducible map" was invented by my father when he was at Princeton in 1912. The four color problem was one of my father's hobbies. I remember that all through the 1920's, my mother was drawing maps that he would then proceed to color. He was always trying to prove the four color theorem.

MP: *There's a story that has been circulating in this area. I should ask you whether it is apocryphal. It concerned a younger colleague of your father's at Harvard, Bernhart, I believe, who also worked on the four color problem. Your mother is reported to have asked Mrs. Bernhart at a tea—the Bernharts had only recently been married—"Did your husband make you color maps on your honeymoon too?"*

Birkhoff: I suspect it is apocryphal, since my parents were married in 1908, and my father's first paper on the four color problem was not published until 1912. A related, definitely true story concerns Lefschetz. He came to Harvard—this must have been around 1942—to give a colloquium talk. After the talk my father asked Lefschetz, "What's new down at Princeton?" Lefschetz gave him a mischievous smile and replied, "Well, one of our visitors solved the four color problem the other day." My father said: "I doubt it, but if it's true I'll go on my hands and knees from the railroad station to

George David Birkhoff in 1941

Fine Hall." He never had to do this; the number of fallacious proofs of the four color problem is, of course, legion.

MP: *Do you believe it has now been solved?*

Birkhoff: Yes. I am not an expert, but there are convincing probabilistic arguments that there exists an algorithm in each of the 1700 cases. There may be some little slips, but you must realize that no such proof is absolutely checked over to the last line. You have to go somewhat on faith. I think that the proof is substantially correct, but that human ingenuity should be able to reduce its bulk by at least an order of magnitude.

MP: *You don't have any philosophical problems with the idea that a proof that you cannot read is not really a proof?*

Birkhoff: There's a very good analogy. Daniel Gorenstein, a Harvard Ph.D., has organized the mathematicians who found all finite simple groups. He was quarterback of his high school football team, and I think this was good training for the task. His view is that there are 11,000 pages of close mathematical reasoning scattered in a lot of journals, and if no mistakes have been made, all finite simple groups have been found. This is a very similar situation. You have to have faith that, on balance, probably each and every one of these people decided where the truth lay, closely enough so that any small gaps in their reasoning can be filled in. Your real problem is to put all this reasoning back on a human scale. If you tell me that you know all the simple groups, why should I believe you?

MP: *No one person can wade through all the details.*

Birkhoff: That's exactly the point. Both human beings and computers are fallible, even if extremely accurate in some ways. So I think the two proofs* should have the same status in the public mind. They are both very elaborate constructs, arising in a civilization with very high technology. The experts believe them.

---

*That of the four color theorem, and that there are no unknown finite simple groups. (Ed.)

Colleagues in Mexico

MP: *In another direction, we are aware of some of your special activities related to Mexico, including an honorary doctorate from the National University of Mexico. We also know that you plan to fly to Mexico City from Monterey before returning to Cambridge. We are curious about your connections there.*

Birkhoff: My father was interested in relativity from 1920 on, as were many others. He wrote two books on it in the 20's and he developed a theory of gravitation about two years before he died, while he was touring South America and Mexico. In Mexico he got great response not only from Vallarta, who was at M.I.T. for many years, but also from a young mathematician, Barajas (whom I shall be seeing tomorrow night), and he invited them to come up to Harvard for the year, but he died that October. It was quite a tragedy for them as well as for his family. The following summer I went down to Mexico and tried to help them to carry on with this theory of gravitation as well as to try to be generally stimulating mathematically. I have been back several times since and have a warm and close relationship with these particular people and some others as well.

MP: *Lefschetz had a Mexico connection too, didn't he?*

Birkhoff: Yes; he went every summer for many years. His role was very different. He was working with the topologists and pure mathematicians and you might say Princeton won out over Harvard as a result. Many of the ablest Mexican graduate students in mathematics went to Princeton for many years, from 1950 on. Wigner also had some influence.

Aesthetic Measure

MP: *I would like to go back to aesthetics briefly. Was your father interested in the theory of music?*

Birkhoff: He was interested in and liked music. He played the piano. He was not very proficient, but he enjoyed it very much. My mother told me that he had ideas about aesthetics, in his early twenties. In the late 1920's he started writing them up. It was quite common in those days for scientists to theorize about other subjects. Simon Newcomb, I think, wrote 500 papers on mathematics, and almost as many on other subjects. It was common for scientists to speculate about the nature of knowledge. Boole, for example, wrote a famous book called *Investigation of the Laws of Thought*, and Helmholtz wrote about cognition theory. My father had his formula $M = E/C$, C being the complexity and E being aesthetic value*. Then he tried to apply this formula to axiomatize our senses of aesthetic attractiveness.

It was in 1927–28, when I was in Lake Placid, that my father went around the world to study Indian and oriental music with help from the Rockefeller Foundation. At the International Congress of Mathematicians in 1928, he gave an invited address on his "aesthetic measure" in the Palazzo Vecchio in Florence. The session was opened by two pages with trumpets in that marvelous medieval setting, which was itself very aesthetic. To develop a theory of aesthetics was regarded as a real plus for mathematics, extending its empire over the realm of thought. This is always the question: should mathematics be inward-looking and concentrate on the integers or should it look out and try to find new things, complex numbers or whatever, and computers. My father's "aesthetic measure" was very much an outward thing. But Hardy did not appreciate it. When I arrived in Cambridge in 1932, he asked, "How is your father coming along with his aesthetic measure?" I replied, "The book is out." He said, "Good, now your father can get back to *real* mathematics!"

MP: *So his aims were broader than those of Descartes, Euler and such who were also interested in the theory of music?*

Birkhoff: I think such comparisons are unreliable. Didn't Mozart write music with some sort of random number generator such as a pair of dice? But I think my father's ideas about aesthetics have withstood the test of time as well as anybody else's.

They have nothing to do with the mathematics of vibrating strings or other musical instruments. His theory was more psychological, in the spirit of Boole and Helmholtz. How does the human mind work?

---

*See Stiny, George, and James Gips, *Algorithmic Aesthetics*, University of California Press, 1978, 155–63. G. H. Hardy's *A Mathematician's Apology* is often quoted by those who like to deprecate "applied" mathematics. But see Norman Levinson's "Coding theory: a counterexample to G. H. Hardy's conception of applied mathematics," *Amer. Math. Monthly* 77 (1970), 249–58.

He also tried to appraise a good many old ideas—the golden rectangle, for example. I remember hearing about them as a teenager and thinking that theories of aesthetics were a lot of hooey.

MP: *Finally, we would like to ask about your plans for the near future.*

Birkhoff: I am presently concentrating on fluid dynamics, and expect to complete another book in this area within the next few months. My current work also includes coordinating the program for the January meeting in Monterey. This conference is being held in connection with my stay at the Naval Postgraduate School next year, for the equivalent of a semester as incumbent of the ONR Mathematics Research Chair. During this visit I plan to do research on numerical weather prediction, teach a course in fluid dynamics, and continue writing a series of books which I hope to publish during the 1980's.

DAVID BLACKWELL

DAVID BLACKWELL

Interviewed by Donald J. Albers

At sixteen, David H. Blackwell enrolled at the University of Illinois to earn a bachelor's degree so that he could get a job as an elementary school teacher. Six years later, he had his Ph.D. in mathematics and a fellowship to the Institute for Advanced Study at Princeton.

Today, he is a much honored professor of statistics at the University of California, Berkeley. He is a theoretician, noted for his rigor and clarity, who has made contributions to Bayesian statistics, probability, game theory, set theory, dynamic programming, and information theory. He says, "I've worked in so many areas—I'm sort of a dilettante. Basically, I'm not interested in doing research and I never have been. I'm interested in *understanding*, which is quite a different thing."

Professor Blackwell was interviewed in his office at the University of California in April of 1983. While being interviewed, he went to the board several times to "share something beautiful with somebody else." In a short time, it became clear that David Blackwell, the theoretician, is also a natural-born teacher.

MP: *You were born in Centralia, Illinois, back in* 1919, *just after World War I.*

Blackwell: That's right, in Centralia, a small town in Southern Illinois with a population of about 12,000.

MP: *You whizzed through elementary and secondary school, graduating at the young age of 16. What remembrances do you have of your childhood in Centralia and the influences on you? Were your mother and father mathematically inclined?*

Blackwell: No, they weren't. My grandfather ran a store. I had an uncle who could add numbers, three columns at a time, and that always impressed me. He never went to school at all; my grandfather taught him.

MP: *Did your family come from Illinois?*

Blackwell: No, my grandfather came from Ohio where he was a schoolteacher and then became a storekeeper.

MP: *To whom do you trace your mathematical abilities?*

Blackwell: To my grandfather, I suppose. I never knew him. Apparently he was a well-educated man—he certainly left a large library of books. The first algebra book I ever saw was in his library. I don't think he graduated from college but I know he was a schoolteacher and, in fact, that's how he met my grandmother. She was a student in his class while he was teaching in Tennessee. The reason that his son, my uncle, never went to school was that my grandfather never let him. He was afraid he would be mistreated because he was black.

MP: *But your grandfather went to school.*

Blackwell: But that was in Ohio, not in Illinois! Southern Illinois was probably fairly racist even when I was growing up there. The school I went to was integrated, but there was also a segregated white school in that same town. There were in fact two segregated schools, one that only blacks could attend and one that only whites could attend. But I was not even aware of these problems—I had no sense of being discriminated against. My parents protected us from it and I didn't encounter enough of it in the schools to notice it.

"Geometry Is a Beautiful Subject!"

MP: *Were there teachers along the way who made a particular difference, who made learning exciting?*

Blackwell: Oh, there were many. A couple of years ago my first-grade teacher came to see me here. She's about 82 now and living in Southern California. I hadn't heard of her in a number of years and of course I was in her class in 1925, but she somehow knew where I was and looked me up here in Berkeley.

But there were a couple of mathematics teachers in particular. My high-school geometry teacher really got me interested in mathematics. I hear it suggested from time to time that geometry might be dropped from the curriculum. I would really hate to see that happen. It is a beautiful subject. Until a year after I had finished calculus it was the only course I had that made me see that mathematics is really beautiful and full of ideas. I still remember the concept of a *helping line*. You have a proposition that looks quite mysterious. Someone draws a line and suddenly it becomes obvious. That's beautiful stuff. I remember the proposition that the exterior angle of a triangle is the sum of the remote interior angles. When you draw that helping line it is completely clear.

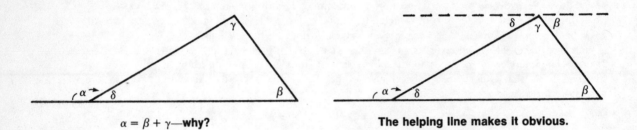

$\alpha = \beta + \gamma$—why?   **The helping line makes it obvious.**

And then there's the river problem where if you want to go from P to the river and then to Q, you have to have equal angles at R to minimize the path. Why is that so?

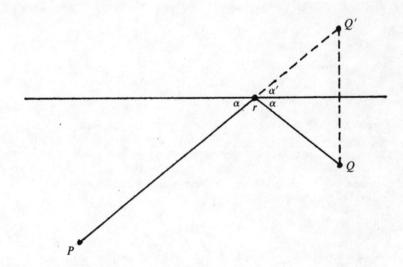

Again, you construct the mirror image Q' and then it's clear what the shortest path is. The construction of this one extra point changes the problem from something that is mysterious to something that is obvious.

MP: *How about your other high-school courses such as Algebra II and Trigonometry?*

Blackwell: I could do it and I could see that it was useful but it wasn't really exciting. When I went to college I knew that I was going to major in mathematics because I liked it and it was easy for me. But through calculus I thought it wasn't particularly interesting. The most interesting thing I remember from calculus was Newton's method for solving equations. That was the only thing in calculus I really liked. The rest of it looked like stuff that was useful for engineers in finding moments of inertia and volumes and such.

MP: *That's curious in a way, because you are generally portrayed as a theoretician who presents his ideas in beautiful and elegant ways. The first two examples that you give from geometry seem to enforce that. But Newton's method, finding approximations to roots of equations, seems inconsistent with the other examples.*

Blackwell: But it's also geometrical. By the way, I just always understood limits. For some students that's a difficult concept.

MP: *Do you mean epsilon-delta proofs?*

Blackwell: No. There probably were epsilons and deltas in the book, but I don't know whether I understood it in that way. The next year I really fell in love with mathematics. I had a course in elementary analysis. We used Hardy's *Pure Mathematics* as a text. That's the first time I knew that serious mathematics was for me. It became clear that it was not simply a few things that I liked. The whole subject was just beautiful.

Planned to Be an Elementary School Teacher

MP: *Was that in your junior year?*

Blackwell: It was my combination junior and senior year; I was an undergraduate only three years. I took some summer courses and proficiency exams. One of the reasons I went into mathematics is that I have never been especially ambitious. When I went to college I expected to be an elementary school teacher.

MP: *But you had gone through high school and found it pretty easy. You still expected to teach in elementary school?*

Blackwell: That's right. I think the reason may have been that my father had a very good friend who was influential on the school board of a town in Southern Illinois and even before I went to college he had told my father that when I finished he could get me a job. It was all laid out that that's what I was going to do. That was about 1935 or 1936 when jobs were scarce. In order to teach you had to take courses in education and I just kept postponing those courses. Before I had to make up my mind, though, it became clear that I was going to get a master's degree in four years, so then I raised my sights a little. I thought I might teach in a college or in high school.

MP: *Was there ever any doubt that you were going to end up in teaching?*

Blackwell: No, never any doubt about that.

MP: *How did your father respond to your postponing the education courses?*

Blackwell: He had so much confidence in me that he thought that whatever I did was right. He himself was not an educated man—he had gone through only the fourth grade in school. He didn't know much about what went on in a university.

MP: *Your father must have been a pretty good guy.*

Blackwell: Oh, he was a great guy. I found out at the end of my freshman year that he had been borrowing money to send me to college. At the start of my sophomore year I told him that he didn't have to send me any more money because I could support myself. At the time I told him that, it wasn't quite clear how I was going to do it, but I just didn't want him to borrow any more money. I had several jobs, as a waiter and as a dishwasher. I had an NYA job, the equivalent for college students of the WPA. I had a job cleaning cases in the entomology lab and filling vials with alcohol. I did that for a couple of years.

MP: *You realized that your university education was a pretty big sacrifice for the rest of the family. What did your mother think of your university experience?*

Blackwell: She was somewhat more concerned about the specifics of what I was doing. She wondered whether I would be able to get a job once I graduated, but she pretty much left it up to me.

MP: *I am interested in your plans to be an elementary school teacher.*

Blackwell: I don't think I could have remained an elementary school teacher but it wouldn't have surprised me at all had I remained a high-school teacher. I think I could very easily have done that. In fact, after I got my master's degree I suspect that if I had gotten a job as a high-school mathematics teacher I would have taken it.

MP: *But you were demonstrating big mathematical talent at that time—you had received a fellowship.*

Blackwell: But I think I would have done it. You go to college for four years and you go out and you get a job. Some people go on. I knew I could do the course work—there's no question about that —but I didn't know whether I could write a thesis. Does anyone really know whether he can write a thesis until he does?

During my first year of graduate work I knew that I could understand mathematics. I could take a graduate mathematics text, read it and do the problems, and with great difficulty I could read a research paper and a journal. I knew I could do that. But whether I could do anything original I didn't know. I didn't mind trying it but it was not the only path in the world for me. I think I would have been perfectly content being a good high-school mathematics teacher.

MP: *I'm sure you would have been active reading mathematics.*

Blackwell: Oh yes, and I would have been active in something like NCTM. Times are different now, too. One of my high-school teachers went on to become a college mathematics teacher after the war and my high-school physics teacher had a Ph.D. in physics. In those days people wanted to get a job they liked.

Graduate School and the Institute for Advanced Study

MP: *You finished your bachelor's degree in 1938 and then stayed on at Illinois for a master's degree?*

Blackwell: I continued working and for the last two years had fellowships from the University.

MP: *Were fellowships commonly awarded to black students at that time?*

Blackwell: If there was any difficulty I never heard a word of it. During my first year of graduate work a couple of my teachers encouraged me to apply for a fellowship. Let me tell you a story about that. Before the fellowships were announced, one of my fellow graduate students told me that I was going to get a fellowship. I said, "How do you know?" He said, "You're good enough to be supported, either with a fellowship or a teaching assistantship, and they're certainly not going to put you in a classroom." That was funny to me because the fellowships were the highest awards; they gave one the same amount of money and one didn't have to work for it. I have no doubt, looking back on it now, that race did enter into it.

MP: *Were there any black faculty members of the department?*

Blackwell: No, not even in the whole university.

MP: *So it turned out to be a lucky break, and you continued through to the Ph.D. as a student of Joe Doob.*

Blackwell: And he was, I would say, clearly the most important mathematical influence on me.

MP: *When did you first come in contact with him?*

Blackwell: My first meeting with him was when I asked him whether I could work with him.

MP: *You hadn't taken a course from him?*

Blackwell: No. Don Kibbey, who was chairman at Syracuse for a number of years, was a teaching assistant at Illinois at that time. One day he asked me whom I was going to work with. I told him I

Blackwell, a distinguished professor at the University of California, Berkeley, went to college expecting to become an elementary school teacher.

didn't know, I hadn't thought much about it, but it was time I started. He said, "Why don't you try to work with Doob? He's a very nice guy." Don was working with Doob. I had a lot of confidence in Don's judgment so I just went up to Doob and asked him if I could work with him and he said yes.

MP: *Was Halmos there at that time?*

Blackwell: Oh yes, he was a year ahead of me. He was also working with Doob.

MP: *Yet you ended up in very different areas.*

Blackwell: Rather different. I have stuck closer to probability than Paul has. I learned a lot from Paul by the way. Many of the papers Doob gave to me, he had given Paul to read a year before. They were on measure theory. Paul had learned them and was anxious to tell someone about them. I was of course anxious to hear them. I suspect I was the first one to hear the first version of Paul's measure theory book.

MP: *Doob took you along with him to the Institute for Advanced Study.*

Blackwell: Now I may not get this story just right, but I think something like this happened. I think it was the custom that members of the Institute would be appointed honorary members of the faculty at Princeton. When I was being considered for membership in the Institute, Princeton University objected to appointing a black man as an honorary member of the faculty. As I understand the story, the Director of the Institute just insisted and threatened, I don't know what, so Princeton withdrew its objections. Apparently there was quite a fuss over this, but I didn't hear a word about it.

MP: *At that time, you were doing measure theoretic work with Doob. But today you're thought of as a statistician. How do you think of yourself? Are you a statistician or a mathematician?*

Blackwell: I don't even try to classify myself and I try not to classify other people. To me that's a fruitless and limiting occupation. In statistics we distinguish between probability and statistics,

David Blackwell

between theoretical statistics and applied statistics, between Bayesian and non-Bayesian statistics, data analysts and other kinds of statisticians. People try to categorize other people and even themselves; they put themselves in pigeonholes. As I say, I think that's stultifying. You can get good mathematics at all levels of abstraction.

Statistical Beginnings

MP: *How did you get into statistics?*

Blackwell: Let me tell you how I got interested in statistics. In 1945, I was teaching at Howard University. The mathematics department there was small and not very lively. I looked around Washington, D.C., to find mathematics wherever I could. I happened to attend a meeting of the Washington Chapter of the American Statistical Association. Abe Girshick gave a lecture on sequential analysis. To me it was a very interesting lecture. The most interesting part was a theorem that he announced that I just didn't believe. Indeed, I went home to see if I could construct a counterexample, and not believing the theorem it was easy for me to think that I had found a counterexample. I wrote it up and sent it to him—he was working for the Department of Agriculture at that time as a statistician. My counterexample was wrong, but instead of just dismissing this counterexample as the misguided effort of somebody who didn't know statistics, he invited me over to his office to talk about it. He didn't tell me it was wrong—he just asked me over to talk about it and this established a personal relationship and collaboration that lasted until his death. And that's how I got started in statistics, just listening to that one lecture by Abe Girshick. In fact, my first paper in sequential analysis was on this very equation that I didn't believe.

MP: *What is sequential analysis?*

Blackwell: It is the analysis of an experiment where the number of trials is not specified in advance. It's the analysis of sequential experiments—that's the only difference between what is called sequential analysis and fixed sample-size analysis. You can either start out with how many subjects you're going to have and how many trials you're going to make, or you can say, I'm going to keep looking until I reach a conclusion. Of course, people had been doing that informally for a long time, but Wald was the first one to formulate that idea and study it systematically. Sequential analysis is no longer considered a distinct branch of statistics. For example, we do not have any course called sequential analysis any more. Its importance at the time was that it led to a re-examination of many things. If concepts for fixed sample-size analysis turn out to be less appropriate for sequential analysis, then that sort of suggests that maybe they weren't appropriate for fixed sample-size analysis either. I'm sure that it's Wald's work in sequential analysis that led to his work in general decision theory and that was very important in the development of statistics.

"I'm Sort of a Dilettante."

MP: *Of the areas you worked in, which do you think are the most significant?*

Blackwell: I've worked in so many areas—I'm sort of a dilettante. Basically, I'm not interested in doing research and I never have been.

MP: *What are you interested in, then?*

Blackwell: I'm interested in *understanding*, which is quite a different thing. And often to understand something you have to work it out yourself because no one else has done it. For example, I have gotten interested in Shannon's information theory. There are many questions that he left unanswered that were just crying out to be answered. The theory was incomplete so I worked on it with a couple of my colleagues because we wanted to know what happens in this case or that case. The drive was not to find something new. It would have been nicer if it had all been done. But since it hasn't been done, you just want to fill out the theory and make it complete. That's what I mean by being a dilettante. When I feel that my understanding of something has been rounded out pretty well, then I'm ready to move on to something else.

MP: *Then maybe that accounts for this long list where you have made contributions: Bayesian statistics, probability theory, game theory, set theory, dynamic programming, information theory.*

Blackwell: But just about everything that I've worked on involves either probability theory or set theory. And of course since the measure theory model for probability involves set theory, I haven't really gone very far away from where I started. I have just looked out in many different directions from it.

Duels

MP: *You are cited as one of the pioneers in the theory of duels. How did you get interested in duels?*

Blackwell: I recall it very well. That happened at the Rand Corporation when I was a consultant for them. One day some of us were talking and this question arose: If two people were advancing on each other and each one has a gun with one bullet, when should you shoot? If you miss, you're required to continue advancing. That's what gives it dramatic interest. If you fire too early your accuracy is less and there's a greater chance of missing. It took us about a day to develop the theory of that duel. I did it and Abe Girshick did it and John Williams did it. Then I got the idea of making each gun silent. With the guns silent, if you fire, the other fellow doesn't know, unless he's been hit. He doesn't know whether you fired and missed or whether you still have the bullet. That turned out to be a very interesting problem mathematically.

MP: *Did you ever go beyond two-person duels?*

Blackwell: I've never gone beyond two-person, zero-sum games at all. They're the only ones I understand. It's regrettable that those are the games for which the theory is clear and beautiful because those are the least important games. One person wins; the other loses. But they're just not the kind of games that are played in the world. For example, the game being played between the United States and the Soviet Union is a much more important game and it is not a zero-sum game. Both sides can win or both sides can lose. But I've never understood those other games. Only the zero-sum games have a clear theory.

MP: *Have you tried to understand those other games?*

Blackwell: I did try for a long time to understand non zero-sum games but I did not succeed and it became clear to me that I was not going to succeed. I was very impressed and I still am impressed by the so-called "sure thing" principle. It was formulated by Jimmie Savage. The "sure thing" principle says this: "If you have to choose between two acts, A and B, and how much you're going to make depends on some unknown situation—it might be S_0, or it might be S_1. Suppose that if you knew it was S_0, then you would choose A over B, and that if you knew it was S_1, you would choose A over B. The "sure thing" principle says that even if you don't know, you should choose A over B. That seems like such a plausible principle, but let me show you what it leads to. The "sure thing" principle leads to the prisoner's dilemma. You and I are playing a game and I can either cooperate with you or double-cross you. And you can either cooperate with me or double-cross me.

		You	
		C	*D*
C		(2, 2)	(0, 5)
Me			
D		(5, 0)	(1, 1)

The first coordinate in each box shows how much I get, and the second coordinate in each box shows how much you get. So I'm wondering, should I cooperate with you or double-cross you? Maybe you're going to cooperate. If I cooperate I get two; if double-cross I get five. So if I knew that what you're going to do is cooperate, then I would double-cross. But maybe you're going to double-cross, then if I cooperate I get zero and if I double-cross I get one. So again it's better for me to double-cross. So I'm going to double-cross. It's actually symmetric. The five is bigger than the two and the one is bigger than the zero so you should double-cross. So we both believe in the "sure thing" principle and we both double-cross. So we each get a dollar, whereas, if we had cooperated, we would each get two dollars. In fact, the situation with the Soviet Union has elements like this in it. To cooperate is to disarm and to double-cross is to re-arm with bigger and bigger weapons. That takes a lot of resources and we would both be better off disarming. But each is afraid that if he throws away his weapons, the other one will not and he will be at a great disadvantage. So when I saw that this "sure thing" principle led to an armaments race, so to speak, I realized I was not the one to come up with a satisfactory theory for non zero-sum games. I keep on encouraging other people to work on it, though.

MP: *Are there parts of your work that have given you particular pleasure? You have a couple of theorems named after you—there must be a certain amount of pleasure there.*

Blackwell: One thing that gave me a good deal of pleasure was finding a game theory proof for a theorem in topology: the Kuratowski Reduction Theorem. I was studying the proof and trying to understand it, when all of a sudden, I recognized the kind of thinking I was doing, exactly the kind of thinking I was doing some years before when I was thinking about games, infinite games. In about three minutes I realized that you could prove this theorem by constructing a certain game. That gave me real joy, connecting these two fields that had not been previously connected.

MP: *That probably surprised a few topologists.*

Blackwell: Actually, some logicians got interested in it. I may have been one of the first to show how infinite games related to set theory. Actually, I was not the first because Banach and Mazur, back in Poland, related infinite games to set theory. *(Blackwell then went to the board and outlined the proof.)*

Blackwell on Teaching

MP: *You're the first person I've interviewed who can't restrain himself, who can't keep from getting up to the board to explain something. You must like to teach. What is it that makes teaching fun for you?*

Blackwell: Why do you want to share something beautiful with somebody else? It's because of the pleasure he will get, and in transmitting it you appreciate its beauty all over again.

MP: *Are you teaching now?*

Blackwell: Yes, I teach at all levels. This quarter I have only a graduate seminar but last quarter I had a senior-level course and a very elementary course. In the fall quarter I had a course for sophomore-level engineers. There is beauty in mathematics at all levels, all levels of sophistication and all levels of abstraction.

MP: *If you were to write down a short list of desirable characteristics in a mathematics teacher, what would be on the list?*

Blackwell: I don't think that one person is a good teacher for all students. There are all kinds of styles of learning and it takes a good teacher to teach in a style that is not the style in which he learns. For example, I love pictures. The first time I had a blind student in my class, though, I realized how inexplicit my teaching is. For example, I put 5 points on the board and say, "Let's try to fit a line through them." The blind student is completely lost because I don't give to the points any coordinates. In another case, I didn't write anything except two letters, *A* and *B*. Some people prefer a somewhat more formal style than that. I think that I'm a good teacher for certain kinds of students, but not necessarily all.

"Formulas and Symbols—I Don't Especially Like Them."

MP: *What I've seen so far reminds me somewhat of Paul Halmos' style. Maybe what we see here is the common influence, Doob.*

Blackwell: My students sometimes complain because I use more than one symbol for the same thing. I forget which symbol I use because the symbols are not very important to me. It's strange to have a mathematician who doesn't especially like formulas and symbols. I remember when von

Neumann and Morgenstern's book on game theory came out. It was a very significant book and it's a big book, because they wrote it twice, once in symbols for mathematicians and once in prose for economists. I read the prose. I found it much easier to read than the symbols. If you are a mathematician it's easy to translate the ideas in prose into symbols if you want to.

MP: *What do you as a mathematician do on a day-to-day basis?*

Blackwell: Well, what I did today was to try to understand two forms of the category 0-1 law, to see if one of them is stronger than the other and to see what each one implies about the existence of what Harvey Friedman calls diagonalizations. It's one of the things I was doing today. That's at a rather high abstract level.

Another thing I was doing today was just playing around with a computer, trying out programs for minimizing a function of five variables, looking at curves and trying various techniques to see which ones work and which ones don't. I would say that the first thing I told you about is a somewhat more serious activity because if I find out the relationship between those two forms, the 0-1 law, I'll probably tell my students about it in a seminar and I may even pursue it further. The other is unlikely to result in anything more than my better understanding the problem of minimizing functions. I play quite a bit.

MP: *What do you mean by play?*

Blackwell: You know the algorithm for calculating the square root. If you want the square root of s and you start out with x, you divide s by x and take the average and that's the new x. Every positive definite matrix has a positive definite square root. It occurred to me that maybe this algorithm would work for positive definite matrices. You take some positive definite X, add to it SX^{-1} and divide by two. The question is: Does this converge to the square root of X? I decided that instead of trying to prove it I would just try it out. If you have an X you can square it and compare it to S and calculate the distance between them. I started out with the identity as my first approximation. In a particular example, the error at first was tremendous. then dropped down to about .003. Then it jumped up a bit to .02, then jumped up quite a bit to .9, and then it exploded. Very unexpected. It is not unusual to have it diverge if you start out far away from the solution, but when you start out close to a solution you expect it to converge to the solution. That's characteristic of Newton's method and this is kind of a Newton-like method. Then I started looking at the theory and it turns out that the algorithm works provided that the matrix you start with commutes with the matrix whose square root you want. You see, it's sort of natural because you have to make a choice between SX^{-1} and $X^{-1}S$, but of course if they commute it doesn't make any difference. Of course I started out with the identity matrix and it should commute with anything. So what happened?

MP: *You must have been having some kind of round-off error.*

Blackwell: Exactly! If the computer had calculated exactly it would have converged. The problem is that the matrix the computer used didn't quite commute.

"I Think Proofs by Contradiction Are a Mistake."

MP: *I have been told that you have a very interesting way of proving a theorem. How do you, in fact, prove a theorem?*

Blackwell: I don't really know. I think proofs by contradiction are a mistake. I've always found that if you start with mutually contradictory hypotheses, you're always working in never-never land. You're saying that in this land $0 = 1$ and you're just trying to show that that's where you're working. Nothing you say is true, and in a way you're not learning anything because everything you say is false. In all the cases I have looked at, you can transform a proof by contradiction into a proof in which everything you're saying is true and you learn more that way. It's not a question of making a big change. Let me give you an example. Take the proof that the set of real numbers is uncountable. The usual proof is this: Suppose you have a list of all the real numbers, the first one, the second one, the third one, and so on. Write out their decimal expansions, then you can write down one where it differs from the first one in the first place, from the second one in the second place, and so on. So you've reached a contradiction. You have claimed to have them all but you've shown one that is not in the list. It's a beautiful proof but I would formulate the theorem this way. Show me any sequence of

Blackwell, the theoretician, says, "I've never been especially interested in research—I'm sort of a dilettante."

numbers and I'll show you one that is not in the list. It's a small change but it's a positive fact. Every proof by contradiction that I have seen or studied can be recast so there's no contradiction at all. You learn something new.

MP: *The first one that we typically see as students is the irrationality of the square root of* 2.

Blackwell: That is more difficult—I would have to think about that.* But I am convinced that imbedded in that somewhere is a positive approach to the proof. In the proof that the number of primes is infinite there is, of course, a construction of a prime that is larger than any of the primes in the list. So that one is clear.

MP: *Thus we should make proofs positive and avoid proofs by contradiction.*

Blackwell: Yes, I learn things that way. It is well known that from contradictory premises you can deduce anything. Well, I'm just not interested in deducing things from contradictory premises.

"I Applied Only to Black Institutions."

MP: *Let's jump back to Howard University for a few minutes. You were there for ten years. In the Neyman biography, Constance Reid mentions that Jerzy Neyman of the University of California first saw you at the University of Illinois where you were president of the Mathematics Club. In 1942 he contacted Doob, hoping to get Doob to join the statistics department, but Doob said, "No, I cannot come but I have*

---

*Later, Blackwell did give a positive approach to the proof.

some good students and Blackwell is the best. But of course he's black and in spite of the fact that we are engaged in a war that's advancing the cause of democracy, it may not have spread throughout our own land." How did that incident affect you?

Blackwell: Let me tell you what happened. This is another one of those cases where I didn't know much about it until much later. Neyman wrote to me and said he wanted to interview me for a possible job in Berkeley. We met in New York and he interviewed me. He said he would let me know. He went back to California and I didn't really expect anything to happen. I had already written 104 letters of application to black colleges.

MP: *You hadn't thought of applying to other institutions?*

Blackwell: Oh no, I applied only to black institutions.

MP: *You mean the door was closed?*

Blackwell: I just assumed that it was, but then Neyman wanted to interview me and I was glad to be interviewed. I would have welcomed the job. I didn't expect anything to happen and eventually I got a letter from him saying something like this: "In view of the war situation and the draft possibilities, they have decided to appoint a woman to this position." It sounded plausible to me and I wasn't expecting anything anyway. It wasn't until I came to Berkeley that I learned that there was more to the story than that. My blackness was a plus for Neyman. He had a tremendous amount of sympathy for anyone who had been oppressed or mistreated in any way. He always favored the underdog. It would have given him a special pleasure to appoint me just because I was black.

MP: *Neyman eventually had that pleasure, although some time elapsed between that first interview and your eventual appointment here.*

Blackwell: Yes, twelve years.

MP: *While you were at Howard you kept going. You did a lot of research.*

Blackwell: I was quite free at Howard. They understood the importance of attending professional meetings, for example. They were quite generous in paying expenses for attending professional meetings. I think everyone had the right to attend one professional meeting per year, more than that if you could show need for it.

MP: *That's better than a lot of colleges today.*

Blackwell: Better than Cal. Now if you are presenting a paper, you can get your expenses paid, but not if you just want to go there and learn something. It causes a lot of incomplete papers to be presented because that's how people get their expenses paid. I was able to maintain mathematical contacts with statisticians in Washington and I went to meetings all over the country.

MP: *Did the employment situation change much for blacks during your twelve years at Howard University?*

Blackwell: Oh yes. The first year I was at Howard one of my colleagues in the economics department got an appointment at the University of Chicago. Looking back on it, I feel that there must have been a big change just resulting from World War II. I think there was a big change between the 1941 attitude and the 1945 attitude.

Berkeley

MP: *You arrived in Berkeley in 1954 and shortly thereafter the Department of Statistics was formed as a separate entity from the Department of Mathematics. Would you have been happier had the division not occurred?*

Blackwell: No, in fact, Neyman had had a separate operation for some years. It was just formalizing something that was in fact already the case. Neyman had had his statistical laboratory and when he wanted appointments in what was thought of as the statistical laboratory, the mathematicians pretty much went along with it. I rather liked being in a smaller group. I think a group loses something when it grows beyond a certain size. I think one would miss that close personal relationship where everybody talks to everybody else. Now our department is a little bit too big.

David Blackwell

MP: *You succeeded Neyman as chairman of the department. At the time, you were still a junior member of the department, at least in years of service.*

Blackwell: I had known all the people of the department before I came here. Mathematical statistics is still not a very large subject and it was considerably smaller at that time. Mathematical statisticians all knew each other and would see each other at meetings of the Institute of Mathematical Statistics.

MP: *You mentioned earlier some people who have influenced you: your geometry teacher and Doob. Was Neyman also an influence on you?*

Blackwell: He was a very good friend. He was not so much a professional influence on me but rather he had a personal influence. His statistical and mathematical ideas did not influence me very much, at least not directly. It was his character as a man—he was a warm, generous, principled man. He regarded himself as a conservative and in some ways he was a conservative person. For example, in dress, he was extremely conservative. He had rather rigid standards about proper behavior.

MP: *Was anyone else a strong influence on you?*

Blackwell: Well, Girshick influenced me. We worked together over many years. He had more good statistical ideas than I had, though I was better trained technically than he was. He would often announce some mathematical idea of his and it would turn out that it was not quite right, but almost right, and what was right was interesting. He was full of ideas and anxious to get other people to work on them.

Blackwell on Leadership

MP: *It seems that since your student days you have been a leader. At the University of Illinois, you were president of the local mathematics club. You have been president of the Institute of Mathematical Statistics and an officer of several other organizations. Do you have a flair for leadership?*

Blackwell: No.

MP: *Do you lead grudgingly?*

Blackwell: No, I don't mind doing it. I have a tendency to figure out what people want done rather than be a leader. When I was department chairman, I soon discovered that my job was not to do what was right but to make people happy. When you set about making up a teaching schedule, you know that *A* can teach it but he won't do a very good job, and *B* will do a better job, but *B* taught it last year so you give it to *A*.

MP: *So you don't miss administrative work.*

Blackwell: Not a bit! In fact, when I gave up being chairman, for about a year my first waking thought in the morning was, "I'm no longer chairman," and it made my day.

MP: *It wasn't too many years after you arrived here in Berkeley that the whole social fabric of the University was torn apart by the free speech movement. You weren't exactly a bystander.*

Blackwell: I was completely sympathetic to the students. I did not like the way they expressed their grievances, but they certainly had grievances. The students had changed but the administration had not recognized the change. You know when Adlai Stevenson was running for President—I believe that was in 1956—he was not allowed to speak on the Berkeley campus. The administration took this line: The University must not get involved in political matters. No candidates were permitted to speak. When I was going to school that would not have bothered me at all. And it wouldn't have bothered most of the other students when I was going to school. But the students in the 1960's were a different breed. A lot of them were very much interested in what was going on and these rules that may have been appropriate 40 years ago were simply completely outmoded. But the administration simply wouldn't move an inch. That's what the students were protesting. They wanted to hear all kinds of ideas discussed on the campus. Looking at it now, it's hard to believe that's the way it was, but it really was that way. I don't like loud noises. There was a lot of violence and destruction in those days, but the students really had something to protest.

Blacks in Mathematics

MP: *At mathematical meetings I still see very few black faces. The list of black mathematicians is short and it does not seem to be growing very rapidly. Do you have any explanation?*

Blackwell: Yes. Black people go in other directions. Black people are going into the professions: law, medicine, and business. I sort of understand that: there's more security. There's more certainty of having a fair income in those areas than there is in mathematics. I don't know if you know J. Ernest Wilkins. He's a black mathematician just a few years younger than I am. He's good. In fact, he's also an engineer and is a member of the National Academy of Engineering. His father was quite a good mathematics student and got his bachelor's degree at the University of Illinois. Then he went into law and became a very distinguished lawyer. He was on some President's cabinet. There are other black people who have had considerable mathematical talent but went into law.

Families and Telephones

MP: *You've been at Berkeley a long time now and you've had a very creative career here. Have you ever thought of going anywhere else?*

Blackwell: Oh no, I've been pretty happy here and a number of our children are living here now, and those who are not plan to come back here.

MP: *You have eight children?*

Blackwell: Yes.

MP: *That's a big family even by the standards of the 'forties and 'fifties. You must like children.*

Blackwell: Yes. I like grandchildren too.

MP: *Have any of your children pursued careers connected with mathematics?*

Blackwell: No, they have no particular mathematical interests at all. And I'm rather glad of that. This may sound immodest, but they probably wouldn't be as good at it as I am. People would inevitably make comparisons. My brother went to the University of Illinois and he was a freshman there about ten years after I was. He joined the same fraternity that I had belonged to. They asked him, when they found he was from Centralia, whether he was related to me. He said: "I think I've heard of him, but there is no connection." Again, he didn't want to be compared to me. He wanted to make it on his own. My name was on some sort of a plaque there. It's hard on the younger one, whether he made a very good record or whether he made a very bad record, for he gets blamed for what his older brother did.

MP: *Some years ago Life Magazine put out a book on mathematics that had your picture in it. You were shown teaching, which I now see was very appropriate. It also mentioned that you did not have a phone in your home. Do you still have no telephone in your home?*

Blackwell: Oh, no, our youngest daughter won that battle years ago. She insisted we have a telephone. But for a long time we did not have a telephone. It wasn't based on any principle at all, but one of our kids ran up an excessive long-distance bill so we decided to have the telephone discontinued for a month. One month went into two and two months went into three and we decided that there were advantages as well as disadvantages to not having a telephone.

I do not have a positive attitude toward telephones, though. During World War II a friend of mine and I were in Washington trying to get a train to New York. There were long lines and trains did not run very frequently. Furthermore, soldiers had priority. We were standing in the ticket line, just waiting to get some information, and my friend said, "Just a minute." He left the line and then I heard the telephone ring—the ticket agent stopped waiting on customers and went over, answered the telephone, and gave my friend the information he wanted to know. That's when my attitude toward the telephone changed. What a rude, impolite instrument that is. It can break in and take priority over all the people who have made the effort of coming in and standing in line.

MP: *Do you have any hobbies?*

Blackwell: No. When I have spare time I listen to music or I go into the country and work. We have some land up in Mendocino County, about 40 acres. It's beautiful—it has a creek and big redwood trees. When we bought it my dream was to go up on weekends, get a martini, sit under a redwood tree, and watch the creek go by. But when I go up there I work from the time I get there until the time I leave, planting trees, repairing fences, cutting weeds, fixing a leak in the barn. So many things go wrong that something always has to be done. When we go up, I'm not the only one who works. My wife also works from the time we get there till we leave for home. It hasn't worked out at all the way we had in mind, but it is a lot of fun.

SHIING-SHEN CHERN

SHIING-SHEN CHERN

by William Chinn and John Lewis

In the spring of 1982, the National Science Foundation announced the creation of two new institutes of research in mathematical sciences. Given the general scarcity of government funding for scientific research, this was remarkable news. Even more remarkable is the unique background of the man who was selected as Director of the Mathematical Sciences Research Institute in Berkeley, Professor Shiing-shen Chern. To get a more precise picture of his background, Professor Chern was interviewed by the authors at the temporary facilities of the new Institute on October 8, 1982. What follows is a product of this interview together with biographical accounts from additional sources.

Shiing-shen Chern was born on October 26, 1911 (a couple of weeks after the overthrow of the Manchu dynasty) in Kashing, Chekiang Province, China. Kashing, located about 50 miles southwest of Shanghai, is noted for its scenic lakes and streams, matching those of nearby Hangzhou. Chekiang is one of the provinces on the eastern coast, known to the Occidental world particularly for its brocades and embroideries. Here, Chern spent his boyhood.

As a young student in high school, Chern studied mathematics from the then heavily used *Algebra* and *Higher Algebra* by Hall and Knight and *Geometry* and *Trigonometry* by Wentworth and Smith, doing a large number of the exercises in the books. At the age of 15, he enrolled in Nankai University in Tientsin (known for one of its graduates, Chou En-lai, the premier under Mao Tse-tung). In the late 1920's, Nankai was a small university with a total enrollment of 300. It was comprised of three schools: letters, science, and commerce, and the school of science had four departments: mathematics, physics, chemistry, and biology. Chern's aptitudes and preferences steered him toward the sciences, and after a year or two, he decided to major in mathematics, although students did not formally choose a major until their third year. One reason for his choice of mathematics was his disinclination for laboratory work. Another reason for selecting mathematics was the presence of Dr. Li-fu Chiang, an excellent professor who had received his Ph.D. from Harvard under Julian Coolidge. Mathematics in China was in a primitive state at that time, and few universities offered a course on complex function theory or linear algebra. Chern was fortunate to have been at a university where such courses were offered along with courses on non-euclidean geometry and circle and sphere geometry (using books by Coolidge). He had four classmates who graduated with him in 1930, and just a few years ago, he had a reunion with two of them in China.

In 1930 there were very few people in China doing mathematical research. One of the few was Dr. Dan Sun, a professor at Tsing Hua University in Peking (then called Peiping). Dan Sun had been a student of E. P. Lane at the University of Chicago, and his research area was projective differential geometry. Through his teacher Li-fu Chiang at Nankai University, Chern had already been attracted to geometry and, upon graduation, he took the entrance examinations for the graduate school of Tsing Hua University. He was the only graduate student in mathematics accepted that year and he was hired as an assistant. His formal graduate training actually began in 1931 and he remained at Tsing Hua until 1934. During this period, Chern read many papers on projective differential geometry and wrote several of his own. In 1934, he received a fellowship from the *Boxer Indemnity Fund*. He was supposed to come to the United States to continue his studies. Instead, he requested and received permission to go to Hamburg, Germany to study under Professor Wilhelm Blaschke, the well-known geometer. This choice was the result of Blaschke's visit to Peking in 1932 when he had lectured on the geometry of webs, an area of mathematics that Chern found very attractive.

Chern entered the University of Hamburg in November 1934 and received his D.Sc. in February 1936. After completing his degree, he had one more year remaining on his fellowship; he sought advice from Blaschke on what to do. Blaschke thought that Chern could either stay in Hamburg and work with Emil Artin on algebra and number theory or go to Paris to work with Elie Cartan on differential geometry. While in Hamburg, Chern had gotten to know Artin quite well. He had listened to more lectures by Artin than any of the others; all of Artin's lectures were beautifully organized. If Chern had stayed in Hamburg, he may well have become an algebraist. Instead, his attraction to differential geometry was so strong that he chose to go to Paris.

Chern's association with Cartan proved to be extremely important for his career. Cartan was the leading differential geometer in the world at that time. Chern had already had some contact with Cartan's work while he was in Hamburg through the lectures and writings of Kähler. Cartan's writings were generally regarded as very difficult, but Chern quickly accustomed himself to Cartan's way of thinking. In retrospect, Chern feels that it was like learning a new language. There is a tendency in mathematics to be abstract and have everything defined, whereas Cartan approached mathematics more intuitively. That is, he approached mathematics from evidence and the phenomena which arise from special cases rather than from a general and abstract viewpoint.

Despite his lofty standing in the mathematical world, Cartan was very responsive to his students. Cartan held his regular office hours on Thursday afternoons, and at first, Chern would line up outside Cartan's office with all the others waiting to see him. After a while, however, Cartan invited Chern to meet with him at his home, and from that time on, they met about once every two weeks at Cartan's apartment (by coincidence, they lived near one another on Boulevard Jourdan). Within a few days of each meeting, Chern would receive a letter from Cartan in which he would discuss further thoughts on ideas and problems that had come up at their meeting.

In the summer of 1937, Chern returned to China to become Professor of Mathematics at Tsing Hua University. He traveled by way of the United States, and in August, embarked on a Canadian boat bound from Vancouver to Shanghai. But while he was on board, the Sino-Japanese war broke out and, as a result, the boat did not stop in Shanghai; it went on to Hongkong instead. Chern never reached Peking on that trip. Because of the war, Tsing Hua University, Peking University, and Nankai

Chern and his family in 1955.

Hail to Chern, a musical tribute, was composed in honor of
Chern and performed as part of the Chern Symposium.

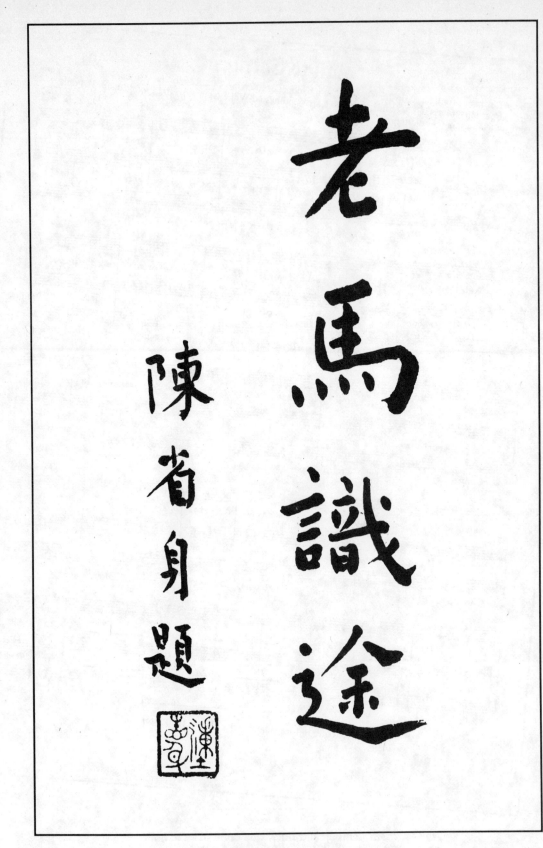

Calligraphy by Chern

Mathematical People

University moved to Changsha and combined to form a new, temporary university. After disembarking in Hongkong, Chern went directly to Changsha. He stayed there for about two months, but the Japanese were getting closer and work conditions became impossible. So in early 1938, the combined university moved to Kunming (where it was named Southwest Associated University) and Chern remained there until 1943. It was during these years that Chern married Shih-Ning Cheng, and his son, Paul, was born. Mathematically, it was a period of isolation, but he had good students and taught courses on advanced topics such as conformal differential geometry and Lie groups.

By 1943, U.S. military aid began to come to China, and the intensity of the Japanese air raids had diminished. That year, Chern received an invitation to come to the Institute for Advanced Study in Princeton; both Oswald Veblen and Hermann Weyl were impressed with his work. Chern accepted, and traveled from Kunming to Princeton by American air transport via Calcutta, Karachi, Aden, Egyptian Sudan, Accra, Ascension Island, Natal, and Miami. The entire trip took about one week. Chern spent the years 1943–45 at the Institute. It was a highly productive period for him; during this time, he completed his famous intrinsic proof of the *Gauss-Bonnet* formula and his work on characteristic classes. He developed close relationships with Hermann Weyl and André Weil during this visit to the Institute. Chern and Weyl spent many hours discussing a wide range of mathematical topics. Chern feels that Weyl had a remarkable ability to predict the future of mathematics; he foresaw the great development in algebraic geometry which has since taken place. Chern also came to know Solomon Lefschetz very well. Under Lefschetz's editorship, Chern served as an associate editor of the *Annals of Mathematics*.

When the war ended in 1945, Chern decided to return to China. After leaving Princeton at the end of 1945, he visited the Lefschetz Institute of Topology in Mexico City (an invitation had been extended at the suggestion of Lefschetz). From there, he traveled to Chicago, and then to San Francisco where he had to wait for more than two months for a boat to Shanghai. The boat, a troop transport, reached Shanghai in April 1946 after about a month at sea. Upon his arrival, he was asked by the Academia Sinica to organize a Mathematics Institute. The Institute was first installed in Shanghai and later moved to Nanking, the capital of the Nationalist Government. Chern decided that the most effective way to start the Institute was to run it like a graduate school. So he invited fresh college graduates and lectured to them on algebraic topology, sometimes for as much as 12 hours a week. Altogether, there were only 20 people at the Institute, and Chern was the sole senior member. This very talented group later developed into distinguished mathematicians.

This period with Academia Sinica lasted about two years. In the fall of 1948, the Civil War in China was approaching Nanking, and Chern was once again invited to the Institute for Advanced Study in Princeton by Veblen and Weyl. Chern accepted the invitation, and at the end of 1948, he and his wife and two children, Paul, and ten-month old daughter, May, left for the United States. In the meantime, he received an appointment as a full professor at the University of Chicago. The appointment was not to take effect until the summer of 1949; so Chern spent the winter term of 1949 at the Institute for Advanced Study and joined the faculty of the University of Chicago in July of 1949. Coincidentally, Chern was the successor to E. P. Lane at Chicago; he took Lane's geometry Chair (as mentioned above, Chern's own teacher, Dan Sun, had been a student of Lane's). Chern remained at the University of Chicago until 1960 when he accepted an appointment at the University of California at Berkeley, with which he has been associated ever since.

Chern is currently working full time as Professor Emeritus at the University of California, where he still teaches, and as Director of the Mathematical Sciences Research Institute in Berkeley. He continues to do research and is integrating research with his administrative duties. Quite a few of the younger members of this Institute have research interests similar to his, and Chern has done work with some of them.

In addition to his work in Berkeley, Chern travels regularly to China, where he has an appointment at Beijing University. He has been back to China seven times since 1972, and prior to that time, he made numerous visits to Taiwan. When he travels to China, he generally gives lectures at various universities and has personal contact with many mathematicians throughout the country. Chern has also helped to organize annual conferences on differential geometry and differential equations in China. The first one was held in Beijing in 1980 and was attended by an American delegation of twelve mathematicians, together with those from other countries. During this past summer (1982), he attended the third DD-Conference, and helped to plan the next one. He also met with government officials in China in an effort to obtain financing for the next conference. Through Chern's insistence, students have been asked to participate in these conferences; the youngest participant at the third

Chern in Berkeley in 1979.

Conference was a very impressive 19-year old. Chern believes that the future looks extremely bright for Chinese mathematics, and that the academic setbacks caused by years of the country's self-imposed isolation from foreign influence are disappearing.

An account of Chern's distinguished career in mathematics would not be complete without further mention of his students, in whom he takes great pride. Chern's former students include Nobel laureate (1957) Chen-Ning Yang and Fields Medalist (1982) Shing-tung Yau. Yang was an undergraduate student of Chern's at Southwest Associated University in Kunming, and Yau was one of Chern's Ph.D. students at U.C. Berkeley; both gave lectures at the June 1979 symposium held on the Berkeley campus to honor Chern. Another of Chern's students was Wu Wen-tsun, who received first science prize in China in the 1950's. Wu was a member of the Institute of Mathematics, Academia Sinica, which Chern organized in Nanking.

Chern's former students are impressively distributed throughout the world. They include the following: Joseph Wolf and Alan Weinstein (Berkeley); A. Rodrigues, P. Simoes, M. do Carmo, and L. Barbosa (Brazil); H. Suzuki (Japan); N. Petridis (Greece). Former students now in remote parts of China indicate the further reaches of Chern's influence. These students include: Chen Chieh, Vice President of the University of Inner Mongolia; Chen Teh-Huang in Urumchi, Xinjiang, China's most northwestern province; Sun Ye-Fon, Kirin University in Changchun, Kirin, China's most northeastern province; and Chu Teh-Hsiang in Kunming.

Reflecting on the account of Chern's life, one is reminded of the age-old question: *Is man a product of history, or is history a product of man?* Chern's career has been indistinguishable from the remarkable growth of differential geometry within the past forty years. In the 1930's, most differential geometers were working on generalizations of general relativity; i.e., they were searching for a unified field theory. Nothing really came of these efforts. When Chern was working on differential geometry in the 1940's, this area of mathematics was at a low point. Global differential geometry was only beginning; even Morse Theory was understood and used by a very small number of people. Today, differential geometry is a major subject in mathematics, and a large share of the credit for this transformation goes to Professor Chern.

Finally, perhaps characteristic of his modesty, when Chern was asked about his schedule for the immediate future (at the end of the interview), he casually mentioned that in a few weeks (November 18) he was going to Switzerland to receive an honorary degree from Eidgenossische Technische Hochschule (ETH), the university where Einstein received his degree, and whose professors included Hermann Weyl and Heinz Hopf.

JOHN HORTON CONWAY

John 'Horned' (Horton) Conway

JOHN HORTON CONWAY

by Richard K. Guy

It's a pleasure to write an obituary while the subject still breathes: of the dead one should speak only good; the living have a chance to reply.

I met John Conway through my son. They were both members of my own Cambridge college, Gonville and Caius. When Mike Guy and his sister, now Anne Scott, became undergraduates in 1960, Conway was in his early years as a graduate student. It's well known that graduate students are a low form of life, well below that of (undergraduate) "scholar," for example. As a scholar, Mike had rooms in college for his three undergraduate years, but graduate students had to find their own "digs." So a useful symbiosis grew up: Conway shared Mike's room until he finished his graduate studies, became a Fellow and regained the privilege of a college room. By then Mike was a graduate student and glad to avail himself of Conway's room.

Tom O'Beirne, who then ran the "Puzzles and Paradoxes" column [14] in the *New Scientist* had visited our family around this time and introduced us to many intriguing things, including Piet Hein's Soma Cube (Figure 1). We found several solutions, but it was Conway and Mike Guy who first found all 240 solutions [1, pp. 802–803]. They did *not*, as stated in the blurb accompanying some commercially produced cubes, use a computer! Another early joint project, in which I believe a computer *may* have featured, was the enumeration of all four-dimensional Archimedean polytopes [3], including a new discovery, the Grand Antiprism.

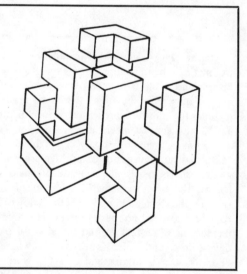

Figure 1.   Piet Hein's Soma Cube comprises the one non-convex piece made from three cubelets and the six non-convex pieces made from four cubelets. John Conway and Mike Guy, in 1961, enumerated the 240 essentially different ways in which these can be assembled to form a $3 \times 3 \times 3$ cube.

Thomas Alva Edison said that genius is one percent inspiration and ninety-nine percent perspiration. If Conway's genius is more than one percent inspiration, then it's because he adds up to more than one hundred percent! He does thousands of calculations, looks at thousands of special cases, until he exposes the hidden pattern and divines the underlying structure. He has drawn (or made out of strings of beads!) tens of thousands of knots, leading to two new ways of looking at their classification and enabling him to push this further than anyone else.

His discovery of the game of Life [1, chap. 25; 7] was effected only after the rejection of many patterns, triangular and hexagonal lattices as well as square ones, and of many other laws of birth and death, including the introduction of two and even three sexes. Acres of squared paper were covered, and he and his admiring entourage of graduate students shuffled poker chips, foreign coins, cowrie shells, Go stones or whatever came to hand, until there was a viable balance between life and death. In the final version (Figure 2), a live cell in a rectangular array survives if there are just 2 or 3 live neighbors (a chess king move away) and a dead cell comes to life if it has exactly 3 live neighbors. The implications of this simple set of rules exceed your wildest guesses: Life can simulate a Minsky machine, so Life is universal!

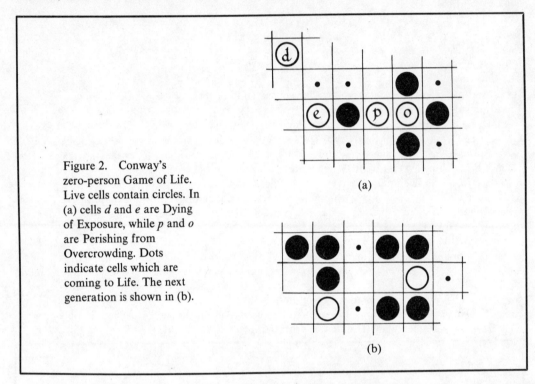

Figure 2. Conway's zero-person Game of Life. Live cells contain circles. In (a) cells d and e are Dying of Exposure, while p and o are Perishing from Overcrowding. Dots indicate cells which are coming to Life. The next generation is shown in (b).

(a)

(b)

Charles Darwin advocated conducting an occasional damned fool experiment, such as blowing a trumpet at a bed of tulips. To his more conventional colleagues, some of Conway's investigations of bizarre and exotic structures seem just about as likely to lead to significant results. By playing a myriad of often quite trivial games he gradually developed their theory to such an extent that it includes the most comprehensive theory of number that we now have [4]: The theories of Dedekind and Cantor are just special cases of the general scheme which includes infinitesimals as well.

Conway is incredibly untidy. The tables in his room at the Department of Pure Mathematics and Mathematical Statistics in Cambridge are heaped high with papers, books, unanswered letters, notes, models, charts, tables, diagrams, dead cups of coffee and an amazing assortment of bric-à-brac, which has overflowed most of the floor and all of the chairs, so that it is hard to take more than a pace or two into the room and impossible to sit down. If you can reach the blackboard there is a wide range of colored chalk, but no space to write. His room in college is in a similar state. In spite of his excellent memory he often fails to find the piece of paper with the important result that he discovered some days before, and which is recorded nowhere else. Even Conway came to see that this was not a desirable state of affairs, and he set to work designing and drawing plans for a device which might induce some order amongst the chaos. He was about to take his idea to someone to get it implemented, when he realized that just what he wanted was standing, empty, in the corner of his room. Conway had invented the filing cabinet!

Claude Elwood Shannon said that he would rather spend two or three days discovering a theorem than two or three hours searching for it in a library. Conway carries the Shannon philosophy to its extreme, often forced by his lack of system to rediscover his own results. With each rebirth, however, the product becomes more complete, more refined, more polished and more translucent. He is never content just to know a topic well. He constantly rearranges it, tries it in different settings, until he gets it into a form in which he can explain it, sometimes literally, to the person-in-the-street.

He has phenomenal powers of concentration. When his four daughters were small, he would often be doing intricate calculations with one or more of them climbing over him. He could either ignore them, or, more likely, integrate them into the process by explaining the simpler parts to them, or by interesting them in the patterns in which he arranged the work.

J. W. S. Cassels, speaking some years ago at a conference in Reading, honoring Richard Rado, described a difficulty he had when working on the Hasse principle for cubic diophantine equations, with Mike Guy. Cassels is diurnal in his habits, but Mike is nocturnal in his. The problem was solved by using Conway, who is irregular in his habits, as a go-between.

A typical Conway uniqueness is exemplified by his lingual calisthenics or tongue gymnastics, or whatever it is called. I forget the details, but the story is roughly this. He read somewhere that about one person in forty (say) was able to make his tongue into a particular shape; that one in forty of those was able to make a second shape (trefoil, rose, or whatever—when you meet him, get him to show you the shapes—he's not unduly shy); one in forty of those can make a third shape, and so on, there being about six shapes altogether, so that only about one person in the whole world can make them all. Well, you've guessed it—after a bit of experimenting in front of a mirror, Conway discovered that he was that person!

But most of his parlor tricks have been achieved only by dint of hard practice. He must have made many hundreds of frogs, peacocks and other objects by folding squares of paper. And how many times has he bent a wire coat hanger into a square, balanced a dime or threepenny bit on the end of the hook, whirled it round many times and finally returned it to the status quo, with the coin still balanced on the hook? He must have twisted a Rubik Cube a million times before most of us reached our first thousand. He was the inventor of "three looks," curing the cube by inspecting it carefully, then holding it under the table while making several moves, bringing it out and examining it a second time, then holding it under the table again for more turns, then taking a third look before finally twiddling it under the table into a perfectly cured cube!

Another illustration of Conway's aura of improbability came when he was contending that when betting on an unlikely event, the exact odds didn't matter. He offered ten shillings against a penny (120 to 1 in those days) against 7 coins all coming up the same (7 heads or 7 tails, 63 to 1 against). Someone was willing to lose a penny, and I got the job of shaking 7 coins. They came out as some mixture, Conway collected his penny and asked if the person wanted to play again. Yes, he did. I threw the coins again; they were all tails!

He has probably supplied Martin Gardner with more material for his *Scientific American* column [8] than anyone else, and *Mathematical Carnival* [9] is dedicated "To John Horton Conway, whose continuing contributions to recreational mathematics are unique in their combination of depth, elegance and humor."

He has made models of dozens of polyhedra, devised and manufactured a computer using ball bearings rolling in grooves, produced lunar calendars, knitted projective planes and twisted wire and string into exasperating topological puzzles. A more recent invention is a remarkable prime producing machine [5, 13] consisting of a row of fourteen rational numbers (Figure 3).

Figure 3. Conway's Prime Producing Machine. **Input** is the whole number 2. A **step** is to multiply by the *earliest* of the fourteen fractions which gives a whole number product. **Output** is the exponent whenever a pure power of 2 occurs.

$$\frac{17}{91} \quad \frac{78}{85} \quad \frac{19}{51} \quad \frac{23}{38} \quad \frac{29}{33} \quad \frac{77}{29} \quad \frac{95}{23} \quad \frac{77}{19} \quad \frac{1}{17} \quad \frac{11}{13} \quad \frac{13}{11} \quad \frac{15}{14} \quad \frac{15}{2} \quad \frac{55}{1}$$

At one time he would be making constant appeals to give him a year, and he would immediately respond with the date of Easter, or to give him a date, so that he could tell you the day of the week or the age of the moon [1, pp. 795–800]. Here are some verses which are essentially Conway's, but true to the oral tradition they've been transcribed by Alf van der Poorten and modified by the present writer.

The Doomsday Algorithm

Months:

It's the last day of Jan. or of Feb. that will do
(Except that in Leap Years it's Jan. thirty-two).
Then in the even months use the month's day
And for odd ones add four, or else take it away,
According to length, or simply remember:
You only subtract for Septem. or November.

Years:

Now to work out your Doomsdays the orthodox way,
Three things you must add to the century day:
Dozens, remainder, and fours in the latter
(If you alter by sevens, of course it won't matter).

Centuries:

In Julian times, lackaday, lackaday,
Zero was Sunday; each hundred went back a day.
But Gregorian four hundreds are always a Tues.
And centuries extra each take us back twos.

Not everything is Conway's unaided work, of course. Michael Stewart Paterson cooperated in the creation of the game of Sprouts [1, pp. 564–569; 6; 16], which continues to defy analysis. Start with some spots (Figure 4). A move is to join two spots, or a spot to itself, by a curve which doesn't meet any previously drawn curve or spot, and to place a new spot on the curve. No spot may have more than three parts of curves ending at it. If you can't move you lose. But a typical Conway twist is Brussels Sprouts, played the same way, but starting with crosses instead of spots (Figure 5), so that up to four curves may meet at a cross. After each curve is drawn, a new crossbar is made somewhere along its length.

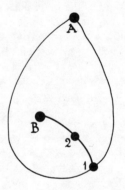

Figure 4. After two moves in a Game of Sprouts. Play started with two spots, *A* and *B*. The first player joined spot *A* to itself and added spot 1. The second player joined 1 to *B* and added spot 2. Spot 1 has now lost its three lives and cannot be used any more.

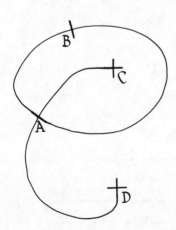

Figure 5. After two moves in a Game of Brussels Sprouts. Play started with two crosses, *C* and *D*. Player *A* joined the west arm of *C* to the South arm of *D* and added a crossbar *A*. Player *B* then joined the two available arms of *A* and added a bar *B*. It's safe to predict that the game will last for another six moves.

Again, it was Roger Penrose who was responsible for the Penrose Pieces [12, 15], the kite and the dart (Conway's names, Figure 6), which were a breakthrough in providing nonperiodic tilings of the plane (Figure 7), but it was Conway who discovered many of their remarkable properties and did much to bring them to the attention of a wide audience [10]. Here the last word is perhaps that of de Bruijn [2].

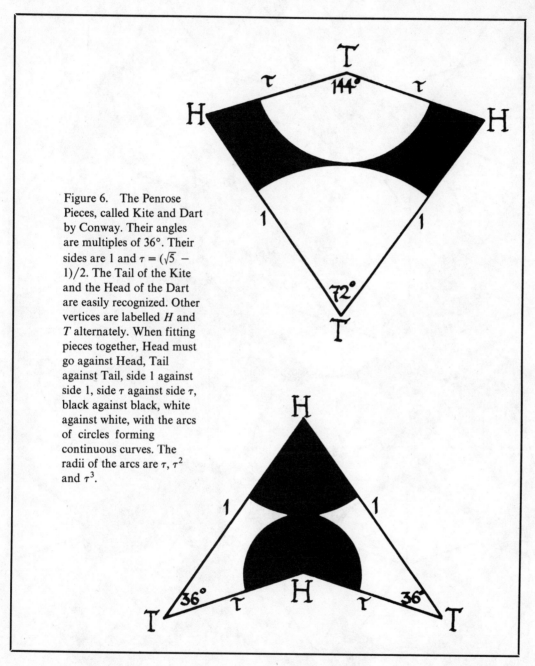

Figure 6. The Penrose Pieces, called Kite and Dart by Conway. Their angles are multiples of 36°. Their sides are 1 and $\tau = (\sqrt{5} - 1)/2$. The Tail of the Kite and the Head of the Dart are easily recognized. Other vertices are labelled H and T alternately. When fitting pieces together, Head must go against Head, Tail against Tail, side 1 against side 1, side τ against side τ, black against black, white against white, with the arcs of circles forming continuous curves. The radii of the arcs are τ, τ^2 and τ^3.

Conway has always been fascinated with language and the *mot juste*. I've spent many hours with him searching through Roget's *Thesaurus* and dictionaries. He's always eager to invent, adapt and pervert in order to achieve the *double entendre* or worse. His extensions of "iff," for example: onnce (once and only once), onne (one and only one) twwo (or was it twoo?), .whenn, and so on [an amusing converse appeared from Germany at the 1981 Banff Symposium on Ordered Sets: a T-shirt with BANF AND ONLY BANF]. His cries against the common misuse of "unique": biunique, triunique, . . . , culminating in the almost meaningless "polyunique"! His names for objects in Life, and his

Figure 7. Part of a tiling with Penrose Pieces. Heads and Tails must not occur at the same vertex.

names of games (star, up, tiny two, all small, ace, deuce, superstar, Col, Snort, ono, oof, sunny, loony, loopy, under, dud, upon, hi, lo, hot, sesqui-up, tis, tisn, . . .), words used in his game of Philosopher's Football (Phutball, shot, pot, tackle, poultry, poison) [1, pp. 688–691]. This is another of his excellent games for two players which can be played on a Go board, with a black Go stone (the ball) and a plentiful supply of white ones (the players). The ball starts at the centre intersection of the board (or nearer to one goal to make a fairer game between unequal players), and a **move** is *either* to place a new player at any unoccupied intersection or to **jump** the ball over one or more players, immediately removing the players jumped over, in any of the eight standard directions, onto the first empty point in that direction (Figure 8). A single move may comprise several consecutive jumps. The object of the game is to jump the ball onto or over your opponent's goal line (edge of the board). The ball may land on a goal line during the course of a move without the game's ending, provided it jumps off the line by the end of the move.

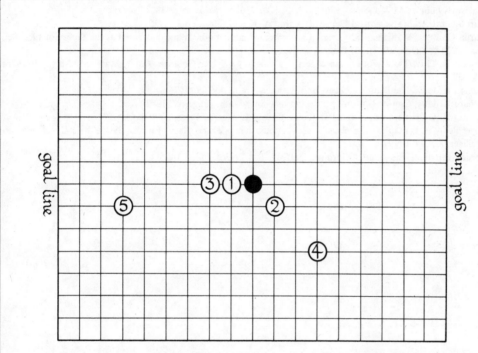

Figure 8. A 15 × 19 Phutball Pitch with the first five moves of a game. The (Black) Ball may now jump over 2 & 4, or over 1 & 3. Remove the (White) Players as they are jumped over.

Another of Conway's games is Sylver Coinage. The two players alternately name different positive integers, but are not allowed to name any number that is a sum of previously named ones, using repetitions if necessary. The player who names 1 is the loser. This seemingly simple game generates many unsolved problems, some of which are probably quite deep [1, chap. 18; 11].

Has all this foolery anything to do with mathematics? Indeed it has! Indeed it *is* mathematics! With Conway it's impossible to draw the line between the trivial leg-pulls and the deep mathematical results. He's made significant contributions to the theory of numbers, the theory of knots, the theory of quadratic forms, the theory of groups and the theory of games. Otherwise he would not have been elected to a Fellowship of that most prestigious scientific body, The Royal Society.

REFERENCES

1. Elwyn R. Berlekamp, John H. Conway and Richard K. Guy, Winning Ways for your Mathematical Plays, Academic Press, London, 1982.
2. N. G. de Bruijn, Sequences of zeros and ones generated by special production rules, Proc. Nederl. Akad.

Wetensch., 84 (= Indag. Math. 43) (1981)27–37; Algebraic theory of Penrose's non-periodic tilings of the plane I, II, *ibid*. 39–52, 53–66.

3. J. H. Conway, Four-dimensional Archimedean polytopes, Proc. Colloq. Convexity, Copenhagen, 1965 (1967)38–39.

4. J. H. Conway, On Numbers and Games, Academic Press, London, 1976.

5. J. H. Conway, Problem 2.4, Math. Intelligencer, 3 (1980)45.

6. Martin Gardner, Mathematical games: of sprouts and Brussels sprouts: games with a topological flavor, Sci. Amer., 217 #1 (July 1967)112–115.

7. Martin Gardner, Mathematical games: The fantastic combinations of John Conway's new solitaire game "life," Sci. Amer., 223 #4 (Oct. 1970)120–123; and see #5 (Nov.)118; #6 (Dec.)114; 224 #1 (Jan. 1971)108; #2 (Feb.)112–117; #3 (Mar.)108–109; #4 (Apr.)116–117; 225 #5 (Nov. 1971)120–121; 226 #1 (Jan. 1972)107; 233 #6 (Dec. 1975)119.

8. Martin Gardner, Mathematical games, Sci. Amer., 215 #3 (Sep. 1966)264 (Mrs. Perkins's Quilt); 221 #1 (Jul. 1969)119 and 227 #3 (Sep. 1972)176 (Soma solutions); 226 #1 (Jan. 1972)104–107 (Hackenbush); 232 #2 (Feb. 1975)98–102 (Conway number system); 232 #6 (June 1975)109 (trominoes); 234 #2 (Feb. 1976) (Conway's cube); 235 #3 (Sep. 1976)206 (ONAG); 236 #5 (May 1977)132–134.

9. Martin Gardner, Mathematical Carnival, Alfred A. Knopf, New York, 1975.

10. Martin Gardner, Mathematical games: Extraordinary nonperiodic tiling that enriches the theory of tiles, Sci. Amer., 236 #1 (Jan. 1977)110–121.

11. Richard K. Guy, Twenty questions concerning Conway's Sylver Coinage, Amer. Math. Monthly, 83 (1976) 634–637.

12. Richard K. Guy, The Penrose Pieces, Bull. London Math. Soc., 8 (1976)9–10.

13. Richard K. Guy, Conway's prime producing machine, Math. Mag., 55 (1982).

14. Thomas H. O'Beirne, Puzzles and Paradoxes, Oxford University Press, New York, 1965.

15. Roger Penrose, Pentaplexity, a class of non-periodic tilings of the plane, Eureka, 39 (1978)16–22; The Math Intelligencer, 2 #1 (1979)32–37.

16. Gordon Pritchett, The game of Sprouts, TYCMJ, 7 #4 (Dec. 1976)21–25.

H. S. M. COXETER

H. S. M. COXETER

Interviewed by Dave Logothetti

Although he has spent the last 44 years at Toronto, Coxeter is a child of England. His name comes from "cock setter"—one who sets cocks in the cockfights that have off and on entertained British rustics. (This is particularly ironic, as Coxeter himself is a pacifist and vegetarian who quietly frowns upon violence and killing.) His boyhood home was a country house in Surrey dating back to the 17th Century. The reluctant proprietor of Coxeter and Son Limited (the "Son" referring to H. S. M.'s grandfather), manufacturers of surgical instruments and compressed gases, his father retired at 50 and sculpted until his death by drowning at 58. One of his father's works, a portrait in bronze of Coxeter himself as a youth, now graces H. S. M.'s piano. Coxeter's mother was also an artist, a professional specializing in portraits and English landscapes.

Coxeter's own creativity ultimately manifested itself in geometry. He is the author or co-author of about 140 articles in periodicals and eleven books: *Introduction to Geometry*, *Projective Geometry*, *The Real Projective Plane*, *Non-Euclidean Geometry*, *Twelve Geometric Essays*, *Regular Polytopes*, *The Fifty-Nine Icosahedra*, *Geometry Revisited*, *Generators and Relations for Discrete Groups*, *Mathematical Recreations and Essays*, and *Regular Complex Polytopes*. His books have been translated into eight languages: French, German, Hungarian, Japanese, Polish, Russian, Serbo-Croatian and Spanish. His vision helped launch the Mathematical Expositions series of the University of Toronto Press, as well as the Canadian Journal of Mathematics, on which he labored as Editor-in-Chief for the first nine years. He also served as President of the International Congress of Mathematicians at Vancouver in 1974. Just last May on his 72nd birthday, he was honored with an international geometry conference at Toronto. Several of the participants at this Toronto conference were former students. Among the names that distinguish Coxeter's students, undergraduate and graduate, are John Coleman, Irving Kaplansky, Nathan Mendelsohn and William Moser. In addition to spreading the geometrical light at Toronto, Coxeter has been a visiting professor at Notre Dame, Columbia, Dartmouth, Florida Atlantic, Amsterdam, Edinburgh, East Anglia, Sussex, Warwick, Utrecht, Bologna and Australian National Universities. He holds honorary doctorates from the Universities of Alberta, Arcadia, Trent, Toronto and Waterloo.

All Those Initials

MP: *Let's get down to brass tacks; how come you have so many initials, and what do they stand for?*

Coxeter: Yes, that was a story. They stand for "Harold Scott MacDonald." Originally my birth certificate said "MacDonald Scott Coxeter"; that's how I came to be known as "Donald" by all my friends and relations. And then some stupid godparent said, "Well, you should have Harold as one of your names because your father's name was Harold." So they stuck the Harold on at the beginning; they made it "Harold MacDonald Scott." However, somebody saw that this was going to be idiotic, because the H. M. S. stands for "His Majesty's Ship" or "Her Majesty's Ship." So they just made a transposition to alter that, put the S before the M, and it became H. S. M.

Early Interest in Mathematics

MP: *How did you first become interested in mathematics in general and in geometry in particular?*

Coxeter: Well now, let me think. I suppose from school I found certain things rather fascinating. I enjoyed learning Euclid, and somebody must have given me a geometry book that had pictures of the regular solids.

MP: *How old were you when you first encountered the regular solids?*

Coxeter: I suppose about twelve or thirteen.

MP: *How un-American! I mean that doesn't happen in the United States.*

Coxeter: No. And somebody lent me an old book called *The Fourth Dimension*, by Howard Hinton, which I suppose one can still see in libraries.

MP: *Which attracted you first, numerical or geometrical mathematics?*

Coxeter: I was always interested in numbers. My mother used to say that when I was very, very young I used to look at the stock market reports in the papers because I was interested in just looking at numbers.

MP: *At what age would that be?*

Coxeter: Oh, about two or three, I suppose.

MP: *That young?—Just you and Gauss!*

Coxeter: I was always fascinated with numbers. And then I became interested in shapes as well—pyramids, cones and spheres; all those things appealed to me at an extremely early age.

MP: *When did you decide that mathematics was going to be your career, rather than just a source of fascination?*

Coxeter: I suppose at about the age of fifteen or sixteen. Before that I thought I might make a career of music. I was interested in composing music.

MP: *Was it hard for you to make a decision at the age of fifteen or sixteen?*

Coxeter: My mother took me to visit some musicians (Gustav Holst and C. V. Stanford). They looked at my stuff and were not too impressed. They said, "Educate him first." They wanted me to concentrate on music. Gradually the inner urge to do something creative transferred itself from music to mathematics.

Adventures at Trinity College, Cambridge

MP: *Having decided on mathematics over music, you had to learn the mathematics prerequisite to entering Cambridge. How did you do it?*

Coxeter: I had to leave school rather early in order to learn enough mathematics to get a scholarship to Cambridge. So I was a paying guest in a house in the same town where there was a famous school (Marlborough) where boys were trained to go to college. And it was while I was there that I had special coaching in German, along with much more intense special coaching in mathematics from teachers in the school. I didn't attend the school properly because it wasn't allowed. They have very strict rules about those big schools in England; if you're over 16 you can't enter. So I was living in the town and went on my bicycle every day to the school for a private session with the mathematics teacher, Alan Robson. Every day he would give me homework to do, and I would bring it along the next day and continue.

MP: *This sounds like an extraordinary circumstance.*

Coxeter: It was. It was quite unusual. But it was the only thing to do because I hadn't learned enough routine mathematics. I was already doing these things on polytopes which later became the climax of Chapter 11 of my Regular Polytopes; but all the standard stuff in algebra, geometry, analysis and even applied mathematics, which I would have to do, I knew scarcely anything about.

MP: *How did you convince the academic powers that they should let you get this special tutoring?*

Coxeter: I suppose it was through the influence of several people. I was introduced to Bertrand Russell, and then he introduced me to E. H. Neville, who wrote about elliptic functions and Farey series. (It was he who persuaded Ramanujan to go to England.) These were sufficiently important people that they just decreed that I should leave the school where I had been, where I'd learned all the mathematics they could teach me, and get on to something better.

MP: *When were you at Trinity College, and in what rôles?*

Coxeter: I was there from 1926 to 1936, off and on, first as a scholar and then as a winner of a Smith's prize, which men who were undergraduates usually put in for. I won a fellowship at Trinity College—a four-year research fellowship. I was an undergraduate from 1926 to 1929, you see; next I got a Ph.D. under H. F. Baker, the great algebraic geometer, and then I became a fellow. I just went on writing about polytopes. I had two intervals going to Princeton from Cambridge and back again.

What is Geometry?

MP: *Let's turn to geometry: What is geometry? I suppose if you had ten years you could answer that.*

Coxeter: It's a very difficult question. You could say it's what geometers do, but then that's only passing the buck. I suppose it's the study of shapes and patterns.

MP: *I know of a professor who interprets geometry primarily as logic, as a study of axioms.*

Coxeter (quickly): No, I've never really wanted that kind of approach too much. I have been interested in developing axiom systems, but on the whole I prefer to plunge right into the subject, rather than to spend too much time on the foundations. I was interested to find that some of the systems of axioms that have been often used could be improved. I mean, lots of people who write about geometry still use the axioms of Hilbert, regardless of the fact that they have been very much improved by later writers.

MP: *What's one of Hilbert's axioms that's particularly weak?*

Coxeter: Well, I would say his treatment of "congruence." He had axioms concerning the congruence of segments and then some other axioms concerning the congruence of angles; later writers such as Moore and Veblen showed that it was quite unnecessary to have separate congruence axioms for angles because they could be deduced from the axioms for distances.

MP: *Where did Henry Forder fit in?*

Coxeter: Forder wrote a very excellent treatment of axiom systems, *The Foundations of Euclidean Geometry*. And then there's another book, called *The Calculus of Extension*, and also a little popular book simply called *Geometry* which is very good.

MP: *Did he work from Hilbert's axioms?*

Coxeter: No, he saw that they had been superseded, and he mentioned the better ones.

MP: *What is your opinion of Forder's axioms?*

Coxeter: Very good! He's still alive, by the way. We're going to celebrate his 90th birthday next year.

The Top Geometers

MP: *Here's another difficult question for you: If we restrict the mathematical world to geometry, who are the top geometers of all time?*

Coxeter: I suppose Archimedes, Apollonius, Gauss and Lobatchevsky.

MP: *Not Euclid?*

Coxeter: Euclid was only a compiler. I don't think he did very much original work. He compiled things that were known in his time. In more recent times I would name Veblen.

MP: *What would you say is Veblen's chief contribution?*

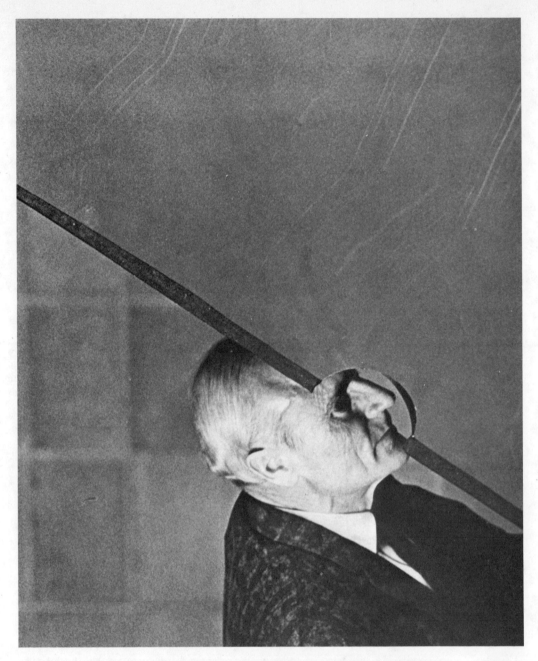

Coxeter in a geometry film studio in 1964, looking through the peephole in one edge of a big tetrahedral kaleidoscope.

Coxeter: I suppose the connections between geometry and topology. And the book he wrote with Young, called Projective Geometry, is a great classic; it is remarkable because it seems just as fresh today as when it was written in 1918.

MP: *Why was that such a valuable contribution to mathematical literature?*

Coxeter: Perhaps chiefly because it put everything so much more clearly and made it much more easy to understand.

MP: *That is a nice virtue. I sometimes get the feeling that many mathematicians would rather not make their work seem too clear; then you won't know what they're doing—or pretending to do.*

Coxeter: Well, that's right. Another name I must add is Von Staudt; I think he was very good. He wrote his big books in 1847 and 1857.

MP: *Von Staudt and many of those other names that you mention are connected with projective geometry. Does that have a special place in your heart?*

Coxeter: Oh, no; it doesn't really. I should also mention Schläfli; he was the man who first discovered regular polytopes.

MP: *Of all those names would you be prompted to pick out one that you thought was the very best geometer?*

Coxeter: No, I don't think I would want to say that, just as I wouldn't really say that any one musical composer is the best.

MP: *The set of geometers is not linearly ordered?*

Coxeter: No.

Elegant Examples

MP: *What are a couple of examples of pieces of geometry which you personally find particularly elegant and beautiful or ingenious?*

Coxeter: Well, I think Klein's enumeration of the finite groups of rotations is a very good example—exactly which are the groups of rotations in three dimensions: the cyclic, dihedral, tetrahedral, octahedral and icosahedral groups.

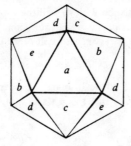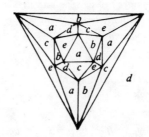

Permuted faces in tetrahedral group.

MP: *Do you have some little pieces of geometry that you just like to show other people, over which your mouth waters a little bit when you get an opportunity to show them?*

Coxeter: Well, I think so, yes. I think the Steiner-Lehmus problem is a good one: Any triangle having two equal internal angle bisectors (each measured from a vertex to the opposite side) is isosceles. There has been an enormous number of proofs of that, I should think over a hundred, but I think the one that is due to Forder is probably the best of all. You'll find it in my *Introduction to Geometry*.

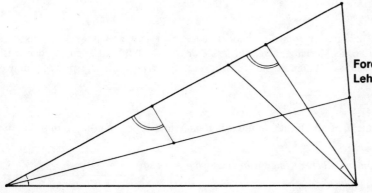

Forder's proof of the Steiner-Lehmus Theorem.

The Four Color Problem

MP: *What is your opinion of the recent Haken-Appel proof of the four color problem?—Is it really completely okay? Do you believe it? What do you think in general of proofs involving computers in this manner?*

Coxeter: I'm certainly happy to accept all of the computer proofs concerned with the theory of numbers, such as finding larger Mersenne primes, $2^p - 1$. I'm just a little bit worried that that seems to be something that could go on more and more; with just a little bit more computer time one could go a little further and get the next one, and so on. I think even more interesting is the other system with Fermat numbers, $2^{2^n} + 1$, and the conjecture that the ones that are already known to be prime are the only ones that *are* prime. Everything that's ever been looked at beyond $n = 4$ seems to be composite, and I think that the use of computers to show that those extremely large numbers are composite is really very striking—$2^{2^{73}} + 1$, for instance.

MP: *Back to the four color problem . . . .*

Coxeter (quickly, again): I don't have the same feeling about that. I have a feeling that that is an untidy kind of use of the computers, and the more you correspond with Haken and Appel, the more shaky you seem to be. They even sometimes seem to mention the word "probability." I've heard recent rumors that somebody has found a flaw, which may be devastating or may not. I don't feel quite happy about that.

MP: *Do you have hopes for a straight-forward proof?*

Coxeter: No, I don't think so. I think it's too difficult for that to be done. It has always seemed to me a different kind of theorem from all other kinds of theorems. When the computing proof has been checked by quite a number of people, and they're all satisfied with it, well then we'll have to accept it. But I think it's very unlikely that anyone can break that proof down into something that one would regard as an ordinary proof. So it's rather in a different category from all other theorems.

MP: *Is it conceivable that some new point of view might crack the problem open?*

Coxeter: I suppose it is. Other problems have seemed to be insolvable, and people have cracked them open, so I wouldn't like to say it couldn't be done. But I don't see anything like that on the horizon.

MP: *My own feeling is that if that's the best we can do with the four color problem, then just aesthetically I wouldn't want to mess with it.*

Coxeter: That's right. I have very much the same feeling. I think the other bits of information that have come from the people struggling with it have been useful—the problems of coloring on other kinds of surfaces, Ringel and Youngs' work, and so on. So I think the problem has stimulated a lot of good mathematics.

Significant Unsolved Problems

MP: *Will you play Hilbert for me? What are some of the most significant unsolved problems in geometry?*

Coxeter: Well, I suppose the classification of three-manifolds is one of the most interesting.

MP: *What is a "three-manifold?"*

Coxeter: If you think of a sphere in four-dimensional space, $x^2 + y^2 + z^2 + w^2 = 1$, for instance, and if you imagine just the surface of that sphere, not inside or outside, but *on* the sphere, then that's an instance of a three-dimensional manifold, just as the surface of an ordinary sphere is a two-dimensional manifold.

MP: *And what is the problem in classification?*

Coxeter: To find invariants to establish that another three-manifold is or is not homeomorphic to a three-sphere.

MP: *So this is a generalization of Pólya's interest in classifying polyhedra?*

Coxeter: Yes, I think one could say that.

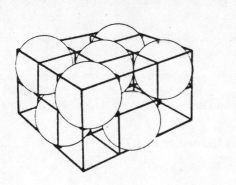 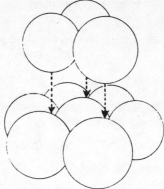

Cubic and hexagonal closepacking.

MP: *What is another significant problem in geometry that you would put on your list?*

Coxeter: There is the tiling problem: What are the possible ways in which you can take a convex polygon and repeat it so as to get congruent replicas of it which when fitted together fill and cover the plane? Even tiling with pentagons, which was supposed to be solved, still isn't. Nobody knows quite what are *all* the possible shapes of pentagon that can be repeated by congruent transformations to cover the plane. People seem to keep on thinking of new ones. Mrs. Doris Schattschneider has just written a very nice essay in praise of amateurs who have contributed to this problem.

And then you can have the same thing in space: Think of a convex polyhedron; can you repeat it to fill and cover the whole three-dimensional space? Of course, there are little things of which one has to be careful. You insist in the planar case that where two tiles meet they have either a vertex or a complete edge of both as their only place of meeting, rather than being staggered, like bricks on a wall. I think the interesting cases are where an edge is just the edge between two points, and it's an edge of both the tiles.

MP: *No vertices of order three?*

Coxeter: No, it's not just that; the tiling of regular hexagons has valence 3 for each vertex. The corresponding problem in three dimensions would be polyhedra—congruent polyhedra—put together with a whole face in common where two meet. Michael Goldberg has done a lot of that and is still writing papers on those. So one unsolved problem would be "Is there essentially only one kind of sixteen-faced polyhedron that's a space filler?" For quite a long time one has been known, and several people have rediscovered this sixteen-faced space filler. Goldberg gave me a reference to the original discovery which was quite a long way back—70 years ago or something of that sort. I have, however, heard rumors that you can have better than 16 faces.

MP: *Any other problems?*

Coxeter: Related to that is the old sphere packing problem, in which you have congruent balls and put them together in space as economically as possible. Are the cubic closepacking and the hexagonal closepacking the best, or could you possibly get anything better? That has been solved if you assume that the centers of the spheres are arranged in straight rows—in a "lattice packing." But if you don't insist on having a lattice, the question is, could you do better than 74% density?

MP: *I have a pedagogical question related to these problems: This is the kind of geometry that I learned only after I escaped from formal courses. Wouldn't it be healthy to have that kind of geometry more accessible to students as undergraduates?*

Coxeter: Exactly. It would be quite good, yes.

MP: *Where would you see that kind of geometry fitting into an undergraduate's program?*

Coxeter: Well, I suppose it's something that's generally known as convexity. I don't see any reason why we shouldn't have undergraduate courses on convex sets.

MP: *I don't either; I taught convex sets to high-school students, and had no particular problem. What are some especially good references—Yaglom and Boltyanskii?*

Coxeter: Yaglom and Boltyanskii would be quite a good one. There's a very good one in German by L. Fejes-Tóth. I think there are quite a number of other books in English, too. Fejes-Tóth himself wrote another very good book and translated it into English; it's called *Regular Figures*. There's quite a lot about convexity in that book.

Who Teaches Geometry Best?

MP: *Which countries are doing the best job of teaching geometry?*

Coxeter: I suppose Russia would be, and then Germany and Austria.

MP: *What kinds of things are they doing that makes their geometrical education superior?*

Coxeter: I think just regarding it as a subject that continues to be worth studying. The English-speaking countries seem to be interested in other things. They give geometry a low priority, which I think is a pity. Two of the Russian geometers whose books I have are Rosenfeld and Yaglom. I think they have done a lot to make things accessible. And then Alexander Alexandrov has written some books on polyhedra and convexity.

MP: *What about the United States and Canada?—What trends do you see? Anything heartening?*

Coxeter: Nothing very heartening, I'm afraid.

MP: *You don't think that maybe the pendulum is swinging back in the right direction?*

Coxeter: I don't see very much sign of it. I must admit my *Introduction to Geometry* seems to have a steady sale, which is rather encouraging to me.

MP: *That's also heartening to me. It seems to me that about twenty years ago I heard lots of people bragging about how they did mathematics without any pictures; the only good mathematics was that which was done without pictures. I do not hear that so much now, and I hear people admitting that in the privacy of their boudoirs they draw a sketch or look at a model. Maybe people are getting a little more interested in geometry.*

Coxeter (quietly): Yes.

MP: *Why is it that English-speaking people have abandoned geometry?*

Coxeter: I think because there was a tradition of dull teaching; perhaps too much emphasis on axiomatics went on for a long time. People thought that the only thing to do in geometry was to build a system of axioms and see how you would go from there. So children got bogged down in this formal stuff and didn't get a lively feel for this subject. That did a lot of harm. And you see if you have a subject badly taught, then the next generation will have the same thing, and so on in perpetuity.

MP: *How about a man like Honsberger, say? He seems to be quite popular now. I'd think that his work would get people interested in geometry.*

Coxeter: Yes, I think that he may do a lot of good. One of his problems appeals to me. It concerns lattice points: Can you draw a circle that has exactly 17, say, lattice points inside it? That, of course, is fairly easy, but if you ask, "Can you draw a circle that has exactly 17 lattice points on its circumference?" it becomes a little more tricky. And that's dealt with by Honsberger in a very nice way, in *Mathematical Gems I*, I think.

MP: *Pólya has favorites along that line, too—finding lattice points on ellipses. He comes with little drawings; he doesn't give me any writing, just a drawing and says, "Think about that."*

Coxeter: Yes, I remember he gave me a nice little problem of that sort: Suppose you are making a model of a polyhedron, and you cut out from cardboard a connected "net" consisting of a lot of polygons and fold the thing up. When you want two edges to be brought together, you have to stick them, and so you have a little tab put on one of the two edges which you then glue to the other. The

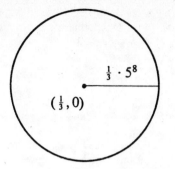

**Circle with exactly
17 = 2×8 + 1 lattice
points on it.**

$\frac{1}{3} \cdot 5^8$

$(\frac{1}{3}, 0)$

question is, for a given polyhedron how many tabs do you need? There's a very clever argument due to Pólya which says it should be just one less than the number of vertices.

Nice Geometry for Colleges

MP: *Okay, you say the prospects for geometrical education are currently rather gloomy. If yours were the hand that moved mathematical education, what geometry would you say should be included in the first two years of college? Right now the first two years for mathematics majors are preempted by calculus and maybe linear algebra. Would you do this differently?*

Coxeter: Certainly. I think that by being careful we could probably do the same amount of calculus and linear algebra in less time and have some time left over for nice geometry.

MP: *What topics would you include in nice geometry?*

Coxeter: I don't see why you couldn't have a revival of Euclid, really. If you don't worry too much about the axioms, I think many of the propositions of Euclid are very interesting.

MP: *What would be in that kind of course that wouldn't be in the standard ninth or tenth grade geometry course?*

Coxeter: Of course, if it's taken in high school first, so much the better. When the students come to college they should have some projective geometry. I had to do a lot of projective geometry before I got to Cambridge, in order to get a scholarship. We had to know many properties of conics. The scholarship candidates going to Oxford and Cambridge also had to know quite a lot about affine and homogeneous coordinates.

MP: *What has happened to the conics in the amalgamation of calculus and analytic geometry?*

Coxeter: I think they have been lost because everyone wants to approach them only by coordinates. An equation of the second degree is a conic; that is what they normally think of. But I think if you can introduce conics other ways, as geometric objects with interesting properties before even introducing coordinates, then students get more feel for them. I'm in favor of the introduction of geometric transformations at the freshman stage, and much earlier too, of course. School children can learn about rotations, reflections, translations and glide reflections; there's no reason why they shouldn't at a very early age. But then at a certain stage they should learn about the transformation called inversion, which is one of the topics I had in mind when I mentioned Euclidean geometry. The geometry of circles, coaxial systems, the fact that inversion preserves angles and transforms circles into circles is all rather fascinating stuff. And then reciprocation comes from inversion; you could define a conic as the figure that you get by reciprocating the circle with respect to a circle that's not concentric with the first circle.

MP: *Going beyond the first two years of college, for a mathematics major now, what geometry is especially desirable?*

Coxeter: To be a realist, you have the kind of geometry that has applications to other subjects. I suppose inversive geometry is that kind, because the inversive plane is the plane of complex numbers with the point at infinity adjoined, which gives you application to analysis straight away.

MP: *Should that geometry be taught as a course unto itself, or should it be taught as part of complex variables?*

Coxeter: I don't see any reason why it shouldn't be geometry as a course. Part of the course could be devoted to inversive geometry and the application to complex numbers. This could serve to illustrate the Erlangen Program, if you like—the group generated by all inversions is an interesting group, and the geometry of that group is quite a rich subject.

MP: *What do you feel about differential geometry? Is it really geometry, or is it really calculus merely using geometrical terms?*

Coxeter: I think it should be included alongside the others. I think that perhaps one should devote more or less equal time to differential geometry and pure geometry. You will see in my *Introduction to Geometry* that I devote one of the four parts to differential geometry. The trouble is that in many places nowadays differential geometry is the only kind the students ever have.

MP: *Do you think that it's really worth vigorously fighting for a separate lower division course, say in inversive geometry?*

Coxeter: I would like to see it, yes. It doesn't have to be called "Inversive Geometry;" it could be called simply "Geometry" and introduce inversive part of the time, projective another part of the time, and a little bit of non-Euclidean. I think that somehow you should fit in at least two courses in geometry, in say the second and third years.

MP: *What are some favorite examples wherein geometry provided crucial insights without which advances in other branches of mathematics would not have taken place?*

Coxeter: Well, I suppose the most famous case is the fundamental theorem of algebra. To prove that every equation has a root would be almost impossible without a geometrical background in the complex plane. Or there's the problem, "For which values of p, q and r will the relations $A^p = B^q = C^r = ABC = I$ give us a finite group?"

MP: *And what kind of geometry comes to the rescue?*

Coxeter: The geometry of rotations of a sphere into itself.

Is Geometry Dead?

MP: *If I or my colleague Jean Pedersen start rhapsodizing about geometry, the reaction that we frequently get is, "Oh well, that's a dead subject; everything is known." What is your reaction to that reaction?*

Coxeter: Oh, I think geometry is developing as fast as any other kind of mathematics. It's just that people are not looking at it.

Coxeter exhuming geometry.

REFERENCES

1. K. Appel and W. Haken, Every planar map is four colorable, Bulletin of the Amer. Math. Soc., 82, 5(September 1976) 711–712.
2. H. F. Baker, Principles of Geometry (4 vol.), Cambridge Univ. Press, London, 1925.
3. W. W. Rouse Ball and H. S. M. Coxeter, Mathematical Recreations and Essays (12th ed.), Univ. of Toronto Press, Toronto, 1974.
4. H. S. M. Coxeter, The Real Projective Plane, Cambridge Univ. Press, London, 1961.
5. ———, Non-Euclidean Geometry (5th ed.), Univ. of Toronto Press, Toronto, 1968.
6. ———, Twelve Geometric Essays, Southern Illinois Press, Carbondale, 1968.
7. ———, Introduction to Geometry (2nd ed.), Wiley, New York, 1964.
8. ———, Regular Polytopes (3rd ed.), Dover, New York, 1973.
9. ———, Regular Complex Polytopes, Cambridge Univ. Press, London, 1973.
10. ———, Projective Geometry (2nd ed), Univ. of Toronto Press, Toronto, 1974.
11. H. S. M. Coxeter, P. Du Val, H. T. Flather, and J. E. Petrie, The Fifty-Nine Icosahedra, Univ. of Toronto Press, Toronto, 1938.
12. H. S. M. Coxeter and S. L. Greitzer, Geometry Revisited, Math. Assoc. of America, Washington D.C., 1967.
13. H. S. M. Coxeter and W. O. J. Moser, Generators and Relations for Discrete Groups, Springer, Berlin.
14. H. G. Forder, The Foundations of Geometry, Dover, New York, 1958.
15. ———, The Calculus of Extension, Chelsea, New York, 1960.
16. ———, Geometry (2nd ed.), Hutchinson's University Library, London, 1960.
17. R. Honsberger, Mathematical Gems I, Math. Assoc. of America, Washington, D.C., 1973.
18. F. Klein, Lectures on the Icosahedron (2nd ed.), Kegan Paul, London, 1913.
19. E. H. Neville, Jacobian Elliptic Functions, Clarendon Press, Oxford, 1944.
20. G. Ringel and J. W. T. Youngs, Solutions of the Heawood map-coloring problem, Proceedings of the National Academy of Sciences of the U.S.A., vol. 60(1968) 438–445.
21. O. Veblen and J. W. Young, Projective Geometry (2 vols.), Ginn, Boston, 1910 & 1918.
22. I. M. Yaglom and V. G. Boltyanskii, Convex Figures, Holt, Rinehart and Winston, New York, 1961.

PERSI DIACONIS

PERSI DIACONIS

Interviewed by Donald J. Albers

Persi Diaconis is only thirty-eight and yet in the midst of a second career at Stanford University, where he is a professor of statistics. His work in statistics is so good that he recently was named a recipient of a MacArthur Foundation Fellowship. As a MacArthur Fellow, Diaconis will receive $192,000 over the five-year period 1982–1987, tax-free and no strings attached. The purpose of the awards, for which applications are neither solicited nor accepted, is to free creative people from economic pressures so they can do work that interests them.

In spite of his mathematical achievements, Diaconis insists that he is better at magic, his first career, than he is at statistics. At fourteen he left his home in New York City to wander the world as a professional magician. After ten years on the road, he decided to try college. At twenty-four, he enrolled as a freshman. Five years later he had earned his Ph.D. from Harvard.

Diaconis applies mathematics to a wide range of real-world problems, claiming that "I can't relate to mathematics abstractly. I need to have a real problem in order to think about it."

His background in magic and statistics has also proven useful in exposing several psychics, including Uri Geller.

Professor Diaconis was interviewed in his Stanford office in April of 1983. Diaconis' love of magic is underscored by one of his closing remarks: "If I could have had a professorship in magic, and if the world recognized magic the way it does mathematics, I probably would be doing magic full-time and never would have done mathematics or statistics."

A Magical Beginning

MP: *After graduating from high school at the age of* 14, *you left home and spent the next ten years on the road practicing magic. What made you do that?*

Diaconis: That's simple. The greatest magician in the United States is a man named Dai Vernon. He called me up one day and said, "How would you like to go on the road with me?" I said, "Great," and he said, "Meet me at the West Side Highway two days from now at two o'clock." So with what money I could pick up and one suitcase, I went on the road. It was simply a question of a magnetic, brilliant expert in the field calling on me, just as a guru calls on a disciple. I was quite honored and excited to do it.

MP: *What did your parents say to your leaving home to practice magic?*

Diaconis: I didn't ask them. I just left home. My parents were upset at my leaving, but somehow they found out that I was okay. For a long time I was the black sheep of the family. Only when I started graduate school at Harvard did my family begin to think that I wasn't terrible.

MP: *So they felt very bad about your going off to practice magic.*

Diaconis: Sure they did. I was being groomed to be a virtuoso musician. I went to Julliard from the ages of 5 to 14. After school and on weekends I played the violin. All of my family members [mother, father, sister, and brother] are professional musicians. They thought I was going to become a violinist and having me desert music for magic was not very appealing to them. I think they have come to accept it all now. They never came to accept the magic, even though I was good at it. I was better at magic than I am at what I do now [statistics].

MP: *When and how did you get into magic?*

Diaconis: When I was five years old I found the book *400 Tricks You Can Do* by Howard Thurston. I picked it up and figured out that I could do a few tricks. I soon did a little magic show at my mother's day camp. I clearly remember that show. I was the center of attention. I wasn't horrible apparently, and magic became a hobby. I sent in my dimes for mail-order catalogs on magic, and for my birthday I would ask for tricks as presents. When I got to public school I met other kids who were magicians and I joined the Magic Club. I threw myself into it with a real fury. All the energy that I didn't put into doing homework or anything else connected with school I put into magic. On many days I would cut school and hang out at the magic store until closing time.

MP: *Who would assemble at the magic store?*

Diaconis: Older magicians and other kids who were interested in magic. In New York City, there was a big, lively magic community. When I was 12, I met Martin Gardner at the cafeteria where magicians used to hang out. He was the kindest, nicest man, and he took time out to show me some lovely, little tricks that I could do. (Gardner, in addition to being a great writer, also is an accomplished magician.) He saw that I was a troubled kid and took a liking to me. He told me to call him if I had any questions. So I used to call him and talk about magic, and he got me interested in working on mathematical tricks because he would warm to that. He was a famous guy, and the things he would warm to were the things I got interested in.

MP: *Did you know that Martin Gardner was a big name?*

Diaconis: Sure. I knew whom the other magicians respected, who was famous and who was not so famous. He was obviously a very special guy, the kind of guy who could go on and on about things and remain interesting and never be pompous, just kind and instructive. He also was genuinely delighted if I showed him a new twist on a trick that he might know. He didn't try to put someone down because it was a trivial twist on something. When I showed him a new little idea, he would make a note of it. Every once in a while he would put something of mine into his "Mathematical Games" column and that was a great thing for me. I found it very inspiring to be recognized and have my name in *Scientific American*.

On the Road

MP: *You went on the road at age 14. What were those years like?*

Diaconis: During the first few years I was in very good company. I was being shepherded around by Dai Vernon, a brilliant man, who is the magician's magician and the best inventor of subtle sleight of hand magic of the century. He was roughly four times my age and he would take me around to hold the curtain. He taught me magic: we talked magic morning, noon, and night. Since he was sort of old, and since I could do the sleight of hand very well, when he would give magic lessons, he would have me demonstrate tricks, and then he would explain them. So my experience was vaguely structured and very colorful—a lot more colorful than I choose to put into any interview. I met all kinds of interesting street people, was often broke, hitchhiked, and so forth. I left Vernon when I was about 16 and was on my own. He went on to Hollywood to found what is now known as the Magic Castle, which is a fabulous magic club, a private, wonderful magic place where movie stars hang out. I decided I didn't want to do that and would stay on my own. So I stayed in Chicago, lived in a theatrical hotel, and played club dates, usually for $50 a night. I did pretty well that way. I eventually drifted back to New York, doing magic and pursuing it as an academic discipline, inventing tricks, giving lessons, and collecting old books on magic, which I still do. It was just my life. I did it with all my energy.

MP: *Magic very often has card tricks associated with it and perhaps card playing. Were you playing cards at the same time?*

Diaconis: No, not at the beginning. Much later somehow I got a copy of Feller's book on probability, and I got interested in probability that way.

MP: *How did that happen?*

Diaconis: It was due to another friend of mine, Charles Radin, who is a mathematical physicist at the University of Texas. He was in college on the straight and narrow while I was still doing magic. We had been kids together in school. One day he went to Barnes and Noble Bookstore to buy a book

The business card of the professional magician Persi Warren (Diaconis), who left home at age fourteen and performed professionally for the next ten years.

and I went along for the ride. He said Feller was the best, most interesting book on probability, and I started to look at it. It looked as if it was filled with real-world problems and interesting insights, and so I said, "I'm going to buy it." He said, "You won't be able to read it," I said, "Oh, I can do anything like that." Well, in fact, I couldn't; I tried pretty hard to read Volume 1 of Feller, and it's one of the big reasons I went to college, for I realized that I needed some tools in order to read it.

MP: *Was this at City College?*

Diaconis: Yes, I started at City College at night. They wouldn't take me during the day because I was something of a strange person, so I went for a couple of years at night taking one or two courses. I discovered that I liked college, and I decided to try for a degree. I finished up in two and a half years. It was a short time after I started college that I dropped magic as a vocation. I went from City College to Harvard.

Persi Diaconis

Martin Gardner and Graduate School

MP: *How did you end up at Harvard?*

Diaconis: It is a humorous story. I graduated in January, and decided to start graduate school in mid-year. It turned out that some places, including Harvard, did accept mid-year applications. Harvard's Mathematics Department hadn't taken anyone from City College in 20 years. All of my teachers said Harvard didn't accept any students from City College, even the really good ones. So, I decided not to apply in mathematics. Instead I applied in statistics; it was the only statistics department I applied to. At the time, I didn't very much care about statistics, but I thought it would be fun to go to Harvard. I thought I would try it for six months and see if I liked it. I did like it, they liked me, and I stayed on to finish a Ph.D.

Because of my strange background I probably wouldn't have gotten into Harvard had it not been for the intervention of Martin Gardner. I was talking to Martin a lot during that time, asking his advice as to where to go, and he was, of course, professing to know nothing about mathematics. I said I was thinking of applying to the Harvard statistics deparatament, and he said that he had a friend there named Fred Mosteller. Now, Fred Mosteller is a great statistician, who in his youth had invented some very good tricks. There is, for example, a trick called the Mosteller Spelling Trick, which is still being used today. Martin wrote a letter in which he said something like, "Dear Fred. I am not a mathematician, but of the ten best card tricks that have been invented in the last five years, this guy Diaconis invented two of them, and he is interested in doing statistics. He really could change the world. Why don't you give him a try?" Fred later told me that I would not have been admitted if it had not been for that letter. Years afterward he wrote to Martin and said: "Dear Martin. You always are being asked to do things. You might want to know if any of it makes any difference. Well, six years ago you wrote me a letter, and on the basis of that letter I let this kid Diaconis in, and he's a real hot shot now." That's the story of how I got into Harvard.

"Statistics is the Physics of Numbers"

MP: *You now spend most of your time doing statistics. What is statistics to you?*

Diaconis: Statistics, somehow, is the physics of numbers. Numbers seem to arise in the world in an orderly fashion. When we examine the world, the same regularities seem to appear again and again.

In more formal terms, statistics is making inferences from data. It is the mathematics associated with the application of probability theory to real-world problems, and deciding which probability measure is actually governing.

MP: *Do you think of statistics as part of mathematics?*

Diaconis: Yes. It is part of applied mathematics. There is something about making inferences that goes beyond mathematics. In mathematics you must have something that is correct and beautiful, and that is enough to qualify as mathematics. In statistics, however, there is the question of trying to decide what is true in the world, and that is somehow going beyond any formal system.

I try to link nearly all of my work with some real-world problem.

The Computer and New-Wave Statistics

MP: *Are computers making much of an impact on statistics and on the way it is being taught?*

Diaconis: I think that the computer is changing mathematics, perhaps slowly, but in statistics it's right at the cutting edge of a real revolution. The usual sharp and tractable mathematical assumptions and approximations grew up in statistics because computing was hard and expensive. Nowadays, computing is fast and cheap, and one can actually use realistic assumptions, go to the computer and obtain the needed numbers. Much of recent statistical work is aimed at making intelligent use of the computer and, of course, this is producing a struggle between the old guard and new-wave statisticians. The old guard grew up with parametric assumptions and the use of things like the normal curve. They tend to say, "Why does anybody need this new-fangled stuff, and what does it all mean anyway?" There is a whole generation of young statisticians that see the power and excitement of the computer in statistics. If you walk around the halls here, you will hear students talking about new, fast algorithms for computing the median of a bunch of numbers, or new hardware for doing computer graphics, whereas ten years ago it would have been sigma algebra and the bread and butter of mathematical statistics.

When I teach a course that has anything computational and real-world about it, I will easily fill the hall. If I teach a course in any sort of esoteric mathematics as applied to statistics or probability, I will be lucky if I get six or eight students. I feel it on a day-to-day basis at the graduate level. Our students at Stanford and the students at Berkeley seem greatly interested in new-wave statistics which really has to do with massive amounts of computing and using graph-theory constructions and all kinds of computer graphics that weren't thought of twenty years ago. It's right at the center of a real revolution, and twenty years from now an old-style statistician from today won't recognize the field. It's really clear—it's not that I am a space-age visionary—that our students are focusing on the computer.

The computer leads you to do a different kind of mathematics—the mathematics of algorithms, the mathematics of numerical approximations, the mathematics of simulation, and all of that can be as subtle and difficult as any of the older mathematics. Indeed, a lot of it is more challenging than the old mathematics.

"I've Got To Have Applications"

MP: *You are now teaching an esoteric course on group theory in statistics. That doesn't sound very real-world to me.*

Diaconis: The course I'm teaching is called "Applications of Group Representations to Statistics." The first day of class I listed twenty applied problems that are easily stated in English. For example, how many times do you have to shuffle a deck of cards until it is close to random? I explained how group theory was the only known way to solve them. Motivating from real problems, I am now going through the theory of group representations in a systematic way, and I'll tackle the problems one at a time.

I'm teaching it the way I think applied math should go. There is a real-world problem, you cast about and find the tool to work with, and very often some mathematician will have created the perfect tool just because it is beautiful. Nothing pleases me more than being able to take something and apply it to solve a problem. But the bottom line for me has to be that I actually get an answer to the problem. In the case of the card shuffling, how many times do I have to shuffle a deck of cards? The answer is seven for real shuffles of a deck with fifty-two cards. Without the number seven at the end, all the group representations wouldn't mean as much to me.

MP: *The group theory is more beautiful for you as a result.*

Diaconis: Absolutely! It comes to life in a way for me. That's a funny feature about me. I can't relate to mathematics abstractly. I need a real problem in order to think about it, but given a real problem I'll learn anything it takes to get a solution. I have taken at least thirty formal courses in very fancy theoretical math, and I got A's and wrote good final papers, and it just never meant anything. It didn't stick at all; that's something about me. I think there are people who can't understand mathematics at the level of applications and problems and can see it functorially. I really do believe there are people like that. For them, the diagram and morphism is everything. It's not that way for me: I've got to have the application.

I am currently working on a beautiful problem that requires precise knowledge of the representations of the two-by-two matrices with entries in the *p*-adics. It is a problem that comes from salmon-fishing. That's a great way to learn about the *p*-adics.

The Art of Finding Real Problems

MP: *How do you find real problems?*

Diaconis: That's probably what I'm best at. What makes somebody a good applied mathematician is a balance between finding an interesting real-world problem and finding an interesting real-world problem which relates to beautiful mathematics. In my case, I browse an awful lot, sit in on courses, and read a lot of mathematics. As a result, I have a rather superficial knowledge of very wide areas of mathematics. Also, I am reasonably good at taking to people and finding out what ails them problemwise. I have a stream of people from all walks of life—psychologists, biologists, mathematicians—coming in saying: "Hey, do you know about this problem, did you ever hear about that?"

"I've got to have applications," says Diaconis. The basketball is used by Diaconis to help illustrate a problem of computer graphics.

MP: *Why would a biologist come to see you about a problem?*

Diaconis: I work at that. I speak to people in English. I am genuinely interested in applied problems, and it really does give me joy to see an applied problem solved by a beautiful mathematical tool. Even if I can't solve a problem I will work at feeding the problem to some other mathematician or statistician who can solve it. If you do it often enough, people get to know that you're a useful resource.

It takes real dedication. I am willing to talk to very strange people, and put up with all kinds of stuff because every once in a while, I hear about some beautiful problem, and in order to solve it, I need to learn some new area of mathematics.

I've actually started to think in a more structured way about how you find problems because I obviously am good at finding them. I'm trying to think of what I do, trying to systematize it so I can explain it to other people. It's still in a very vague state, but it does have to do with not being afraid to dip into an area and not to learn it in real depth. It's important to know even vaguely that there are mathematical tools which might enable you to say: "Aha, this problem might yield to this kind of tool," and then if you have enough mathematical background you can read more carefully and then

Mathematical People

try it out. It involves reading surveys and talking to people. I am always saying to mathematicians, "Tell me what you do in English in 20 minutes." Also, I am good at browsing and, in fact, I'm working on the mathematics of browsing.

MP: *What do you mean that you're good at browsing?*

Diaconis: Whenever I look in a volume of a journal I look at the table of contents, and I just scan titles, and if it looks interesting, I'll open it up and spend a minute. Part of browsing has to do with being able to decide whether something is interesting. It also involves the decision to spare the minute to look. I've had marvelous luck finding things that you ordinarily would never find.

I was working with a mathematician on a problem that required formulas for the characters of a certain representation, and they weren't in any of the standard literature. Then I went to the Reviews in Group Theory—and browsed in it for about an hour. I couldn't just look up what we wanted—I had to browse, free associate, read, and turn a page here and pick up phrases, and locally try to build some structure, and I managed to do that and was led to papers of Frobenius that I hadn't seen before. Sure enough, Frobenius had written out explicit formulas for the characters that we wanted, and they weren't in any modern literature. Knowing how to use the literature is a kind of art.

There really are two different kinds of researchers—those who look things up and those who derive things. I have co-authors, marvelous mathematicians and statisticans, who never read the literature and do everything from first principles. We might need a particular asymptotic expansion and I'll say, "Let's go look it up," and they'll say, "Let's derive it, and then we'll understand it better." It's also worth thinking about how much time you should spend browsing. We all know dilettantes who spend all of their time browsing and never get any work done. There is some trade-off in these two approaches that has to be fine-tuned. Anyway, I'm trying to build a theory of browsing that ties search theory into the problem of browsing, because browsing is looking for something.

What is Teaching?

MP: *You seem to enjoy very much explaining ideas to others which suggests that you like teaching.*

Diaconis: I'm not sure about that. I love to explain and understand at a personal level. People can explain things to you in a few sentences at the board or over lunch in a way that is impossible to get from a more formal lecture or a paper, and I love those kinds of insights. That gives me enormous pleasure and in fact drives much of what I do. On the other hand, teaching is different from that in some ways. You have to be more careful—you're not only trying to give the essence but actually to teach the skills. I have never been able to figure out what I am supposed to be doing when I am teaching. I mostly teach graduate students. The students I teach are actually here because they want to be here, but what am I supposed to teach them? Am I supposed to give them a good set of notes? Am I supposed to teach them the basics really clearly and deeply? Am I supposed to teach them how to do research? Am I supposed to give them an overview of the subject? I don't think you can do all of those things in one course. They are all different tasks, and I don't think you can do all of them. As a result, I have never been able to figure out teaching.

MP: *What is your teaching strategy then?*

Diaconis: Chaos! I am not a bad teacher, but I'm probably not a very good teacher. I like instilling in people the richness of mathematics and statistics, and so I very often will do lots of special cases of a general theorem. I try to instill a feeling for the many different cases that the theory will encounter.

MP: *You had previously said that you had been driven by your interest in applications—you wanted to see the mathematics that you worked with applied to the real world. Yet, you have a powerful interest in number theory. You wrote your thesis in number theory. How do you reconcile your interest in number theory, which to most people is regarded as a very non-applied area, with your previous statement about being driven by applications?*

Diaconis: (Laughing) The integers are very real. There are all kinds of number theory, and there really is a difference between very fancy, modern Galois theory and modular forms theory and Erdös-style problems that you can explain to your grandmother. Number theory is just beautiful. It's all mathematics. But, at the heart of my thesis, which was about analytic number theory, was a very concrete problem, namely the crazy first digit phenomenon. If you look on the front page of *The New York Times*, and observe all of the numbers which appear there, how many of them do you think will begin with one? Some people think about a ninth. It turns out empirically that more numbers begin with one, and in fact it is a very exact proportion of numbers that seem to begin with one; it is .301. (The more you look at it, it's the logarithm of 2 to the base 10.) Now that's an empirical fact, and it's sort of surprising. It comes up in all kinds of real data. If you open a book of tables, and look at all of the numbers on the page, about 30% of them begin with one. Why should that be? I had a funny explanation for it, and the explanation involved doing some computations with the zeta function. So then I started reading about the zeta function. Then I pushed my little explanation through, and it was sort of an okay explanation, and in the course of doing that I had to learn certain things. That always seems to happen in my research. Bott once said to me that research is just understanding part of mathematics in your own language. In the course of understanding these tools in my own language, some theorems came out and I learned the machine of classical and analytic number theory and the theory of complex variables, *but* I learned it to solve a problem. It's always been that way for me. There is some question and some set of tools, and often the question has been asked several times, and eventually the question drives you on to understand the set of tools, and then for me the game isn't finished until the set of tools yields the answer. This can take years. There are questions I have worked on for 30 years. Until I get the right answer, I don't stop.

As an undergraduate, I had a very good teacher named Onishi, who loved analytic number theory. I took a course from him, and he was nice and seemed to think I was smart, and that probably made some difference to me, too. We wrote a funny paper together in my first year in graduate school. When I interviewed at Harvard, Fred Mosteller said, "It says on your City College transcript, you're interested in number theory. Well, here is a problem I've always wondered about." He asked me a problem about the distribution of prime divisors of an integer chosen at random; how many different primes divide it. It was a beautiful question, and he had done a lot of numerical computing in searching for hints to a solution. I went back to New York, and I started to think about his problem. I got started on it, and then I got Onishi interested. He taught me some stuff, and the two of us hacked out a solution. Then we went back to Fred, who had more questions, and eventually it wound up as a triple paper. It was my first paper, and it appeared in the *Journal of Number Theory*. That happened when I was more or less an undergraduate. It was due to a bright, friendly teacher and an older guy who asked a really good question and had done a lot of work leading up to the question.

MP: *Who, in addition to Onishi and Mosteller, have been important influences on you?*

Diaconis: Dai Vernon was a very important influence on me as was Martin Gardner. I had a marvelous statistics teacher at City College named Leonard Cohen. When I said to my math teachers at City College that I was going into statistics, they all thought I was crazy and said, "Why are you going into statistics when you can actually do math?" Cohen was the only one who said that it was okay to be interested in statistics. He was a superb, clear teacher who also took the trouble to give me a reading course and just talk to me for a couple of hours between classes and give me a feeling for what the subject was about.

Other prominent influences on me have been my co-authors. My main co-author is David Freedman, who is chairman of the Statistics Department at Berkeley. We've written 25 papers together, and it has been a very fruitful, on-going relationship. David is a first-rate probabilist, and he has taught me a tremendous amount of modern probability. Ron Graham and I have co-authored half a dozen papers. They all start with applied problems. At present, we're working hard on finite Radon transforms.

MP: *Do you think your instincts are such that you are more inclined to collaborate than to work alone?*

Diaconis: I have done a lot of joint work. I have been thinking about that recently because when you get into a pattern with somebody you have a certain set of problems that you enjoy talking about, and so whenever a new question comes up there is a tendency to put it into that mold, and it has bothered me enough that recently I have written a year or two's papers alone. I wanted to shape the direction more independently than I would in a joint project. There is a great advantage in working with a great co-author. There is excitement and fun, and it's something I notice happening more and more in mathematics. Mathematical people enjoy talking to each other. Teaching someone else your tricks is as much fun as doing the new math. It's a great way to learn stuff and solve problems and also to have somebody to compete with in a friendly style. Collaboration forces you to work beyond your normal level. Ron Graham has a nice way to put it. He says when you've done a joint paper, both co-authors do 75% of the work, and that's about right; if it's not done that way, it's not much fun to collaborate.

MP: *What do you offer as the main reason for increased collaboration among mathematicians?*

Diaconis: I think it's due to the huge volume of tools and techniques that are available. In the case of working with Ron Graham, I know the probability machine, and he knows the combinatorics machine, and I don't think anybody knows both of them. In the case of David Freedman, it was my having questions that arose from applied statistics and his being a superb problem solver. Collaboration for me means enjoying talking and explaining, false starts, and the interaction of personalities. It's a great, great joy to me.

Psychics and ESP

MP: *How did you become involved with psychics like Uri Geller and ESP research?*

Diaconis: ESP is a nice example of an area where my background as a magician and my interest in statistics come together. It's a marvelous, clear example of a nice applied math problem. Any respectable proof of parapsychology by the standards of today is statistical in nature, and therefore in order to be a good investigator you have to know about statistics. One of the big problems for parapsychology investigators is that sometimes they work with people who cheat, deliberately or subconsciously, or both.

My involvement began when *Scientific American* reviewed a book that contained a report of a psychic in Denver who purported to make psychic photographs with his mind. Investigators would bring their Polaroid cameras and film and snap a picture of this guy's head, and usually they would get a picture of his head; but once in a while the photographs would look something like a fork, or a biplane, or Cromagnon Man or something like that. Martin Gardner arranged for me to go to Denver from New York—paid for by *Scientific American* and later by *Popular Photography*—to investigate him; and while I was there I caught him cheating unquestionably.

During the course of preparing for it, I began to read some of the literature on psychic phenomena. I always have been a complete skeptic about psychic phenomena, but I also think you can't dismiss something without thinking about it a little bit. So I read a fair amount. My reading on psychic phenomena suggested many nice mathematics problems. Years later when I was a graduate student at Harvard somebody asked me to give a talk about psychic phenomena. In preparing for the talk, I was going to tell some stories about psychics, and I was also going to solve some related, little math problems. I turned out that the little math problems weren't so little. They're right on the boundaries of what one can do.

Suppose you and I are doing a card-guessing experiment. We are in the same room, and I'm shuffling the cards and looking at them and concentrating at them. You are trying to guess what the cards are. We can see each other as we proceed. Well, to somebody with a little training in deception, it's obvious that there could be a good deal of cueing going on; it could very well be subconscious. That is, when I look at a card, and you get it right, I smile, and I'm happy, and when you get it wrong, I'm unhappy and I frown. You can pick it up and get an idea of what the cards are that have gone by and can improve your guesses and do better that way. That's a typical example of a problem that I heard about when I was doing investigations of psychics as a magician. Ron Graham and I wrote three or four papers on solving some cases of those problems.

For example, suppose we are in a card-guessing experiment with a deck of 52 cards. The simplest experiment goes like this: I shuffle the cards, I look at the top card, you try to guess it, and I don't tell you anything. I just put it to the side. Then I look at the next card and put it to the side. We keep going throught the deck that way. If you have no ESP, and I tell you nothing, your chance of getting the top card right is one in 52. Your chance of getting the second card right is one in 52, no matter what you guess. So, you have one chance in 52, 52 times, and so you get one card right on the average. (The distribution is a Poisson distribution, approximately, with parameter $1/52$.)

Well, then you could ask, suppose you get feedback. That is, suppose I shuffle the cards, and look at the top card. You start by guessing the ace of spades; and I say, "No, it was the six of hearts." Then you look at the second card, and you guess again. If you are trying to get the cards right, you probably will not guess the six of hearts again. As long as you don't, it doesn't matter what you guess. Well, that changes your odds. The chance of being correct is one in 52 on the first guess, one in 51 on the second, one in 50 on the third, and so forth. The expected number correct is the sum of the reciprocals of the integers from 1 to 52, and that's about 4.5, which is about log 52. So, instead of one in 52 on the average, with that kind of feedback, you get about $4\frac{1}{2}$ right on the average. (You can work out the associated distribution of correct guesses; it is approximately normal.)

A third version of the problem is actually a lot harder. Suppose instead of my showing you the card, I just tell you if your guess—let's say, the ace of spades—is right or wrong. If you're right, I'll tell you, but if you are wrong, I just say, "No, you're wrong." First of all, what is the optimal strategy. Suppose you want to get as many cards right as possible and that you don't have ESP; you just want to get a high score. You guess the top card as the ace of spades, and I say you're wrong and put the card down on the table. Well, all you know is that the ace of spades is still in the deck. The best thing to do is to guess the ace of spades again and to keep guessing ace of spades until you're told you're right, then guess something else. That is the optimal strategy. Strangely that is quite hard to prove. It sounds as if the optimality should be easy to prove. It turns out that the expected number of correct guesses is about $e - 1 \simeq 1.7$

That kind of problem is clearly a bread-and-butter problem if you try to investigate an experiment where feedback was given to the subject. You want to know what's high score in this experiment. It turned out to be a hard math problem. As usual, when you do some honest mathematics, the results are useful to people in a variety of disciplines. This work interests people who do taste testing. They do essentially the same kind of experiment to see if somebody can taste the difference between one kind of scotch and another. Our results are also used in medical trials.

After doing this research, I was in a situation where I had accumulated some good stories about psychics and I had done some mathematics related to psychic phenomena. As a result, I could give talks that were popular but had some science in them. That got me a fair amount of publicity, and as a result, I was asked to help investigate many other ESP projects. Somebody would hear about a great psychic and call up and say, "Do you want to investigate somebody like Uri Geller, who can bend keys or move a scale with his mind and so forth?" I have done such investigations on and off as a kind of hobby and also as a source of interesting problems. I guess it's also a service to the scientific community. It's hard for ordinary scientists to do a good job at debunking psychics. We may all feel that it is baloney, but it's very hard to determine why. I'm tired of it now.

Debunking

MP: *Why is it hard for scientists to debunk psychics?*

Diaconis: It's because most people (a) don't know the tricks, and (b) don't have the statistical background. It is very easy for the tricks to be concealed in poor statistics. A combination of (a) and (b) can be devastating. You can be a terrific physicist or mathematician, but if you don't have experience in running experiments with human subjects and with cueing, etc., you may have a very tough time. Having the experience often makes it very obvious what's wrong, and when you point out the trick or statistical fallacy to somebody else, they say aha. It's hard for people to spot it on their own.

MP: *The public's interest in ESP, astrology, and numerology remains very high. How do you explain the public's fondness for it?*

Diaconis: It is a basic human reaction to wonder at something surprising such as an unusual coincidence. That seems to be a hard-wired reaction in people. Perhaps it is wired in there for protection. Say you observe some pattern in the background. You say: "Aha! Something is different from what it usually is." I think it is unquestionable that we have a pattern-detecting mechanism that works and is alerted and delighted by surprising coincidences.

I don't think that there has been a reasonable explanation of the psychology of the appeal of psychics. When I was a performer, I learned that it is much easier to entertain people by pretending that your tricks are real magic, than to do wonderful tricks and just present them as tricks. People, if you let them, are quite willing to believe the most outlandish things, and the fact that you can do a little sleight of hand and actually make something happen in addition to creating a spell of wonder makes it all the more believable. It seems to be true that there is a growing interest in parapsychology and a growing disenchantment with ordinary science. Large proportions of our undergraduates believe that parapsychology is a demonstrated fact.

I have read very thoroughly for ten years all of the refereed, serious parapsychology literature. There is not a single, repeatable experiment in that literature. Most people don't seem to know that. I guess it is useful to go on record and to say that loud and clear.

MP: *You've done a service to the scientific community working to expose some of these individuals. Frustration must set in after you've exposed a charlatan or two and see that the general public does not welcome your work.*

Diaconis: I do see things changing. The changes have implications far from ESP. I think that math and the sciences are going to capture a larger share of the audience than they have in past years because of a new seriousness in students that I haven't seen for the past ten years. They seem more interested in beginning at earlier stages to become serious about careers and not to take five years off and discover the earth.

MP: *Sometimes I wonder how mathematicians are going to live through all the changes brought on by computers. I've heard a few serious applied mathematicians make the following accusation about those who teach mathematics: "A lot of them have simply run away from reality. They're doing mathematics because reality is not always that pleasant."*

Diaconis: I don't accept that. Historically there is a certain sense in which mathematicians have followed their noses and have said: "Here is an interesting mathematical object. My interest is to bring that structure into a clear, definite perspective so that someone else can understand what is going on in an applications sense." They do it because of the inner beauty of things. Time and time again something was done with absolutely no applications motivation, even to other areas of pure mathematics, and then later it turns out that it is wonderfully useful in an applied problem.

Take group representations, for example. Much of the mathematics that is going on today is concerned with representations of infinite groups and figuring out what they are, what the structure is. It's being done for its inner beauty and not in any sense because it's good for anything. But physicists, chemists, and now statisticians have found lots of applications for group representations.

I have friends with whom I work who absolutely can't focus on the real part of problems. When I start explaining why this or that is the right choice or this or that must be true from an applications point of view, they draw some diagram. That's the way they think. For them, there is some functorial reason why it should be true, but there is nothing that makes me right and them wrong. Often they can prove things that intuition only suggests. I'd hate to have pure mathematicians stop doing whatever they please.

The MacArthur Prize

MP: *You recently received a big award, namely the MacArthur Prize. What does the MacArthur Prize mean to you?*

Diaconis: I have actually figured that out. The money doesn't make a huge difference. Stanford University and the academic community have taken reasonably good care of those of us who work hard and are lucky. However, the recognition and the associated prestige mean a great deal. It means that I'm really encouraged to take ideas of mine that are a little bit fringy and pursue them. I really say to myself, "The world is giving you a pat on the back, and in fact, they want you to do it, so do it. Forget about pleasing the old-timers and dotting another 'i' in some theorem."

The key difference about the MacArthur Prize is that without the money the recognition isn't believable. That is, if somebody gives you a prize and says that you're terrific, well, that's nice. But if someone gives you a prize and says here is $200,000 to show you that we mean it, maybe they really mean it. For me, it's a really wonderful thing to have, but the impact of the money is to make me and the community around me take it somewhat more seriously as an indication of good work. It doesn't seem to change my life very much.

MP: *But you do feel more free. You said you can play a little bit more with ideas that you previously referred to as fringe ideas.*

Diaconis: I do feel more free intellectually. For example, I haven't figured out any way to use it to go on a cruise. I go on a sabbatical leave or take a year off every three years anyway. I'm at the center of mathematical statistics in the world, namely the Berkeley-Stanford axis. If I want to give any crazy course at all—like Choquet theory in statistics, or group theory in statistics—I put up a sign and I get twenty smart graduate students, and five visiting faculty to attend, and I get to talk about my research. It couldn't be more idyllic, and so there is no motivation for me to go elsewhere.

Professor Diaconis being interviewed by the media after being awarded a MacArthur Fellowship in 1982.

"Teaching is Terrific"

MP: *Your last comment suggests that teaching is very important to you.*

Diaconis: Teaching is terrific. It's the thing I missed most when I was in industry. It is tremendously exciting to structure your ideas and explain them to a bunch of bright, eager graduate students, who want to catch you when you're wrong and do problems when you're right. I remember the first day I came back after a year away and there are my students. They listen to every word you say, and your views and values become their views and values. That's a very good feeling. It drives me to do work that I wouldn't normally do. I'm always running seminars and that gets me to read papers that I ordinarily wouldn't read. You know there is a big difference between reading a paper and understanding it for yourself and reading a paper and digesting it and explaining it to somebody. You actually have to figure it out if you're going to explain it to someone else. I use teaching to trick myself into understanding.

MP: *What do you want to do with the rest of your life?*

Diaconis: I have already had two careers. I was really good at magic; I still love it, and I love and enjoy doing full-time statistics. My friend, Brad Efron, another MacArthur winner at Stanford's statistics department, said to me last year that he had a premonition that I was going to have yet another career. It's puzzled me; I can't imagine what it would be. At the moment anyway, I like the academic life—thinking about problems and talking to people about them. I could see that I could get tired of it or slow down—enough to make it painful. It's possible to reach a point where you know so many things and have so many directions to go in that you stop being able to do any one of them very well. There is tremendous pressure in society to make that happen. I get lots of offers to be editor of something or head of some committee. All of that cuts into what I am really good at doing and want to do, which is doing research and explaining my ideas to other people. I hope I can fight off the pressures for another ten or twenty years.

MP: *Earlier you said, "If you're talented, the world takes care of you. It took care of me."*

Diaconis: I think I'm good at asking questions that lead to interesting mathematics and useful statistics. That's a talent. Here I am really able to exercise it in the most idyllic circumstances. The academic community has figured out that I'm good at what I do, and they've managed to get me a good job and get me some money and let me do it full time, and I'm very happy for that. I think about it almost every day. I work from seven to midnight each day. I'm just always going, and I'm always doing mathematics.

MP: *Is there life outside of mathematics?*

Diaconis: Well, it's all one world.

MP: *How about music?*

Diaconis: I don't do music anymore, but I still do magic. The way I do magic is very similar to mathematics. I do it seriously as an academic discipline. I study its history, I invent tricks, and I write material for other magicians. I meet with them, do tricks occasionally and practice. That's an activity that is not very different from mathematics for me. I subscribe to 20 magic journals. You might say I do magic as a hobby, but for me it's quite close to math.

That's a nice point. Inventing a magic trick and inventing a theorem are very, very similar activities in the following sense. In both subjects you have a problem that you're trying to solve with constraints. In mathematics, it's the limitations of a reasoned argument with the tools you have available, and with magic it's to use your tools and sleight of hand to bring about a certain effect without the audience knowing what you're doing. The intellectual process of solving problems in the two areas is almost the same. When you're inventing a trick, it's always possible to have an elephant walk on stage, and while the elephant is in front of you, sneak something under your coat, but that's not a good trick. Similarly with mathematical proof, it is always possible to bring out the big guns, but then you lose elegance, or your conclusions aren't very different from your hypotheses, and it's not a very interesting theorem.

One difference between magic and mathematics is the competition. Somehow the competition in mathematics is a lot stiffer than in magic. When I was doing magic, all those bright, young kids were learning calculus. Now, here they are all competing for the same pot of theorems.

A Professorship in Magic

MP: *Why did you leave magic as an occupation?*

Diaconis: I left the performing part of it. Show business is very different from being a creative magician. In fact, the reason I left it is because you can't be too creative. There is tremendous pressure to do the same 17-minute act: it works and it gets laughs. I can remember very clearly changing the closing trick of my act, a trick with butterflies. I took the butterfly trick out to do something else. After my performance, my agent rushed up to me backstage, and said I couldn't take the butterfly trick out of my act. He said, "That's what I book you on." At that point, I wondered if I was going to end up doing the same seventeen minutes for the next twenty years?

If I could have had a professorship in magic, and if the world recognized magic the way it does mathematics, I probably would be doing magic full-time and never would have done mathematics or statistics. Magic can be done as a very academic and creative discipline; it's very similar to doing mathematics, except for the fact that the world treats you more seriously if you're a mathematician. If you say that you're a professor at Stanford, people treat you respectfully. If you say that you invent magic tricks, they don't want to introduce you to their daughter.

MP: *When you were doing magic, you said that you were following the wind. Are you still following the wind?*

Diaconis: When I was young and doing magic, if I heard that an Eskimo had a new way of dealing a second card using snowshoes, I'd be off to Alaska. I spent ten years doing that, traveling around the world, chasing down the exclusive, interesting secrets of magic. An analog of that in my second career is not just doing any one thing. For example, my thesis was in number theory, and some might think that I do that kind of mathematics. I've done a fair amount of classical mathematical statistics, so you might think I do that. I have worked in philosophy of statistics, psychology of vision and pure group theory. What happens now is that if I hear about a beautiful problem, and if that means learning some beautiful math machine, then, boy, I'm off in a second to learn the secrets of the new machine. I'm just following the mathematical wind.

PAUL ERDÖS

PAUL ERDÖS

Interviewed by G. L. Alexanderson

Paul Erdös, of the Mathematical Institute of the Hungarian Academy of Science, is one of the best known, most prolific and widely traveled of mathematicians. The following conversation was held in Santa Clara in December, 1979, where he had just arrived after visits to Montreal, Winnipeg, and Mexico City on his way to the Number Theory Conference in Asilomar, and then to Texas, Florida, Memphis, Zürich, Budapest, and India.

The Peripatetic Mathematician

MP: *You are certainly among the most renowned and prolific of mathematicians, but you are also, undoubtedly, the most widely traveled.*

Erdös: Yes, but Marshall Stone traveled a great deal, not perhaps now, he's too old, but he used to. He also went to some places like Indonesia and the Fireland (Tierra del Fuego) where there is not much mathematics. I rarely go any place where there is no mathematics.

MP: *It has been pointed out that one can travel like this in mathematics, but one couldn't do it so easily in one of the laboratory sciences.*

Erdös: I am also accustomed to working while traveling. I can work in an airplane very easily. I can adjust very fast to surroundings. Also, I can jump from problem to problem easily. It has been commented that sometimes, without changing my voice, I change the topic completely, something that causes confusion sometimes.

MP: *I personally find travel very annoying—passports, customs, visas, and so on.*

Erdös: It is worse with me than with you because I need a visa everywhere.

MP: *Doesn't that mean a lot of standing in line and waiting?*

Erdös: Yes, it is a bit of a nuisance, but it has gotten easier now. But I cannot travel at a moment's notice.

MP: *Does jet lag bother you?*

Erdös: That doesn't seem to bother me. You see my record is this: There was a Saturday meeting in Winnipeg on number theory and computing. On Saturday evening we had a dinner in a Hungarian restaurant, a farewell dinner for the speaker; then on Sunday morning I flew to Toronto. I was met at the airport, and we went to Waterloo to a picnic. In the evening I was taken back to Toronto, and I flew to London where I lectured at 11 o'clock at Imperial College. It didn't bother me that much.

MP: *How many continents have you visited this year?*

Erdös: This year, I crossed the Atlantic five times. I was in Israel too, so that makes only three continents this year. I have never been in Africa.

MP: *Not even in Cairo?*

Erdös: No, but I will go now, if I live. It was difficult, since my passport says that I am a resident in Israel. I was never in Africa or in South America. The most important mathematical country I have missed is undoubtedly Japan. Once I was asked: "Do you have a prejudice against Japan?" So I said, as a joke: "Yes! They are too interested in algebraic geometry and algebraic topology." (I don't know much about those.) It is really an accident. Before I leave I hope to visit Japan. Mathematically it is very interesting.

MP: *When did you start traveling so extensively?*

Erdös: I liked to travel already as a child. When I went to England first in 1934, I went three times back and forth between England and Hungary in a year. And then when I got to America my travels increased. I traveled more and more. On a world scale I have traveled since 1954 when I left the U.S. They didn't want to give me a re-entry permit. I remember as a joke, when I decided to leave without a re-entry permit, I drew a picture: here is S (Sam) and here is J (Joe).*

$$\textcircled{S} \quad \textcircled{J}$$

And I said I will be in the middle. Someone said, "Yes, but your picture is unrealistic. Let me draw it realistically."

$$S \mid J$$

That seemed more realistic.

In Europe the Franco-German frontier is much better than the American-Canadian frontier—no control, or practically no control. They couldn't stamp my passport, because they had no stamp. In a car they don't even stop you.

A Mathematical Prodigy

MP: *This brings me to another question about your childhood, aside from travel. Now, both of your parents were mathematicians. They both taught?*

Erdös: Yes, in high schools. So I learned a great deal from them.

MP: *On being asked, "At what age did you realize you were a genius?" Noel Coward said, "At two." At what age did you or your parents realize you were a genius?*

Erdös: It was fairly obvious that I could calculate very well when I was four. At that age I told my mother that if you take away 250 from 100 you have 150 below 0. But you see my first paper was not published as phenomenally early as some by other, later Hungarian child prodigies. At eighteen I gave a new proof that there is always a prime number between n and $2n$.

Issai Schur was one of the first foreigners with whom I corresponded. I proved that there is a prime of the form $4k + 1$ and $4k + 3$ between n and $2n$. Schur was very impressed and he wrote me a nice letter in German. It pleased my parents. Another such story is this: You know that Brouwer, the great Dutch mathematician, was also a businessman, as was an uncle of mine. Brouwer asked him, when he (my uncle) introduced himself, "Are you a relative of Paul Erdös, the young Hungarian mathematician?" This was later, though, when I was a little over twenty.

MP: *Where was Schur then?*

Erdös: In Berlin. This was before the Nazis, in 1932. Later I visited Schur. I saw him in 1936 and I saw him in 1938 for the last time. My mother and I visited Mrs. Schur in Israel just before her death. I visited her every year; but my mother first came to Israel in 1964–65, and when we were in Tel Aviv we visited her.

MP: *One thing I have observed in you is your intense curiosity. Is that common to all first-rate mathematicians?*

Erdös: No, I don't think so. I was always interested in many things. My father taught me, when I was a small child, all kinds of things—physics, chemistry. I was always interested in history. This is also because I have a good memory for facts. Not for faces.

MP: *I remember one of the questions that you asked on your first visit to Santa Clara. You asked me the temperature of this valley during the Ice Age. I have never been asked that before or since.*

Erdös: I was always interested in geology. Actively I was not interested, not as a research subject, but I read a great deal. I read the *Scientific American* regularly, as much as I can. It is getting harder,

---

*S = U.S.A.
J = U.S.S.R.

even in mathematics, to follow everything. Fejér used to say in the 1930's, "Everybody writes and nobody reads." This was true even then. Reviewing has improved, but even so it is very hard.

MP: *You mentioned Fejér. Your parents were contemporaries of Fejér and von Kármán, weren't they?*

Erdös: They went to the University at the same time.

MP: *You studied later with Fejér.*

Erdös: Yes, in a way. I mean, I took his courses. Fejér was formally the advisor for the thesis, but he was not in number theory so he didn't have very much to do with it. My work on interpolation with Turán was certainly influenced by Fejér.

MP: *Fejér was renowned as an inspiring teacher, I believe.*

Erdös: Yes, he explained very clearly. He had many inspiring ideas, but he didn't give very good lectures in the following sense: he didn't carry through. He once told Turán: "You know, I feel I was burned out by thirty." He still did very good things, but he felt that he didn't have any significant new ideas, though he did very, very beautiful things afterward. When he was sixty, he had a prostate operation and after that he didn't do very much work. Then he was on an even keel for fifteen or sixteen years, and then he became senile. There was some disturbance of the circulation. It was very sad because he knew that he was senile and he said things like "Since I became a complete idiot..." and such things. He always recognized me and my mother. He was happy when he didn't think of it. He was very well kept in the hospital but died in 1959 of a stroke.

MP: *Did you have other teachers who were particularly influential?*

Erdös: Kalmár. He rewrote my first paper. I learned a great deal from him in conversation. We made excursions nearly every Sunday and we talked mathematics while walking. I got accustomed to working without paper.

MP: *The great flowering of Hungarian mathematics—to what do you attribute this?*

Erdös: There must be many factors. There was a mathematical paper for high schools already and the contests which started already before Fejér. And once it started it was self-perpetuating to some extent. Hungary was a poor country—the natural sciences were harder to pursue because of cost, so the clever people went into mathematics. But probably such things have more than one reason. It would be very hard to pin it down.

**Paul Erdös, peripatetic mathematician, lecturing at
the California Institute of Technology in 1983.**

Changes in the Teaching of Mathematics

MP: *Have you in your travels observed any trends in the teaching of mathematics? I know that you have been interested in geometry, and yet, at least in the United States, geometry seems to be de-emphasized.*

Erdös: They don't teach it much now. They do different things from what was done fifty years ago. That's true in all countries. Even in Hungary, they don't teach geometry that much now. Turán felt that perhaps combinatorial analysis could replace geometry for encouraging young people in problem solving. The great advantage of geometry is that a young person can see very nice problems. Now in Euclidean geometry, not much new has been done, but these combinatorial geometry problems are different. There are many unsolved problems in combinatorics. So this move away from geometry is to some extent worthwhile, but they may overdo it a bit here. Also perhaps number theory is not so much taught now. Combinatorics is, on the other hand, taught much more.

MP: *All these subjects are accessible without lots and lots of definitions.*

Erdös: That makes it easier. That is why many of the prodigies do number theory and combinatorics, because they are more easily accessible. There is some danger in that, as Sabidussi, who is a graph theorist, told me. He doesn't like it if his graduate students start with these subjects, because he wants them to know also other types of mathematics and that is a good idea.

MP: *Now your dissertation was in number theory and many of your most famous results, your proof with Selberg of the prime number theorem and your asymptotic expression for $p(n)$,* for example, were in number theory. But these days I seem to be aware of more combinatorial results than number theoretic results. Have your interests shifted at all over the years?*

Erdös: No, not really. I think that, as Shelah, a very clever young Israeli mathematician, once said, "I am an opportunist, I do what I can do." If there is anything in number theory I can do, I certainly do it. But you see some of the problems in number theory are enormously difficult and many of these classic problems are very, very hard to make any progress in. Combinatorics is a much newer field, and there are many more problems that are still accessible.

Erdös's Mathematical Contributions and "The Book"

MP: *Is there any piece of mathematics or theorem of which you are especially proud?*

Erdös: It's hard to tell. Certainly the proof with Selberg of the prime number theorem is good, and I like the Erdös-Kac and the Erdös-Wintner theorems in additive functions.† I like my work on applications of probability to combinatorics and number theory, also some things in set theory, you know, the partition calculus. And some things in elementary geometry I like. A theorem of Anning and myself that, if you have an infinite set of points in the plane and all distances are integers, this is only possible if the points all lie on a straight line.‡ The first proof with Anning was a bit messy, and Kaplansky said I should look for a simple proof. And I found a proof of a few lines that is rather pretty. It is straight from the book. You know my joke about the book.

MP: *Yes, I wanted to ask you about that. Would you explain?*

Erdös: Once I wrote I will go to Poland if my cold gets better, which the S.F. (God) saw fit to send me. I didn't realize it would be published. I got a letter from my mother and a Communist friend of mine saying that it was improper of me to say such things. It may offend many people. And I said, "Yes, they are right, but it wasn't my fault." Gordon Walker said he just didn't want to alter the letter.

---

*The function $p(n)$ gives the number of partitions of n, a positive integer, that is, the number of ways it can be written as a sum of positive integers, disregarding order.

†These are two probabilistic theorems on additive functions. For an elementary statement of the Erdös-Kac theorem, see M. Kac, *Statistical Independence in Probability, Analysis and Number Theory*, Mathematical Association of America, 1959 (Carus Monograph #12). For a more technical treatment of both theorems, see J. Kubilius, *Probabilistic Methods in the Theory of Numbers*, American Mathematical Society, 1964 (Translation of Mathematical Monographs, #11).

‡See R. Honsberger, Stories in Combinatorial Geometry: A Theorem of Erdös, *Two-Year College Mathematics Journal* 10, (1979), 344–7.

But anyway, the S.F. has a transfinite book of theorems in which the best proofs are written. And if he is well intentioned, he gives us the book for a moment.

MP: *The best proof being defined as?*

Erdös: The simplest. In some cases it's not really clearly defined, and in some cases it is obvious. It is the simplest and the most elegant. Sometimes there is no unique best proof.

Erdös Numbers and Erdös Lore

MP: *How many people have an Erdös number of one?*[§]

Erdös: You know, I don't keep an exact count of them. My mother kept records. I think it is over two hundred. There will be a more complete record out. Ron Graham has a collection of my reprints and there is one in Hungary where someone is preparing a list which should be as complete as possible.

Professor Vladimir Drobot (right) discusses his Erdös Number with Paul Erdös.

MP: *How many papers are there?*

Erdös: On the order of magnitude of nine hundred, but it could be off by thirty or forty. The record so far is Cayley with nine hundred twenty-seven. There is an old Hungarian saying—*Non numerantur, sed ponderantur* (they are not counted but weighed). In the old parliament of noblemen, they didn't count the votes: they weighed them. And this is true of papers. You know, Riemann had a very short list of papers, Gödel had a short list. Gauss was very prolific, as was Euler, of course.

MP: *Also Cauchy, I believe.*

Erdös: Yes, he was very prolific. Riemann and Gödel are typical of those with very short lists. And Gödel lived fairly long, too, but he didn't publish much.

[§]A person has an Erdös number of one if he has written a paper with Erdös, an Erdös number of two if he has written a paper with someone who has, in turn, written a paper with Erdös, and so on. See C. Goffman, "And What Is Your Erdös Number?" *Amer. Math. Monthly* 76 (1969), 791.

MP: *Now I know that you offer monetary rewards for solutions of problems.*

Erdös: I have to restrict myself to my own problems, with one or two exceptions. In Turán's memory I offered $1000 for the solution of a problem of his in graph theory. The reason I have to restrict myself to my problems is that otherwise I would be completely broke very soon.

MP: *What is the range of these rewards?*

Erdös: There is a maximum of $10,000, but it is a hopeless problem in number theory. A problem of $3000 is: If you have an infinite sequence of integers, the sum of whose reciprocals diverges, then the sequence contains arbitrarily long arithmetic progressions. This would imply that the primes contain arbitrarily long arithmetic progressions. This would be really very nice. And I don't expect to have to pay this money but I should leave some money for it in case I leave. (There I mean leave on the trip for which one doesn't need a passport.) The next amount I had to pay was $1000 to Szemerédi. Turán and I had a conjecture that goes back to the early thirties, which is a generalization of van der Waerden's theorem: If you have a sequence of positive density, then it contains arbitrarily long arithmetic progressions. It was actually very surprising that this was so difficult. We didn't realize it when we proposed it. It was enormously difficult. You can't always recognize the difficulties.

MP: *What's an example of a problem at the low end of the scale?*

Erdös: I made a fool of myself. I offered $25 for a problem during a lecture. I thought it wouldn't be very difficult, but it was disappointingly easy. You know, Herzog and Stewart had studied the following situation: You call a lattice point visible if the coordinates are relatively prime. Now you join two visible lattice points if they are neighbors: one coordinate coincides and the other differs by one. Pólya studied these points too. Herzog and Stewart proved that there is only one infinite component of this graph. Strangely enough the density of these is not yet known. I don't think it has been firmly established. Anyway, I asked this question during my lecture at Michigan State. Stewart and Herzog were there. There is an infinite path going to infinity in this infinite component. But what Herzog asserted: you go on the line l parallel to the x-axis. But I asked for another infinite path that doesn't touch the line l. Stewart found a very simple way: you take the point with coordinates (P_k, P_{k+1}) which you can connect to (P_{k+1}, P_{k+2}) and then you go off to infinity. I should have thought of that. So I said that if I continue this I should be put under tutelage. It was customary. Hungary used to be sort of a semi-feudal country and if a rich aristocrat became old and spent all his money on poison, noise, and bosses (wine, women, and song) his family put him under tutelage. He was given a large sum of money, but he couldn't touch his main capital.

I didn't realize it was so simple. There are still interesting questions: for example, can you find an infinite path that doesn't touch any point where both coordinates are primes? That is still unsolved. I still offer another $25 for that. Maybe it is also trivial. Or is there a monotone path where each vertical and horizontal strip is a bounded line and which goes through these lattice points? Maybe there isn't any. And maybe this could be also simple. It is not always easy to tell. It was a clever idea of Stewart, but it was disappointingly simple.

MP: *Now you have established prizes for young mathematicians in Hungary and Israel.*

Erdös: I have two such prizes. I left in memory of my parents a prize in Hungary which was 15,000 florins and I have a similar prize in Israel. The first man who won it in Israel was Shelah. And now probably a young group theorist will get the second one. In Hungary it has been going for five or six years. And Mrs. Turán and I established the Turán Memorial Lectureship. The first Turán lecturer was Baker, and K. F. Roth will be the second.

Nobel Prizes: Appropriate for Mathematics?

MP: *You mentioned the other day reasons why a Nobel prize might not be appropriate for mathematics.*

Erdös: Yes, there would be some difficulty for many different fields. Bers told me this story about twenty years ago: When money was very available, someone told him we should have 10, 15, or 20 Fields Medals. Someone said, "I don't think we could get the money." Bers said, "Yes, we could. But I'm not quite happy about it, because then it might cause real fights." As long as there are two or at most four Fields Medals (every four years), nobody will be offended seriously if he doesn't get it, as long as good people get it. Sometimes the committee just doesn't like a field. For example, in combinatorics, Szemerédi should have gotten a Fields Medal. The people who decide are not that interested in combinatorics.

Shelah should certainly get it. But whether he will get it is doubtful. Cohen got it but that was such a sensational result, it couldn't be ignored. They slightly lean too much toward algebraic geometry. But there's no doubt that all the people who got it are good. Nobody claims that you can judge who is best. If there were 20 Fields Medals, you might say you could give awards for all the important results.

MP: *Then not to get it would look bad.*

Erdös: Yes, it might cause more antagonism. For a Nobel prize, the same could apply. How can they decide? In some cases it is clear, but overall it must be very hard to decide.

Von Neumann and Gödel

MP: *Who have been the most impressive mathematicians you have known?*

Erdös: Von Neumann was very impressive to talk with. He was very quick, but I don't know his work that much. Among the younger people with whom I have had more contact, I would have to mention Szemerédi and Shelah. There are some very good ones I don't know so well, like Fefferman, but I never had much contact with him so I can't compare him to others in any way.

In speed and understanding Von Neumann was certainly phenomenal. He could understand a proof even far from his own subject very fast. I remember once in Cambridge I told him a proof of interpolation that was not quite correct. By the time we met again I had a correct proof. Von Neumann told me, "Something seems to be wrong in that proof." And it was really not his subject. He wasn't that interested in it, but he was quite right. Gödel I talked with a great deal. He was certainly a remarkable intellect. He understood everything, even what he didn't work with. It is strange how little he published. He could have certainly done more things.

I always argued with him. We studied Leibniz a great deal and I told him, "You became a mathematician so that people should study you, not that you should study Leibniz."

MP: *How did he happen not to publish more? Was it intentional? Rossini stopped writing operas at thirty-six and lived on another thirty-nine years, apparently quite happy.*

Erdös: No, he continued to work. He had a proof that the axiom of choice is independent. And there was a rumor that he had a proof of the independence of the continuum hypothesis before Cohen. I asked him and he said, "No, it is not true." He had a proof for the independence of the axiom of choice, but he didn't like the proof.

Child Prodigies

MP: *There is a stereotyped view of prodigies, that they are not socially well-adjusted.*

Erdös: Well, I don't think this is true. My experience with Hungarian child prodigies doesn't point to this. You see, all of them seem to be reasonably well-adjusted. Fefferman seems to have been a perfectly well-adjusted youth. One would at least have to have some statistical evidence.

MP: *Now Wiener was a child prodigy.*

Erdös: Yes, he wasn't very well-adjusted but he was certainly pushed very, very strongly. And he had an unhappy childhood. He wrote about it.*

MP: *He apparently, even in later years, suffered some doubts about his talent.*

Erdös: Yes. He wanted to be praised in a childish way. So he [Wiener] remained a child prodigy all his life. He had some emotional difficulties.

MP: *Now, I won't go into the cases of the prodigies you have worked with. That's fairly well documented, for example, in the chapter on Pósa in Honsberger's book.†*

---

*Norbert Wiener, *I Am a Mathematician* (1977) and *Ex-Prodigy* (1973), MIT Press.
†R. Honsberger, *Mathematical Gems from Elementary Combinatorics, Number Theory, and Geometry*, Mathematical Association of America, 1973 (Dolciani Exposition Series #1).

Erdös: Yes, it appeared in the proceedings of a meeting in Carbondale. It appeared also in a paper from the number theory conference in 1971 in Missoula.

MP: *Is there something special about Hungarians?*

Erdös: I doubt it. Now that child prodigies are more encouraged and recognized in this country, you have quite a lot of them. Fefferman, Friedman, ... In the past they would have had a difficult time when they were told they had to stay with their peers. You know, all of these rules that may be completely correct for normal people, make no sense for prodigies. To say that Bach should pay any attention to how he was socially adjusted is just a bad joke. It is obvious that this is secondary.

MP: *Your mention of Bach brings up another question. The Bachs were a remarkable family and to some extent their mathematical analogue is the family of . . .*

Erdös: Bernoullis.

MP: *What happens in a situation of that sort, where one family carries on an intellectual tradition for generations?*

Erdös: It's hard to tell because human beings are so complicated. It's hard to decide how much is heredity and how much is environment. Clearly with the Bachs, there was inherited genius. But the fact that they were encouraged and that they grew up in a musical atmosphere, if they had any ability at the start it was increased by these influences.

It seems that it is not demonstrated that mathematicians are better musicians. There was a Hungarian psychologist, Revesz (uncle of Rényi), who made a study of this. He compared mathematicians with other academics, doctors and physicists, and he didn't find any significant difference, but it is hard to prove a negative statement. He wrote a little paper saying that he did not find any evidence that mathematicians are more musical than others.

MP: *It is certainly a very common view that they are.*

Erdös: But mathematicians have more time, to some extent, and they can often work at leisure. I can certainly work while listening to the radio. Perhaps if I were really musical I couldn't do it, because I would be too much occupied with the music.

Erdös's First Two and a Half Billion Years in Mathematics

MP: *Let's move on now, a little, to Erdös lore. I once heard you speak in Albuquerque with the title, I believe, "My First Two and a Half Billion Years in Mathematics."*

Erdös: That was a joke. I complain often about my old age. Now I have stopped complaining because I can't get old anymore, the process is finished. I said I'm two and a half billion years old because when I was young the earth was two billion years old and now it is four and a half billion years old so I must be two and a half billion years old. Once at a party a lady asked me: "How were the dinosaurs?" If I would have been clever I would have answered: "I don't remember, because an old man remembers only events of his youth and for me the dinosaurs were only yesterday." But I didn't think of it. Another such question is this: You know I have this language of my own and I call children "ϵ's." Women are bosses and men slaves but children are bosses *per se*. So I was asked once: "When does a slave child become a slave?" If he is a boss originally, when does he become a slave? And here I answered immediately: "When he starts running after bosses."

MP: *How extensive is this special language: poison (alcohol) and such?*

Erdös: Just a few words, not very extensive. Another was a joke, something they called Erdese. I pronounced English words as if they were Hungarian. When I started to learn English with my father I knew German already, so I thought English also could be read phonetically. Hungarian is pronounced phonetically. Once you have this in mind, the only problem is to learn to speak it fast. It's like this funny language called pig Latin. Once you know the principle, with only a little practice, you can understand it. When I was away in '54 it was practically forgotten.

MP: *I know that there is at least one limerick that refers to you.*

Professors Alexanderson and deBouvère discuss a problem with Erdös following his lecture at the University of Santa Clara in 1980.

Erdös: Yes, Moser has one:

> A conjecture both deep and profound
> Is whether a circle is round.
> In a paper of Erdös
> Written in Kurdish
> A counterexample is found.

I tried to publish a paper in Kurdish; there is no journal.

MP: *What is the largest known Erdös number?*

Erdös: I don't know. Ron Graham wrote a joke paper and said it filled a much-needed gap. He investigated some of these questions. It is difficult to know; you would have to have very good records. He tried to figure out whether Gauss has an Erdös number. It is not yet known. Einstein had an Erdös number of two because I have papers with Straus and Straus had papers with Einstein. Now Gauss wrote almost no joint papers because in those days people didn't do it as much. But there was a physicist, Weber, with whom Gauss collaborated. There is even a Gauss-Weber statue. It seems they came close to inventing the telegraph. Weber can be joined to Helmholtz, but Helmholtz could not be connected to Einstein.

NOTE:

For additional Erdös lore, see
    Gina Bari Kolata, "Mathematician Paul Erdös: Total Devotion to the Subject," *Science*, April 8, 1977,
and
    S. Ulam, *Adventures of a Mathematician*, Scribner's, 1976.

MARTIN GARDNER

MARTIN GARDNER: DEFENDING THE HONOR OF THE HUMAN MIND

by Irving Joshua Matrix*

We expect Martin Gardner to amuse and delight us, but he does more. He teaches us to be critical. We must be on our toes with him, so as not to be fooled. He raises many questions that do not have pat answers. He does this for the casual reader of his books and columns as well as for those who follow his work closely.

Martin is an accomplished conjurer. In his hands common objects may take on magical properties. He works this same magic with ideas—drawing something from nothing (February 1975 column†) and even successfully taking on everything in a single column (May 1976 column). He has a conjurer's eye for hidden causes and has mastered the art of transforming base metal into gold, not using a philosopher's stone, but using his philosopher's mind.

The questions Martin asks are often unusual and revealing. When the subject of extraterrestrial life came up, he inquired what I knew of the physicist John A. Wheeler's views on the subject. Martin had read that Wheeler believed life exists only on Earth. To Martin a flat statement on what seems a problematical subject suggests reasons not in evidence.

This is a Gardner twist. While I can't do anything directly to demonstrate or disprove the existence of extraterrestrial life, I certainly can find out what people think about the possibilities of extraterrestrial life and, to an extent, why they think as they do.

Martin sets interesting and useful questions, questions to sharpen our wits and our critical ability, questions that celebrate our abilities to solve problems. Like Wittgenstein, Martin seeks to show the trapped fly how to find its way out of the fly bottle.

That brings me to a problem you will not find in Martin's columns. Who is Martin Gardner, and how did he come to write the "Mathematical Games" column in *Scientific American*? I will answer these questions as best I can, but though I am a creation of Martin's mind I am not privy to his innermost thoughts, nor am I infallible in my opinions about him.

Martin Gardner was born in 1914, the first of three children of Dr. James Henry Gardner and Willie Wilkerson Spiers Gardner. His father was a geologist, first with various state geological surveys and later as a consultant and as president of his own oil company. The Gardners were of Methodist stock. Dr. Gardner was a director of the Tulsa Chamber of Commerce, active in the Audubon Society, a thirty-second-degree Mason, and a Democrat.

Martin's background combined fundamentalist faith with a strong commitment to science. His interest in magic began early, when his father showed him his first trick, the papers-on-knife trick. By his high-school years he was contributing regularly to *The Sphinx*, a magazine devoted to magic. His high-school mathematics teacher, Pauline Baker Perry, first stirred what later became a lifelong interest in mathematics. He decided he wanted to study physics at Caltech. In 1932 Caltech did not have a program for the freshman and sophomore years; so Martin set off for two years at the University of Chicago, intending to transfer to Caltech later. This was a fateful choice. At the time Robert Maynard Hutchins had recently become president of the university and a general education in humanities was required for the student's first two years. Thus it happened that Martin did not take a single college mathematics course in his first years at Chicago. By the time he was an upperclassman,

---
*See note on authorship at the end of this article.
† "Column" refers to Martin Gardner's "Mathematical Games" column in *Scientific American*.

he was caught up in the excitement of the philosophy department at Chicago with such teachers as Charles W. Morris, Rudolf Carnap, and Charles Hartshorne.

During his undergraduate years at Chicago he struggled to reconcile the Methodist fundamentalism he was raised in with the rational scientific philosophy he found at the university. He also made lasting friends at the university and among Chicago's magicians. His long spiritual and philosophical struggle is fictionalized in his novel, *The Flight of Peter Fromm* (William Kaufmann, Inc., 1973), which he first roughed out in 1946 and 1947. The skeptical rationality he gained in this struggle served him well.

In 1936 he graduated a Phi Beta Kappa in philosophy and, after a brief stint as a reporter for the *Tulsa Tribune*, he went to work in public relations for the University of Chicago. From 1936 to 1939 Martin also pursued graduate work in the philosophy of science at Chicago. In 1941 he enlisted in the United States Navy and served as a yeoman on a destroyer escort in the North Atlantic until the end of the war.

After World War II, Martin returned to Chicago and, aided by the G.I. Bill, resumed his studies at the university. He worked with Rudolf Carnap in the philosophy of science, and he began his career as a free-lance writer. He was no stranger to writing. As mentioned, he began publishing in magazines devoted to magic in high school; he had worked as a reporter; in his work for the University of Chicago he wrote publicity material. He published his first book (on magic) in 1935. He began to write fiction for a number of magazines. Mathematicians will recall the story of the "No-Sided Professor," which first appeared in *Esquire* and has been reprinted since in Clifton Fadiman's *Fantasia Mathematica*. This is the only one of Martin's stories from these years with a mathematical theme.

About 1947 he moved to New York and found congenial friends among the magicians and writers there. Among these friends were the magicians Persi Diaconis and Bill Simon and the writer Gershon Legman. It was Simon who introduced Martin to Charlotte Greenwald and later served as best man when Martin and Charlotte were married in 1952. Persi Diaconis is one friend who, like Raymond Smullyan, shares Martin's interest in magic and mathematics. New York offers a free-lance writer the advantages of a superb research collection in its main public library, where Martin did much of the research for his *Fads and Fallacies in the Name of Science* (G. P. Putnam's Sons, 1952; Dover, 1957). In New York Martin also worked as a staff writer for *Humpty Dumpty's Magazine* for eight years and only reliquished this job after he began his "Mathematical Games" column for *Scientific American* in 1957—but I get a bit ahead of myself in mentioning the column.

At this point the stage was set for the appearance of the column. Readers acquainted with Martin's tough, brilliant, and amusing attack on pseudoscience, *Fads and Fallacies in the Name of Science*, will know him as an armed and dangerous skeptic. Those who read his novel, *The Flight of Peter Fromm*, will know that Martin, having laid his own spiritual ghosts to rest, was ready to go forth without qualms to dispatch ghosts or dragons of others to oblivion. Those who know his short stories can have no doubts about his talents as an author. Finally, after seven years on staff at *Humpty Dumpty's*, his ability to produce creative material on schedule was an established fact.

However, important elements are still missing from our story. Although the "Mathematical Games" column in *Scientific American* is about to be launched, it seems that the main elements—*Scientific American* and mathematics—have not yet appeared. Martin's last formal class in mathematics was in high school. His training was in philosophy of science, and he was a professional writer. Martin had done an article titled "Logic Machines" for *Scientific American* in 1952, but that was ancient history. The turning point came with his December 1956 *Scientific American* article "Flexagons." Perhaps we can trace a thread that leads from his work in philosophy to his article "Logic Machines." Perhaps there are links between the cut-and-fold features he did for *Humpty Dumpty's* and the article "Flexagons." The flexagon article brought Martin into contact with mathematicians John Tukey, Bryant Tuckerman, and A. H. Stone and with physicist Richard Feynman—an indication of the sort of person he would meet more often once he began his column. Whatever these continuities may have been, in 1957 there was a major break. Gerard Piel of *Scientific American* was impressed by Martin's article and the interest in it and by the popularity of James R. Newman's four-volume *World of Mathematics* (Simon & Schuster, 1956).

Gerard Piel asked Gardner if he felt there was enough material on recreational mathematics to fuel a monthly column. Martin said yes and took on the job. The rest is history.

Thus men of action seized the day and shaped events to come. Piel had initiated one of *Scientific American's* most successful features, and Martin Gardner set out on a new phase of his life. He quickly assembled a library of recreational mathematics classics, including Ball's *Mathematical*

Recreations and Essays and Kraitchik's *Mathematical Recreations*, and subscribed to a dozen journals related to mathematics. These resources would have availed him little had he not had a passionate interest in understanding things and the ability to write with clarity and humor even on so dry a subject as mathematics.

Of course in 1979 his library and files are much more extensive than they were when he started the column. However, neither a library nor files nor a network of informants can write an interesting column. Each month Martin still faces the same difficult task he did at the start, but with more material to choose from. Gathering material is now less a problem, but the choices are harder to make. He handles most of the work himself, with help from his wife on checking and proofing.

To explain arcane science or mathematics one must first understand it. Because Martin is neither a mathematician nor a physicist, he must first learn the material before he writes a column or a book. The readers of these books benefit from Martin's labor at understanding a subject, whether it be in mathematics or physics or philosophy of science. They can be sure that Martin, the author of *Fads and Fallacies*, will sift out any nonsense—there is some nonsense and careless thinking even in legitimate science. Finally, Martin takes care to make things clear, logical, and understandable, because these are essential qualities that distinguish scientific knowledge from pseudoscience.

Martin Gardner with one of his great loves, *Alice in Wonderland*.

It may seem that all this discussion leads only to "Mathematical Games." But one should not shortchange Martin's other interests. In 1979 he has about thirty books in print. His *Annotated Alice* sold 40,000 copies in the fifteenth year after its publication. His writings range from articles in mathematics journals to books on science, philosophy, mathematics, literary criticism, and magic. He has also written children's books, not to mention his many books on mathematical games. He is active in the Committee for the Scientific Investigation of Claims of the Paranormal, which publishes the *Skeptical Inquirer* to fight the rise of pseudoscience.

His columns are wide ranging and substantial. The kindred spirit between his October 1975 article on extrasensory perception and his interests in magic and his classic *Fads and Fallacies* is easy to trace. Many of his articles are as much about philosophy as about mathematics. Despite our friendship and the occasional help I am able to give Martin, I sometimes have the feeling that he may secretly harbor doubts about the science of numerology. Of course, as the numerologist, I must note that he simply does not give enough coverage to the important field. Otherwise, I cannot fault his work.

By happy chance Martin lives on Euclid Avenue in a small town up the Hudson from New York with his wife, Charlotte. One of their two sons is at Bard College and the other is married and lives in Bowling Green.

I offer these few facts and thoughts about Martin Gardner so that his readers can know something of how the "Mathematical Games" column came to be, and how it is that Martin Gardner was able to carry it off with no special training in the subject.

It is the search for clarity, understanding, and pattern that drives mathematics. Martin is firmly committed to this search. He adds to these qualities wit, humor, and a relentless devotion to the truth. Why, if there had been no such thing as recreational mathematics when he started, he might simply have invented the whole field just as he invented me.

Note on the Source of This Text

The above text is a faithful copy of output from the LOGOS XII text generator running on the HARE 6340 computer at the Academy of Lagado early in 1979. This device generates material on every subject, selecting, editing, and cross-checking it for accuracy and consistency. It is the modern scientific realization of a crude project first described in the western literature by Lemuel Gulliver in his 1727 account of his voyages—in particular, his 1707 voyage to Laputa and Lagado. Happily, jet travel and modern technology have brought Lagado to our very doorstep, and we no longer have to suffer the inconveniences Gulliver did to admire the works of the Academy of Lagado and its successors.

LOGOS XII has a large, accurate data base, and its internal checks for consistency were hitherto believed to be totally effective. Thus it should be possible to assure the reader of the absolute truth of the above account. However, this leaves us with a paradox, since the text is alleged to be the work of one Irving Joshua Matrix, whereas it is actually output from LOGOS XII.

In an effort to discover if the LOGOS XII program was malfunctioning, I have checked the dates and other facts presented against other sources and found them to be completely accurate. Thus I must leave this paradox to the reader to resolve. Perhaps the provenance of this document will never be known.

PETER RENZ

MARTIN GARDNER

Interviewed by Anthony Barcellos

The interviewer was assisted by a panel consisting of the *Two-Year College Mathematics Journal* editor, Donald Albers; W. H. Freeman mathematics editor, Peter Renz; and mathematicians Ronald Graham, of Bell Laboratories, and Stanislaw M. Ulam, of the University of Florida at Gainesville. Gardner looked back over his twenty-two years as the mathematical games columnist for *Scientific American* and forward to his sixty-fifth birthday and possible retirement.

Origin of "Mathematical Games"

MP: *How did a philosophy student at the University of Chicago eventually find himself writing about mathematical games and recreations for "Scientific American"?*

Gardner: It was a combination of a lifelong interest in mathematics, without any formal training, and just a series of accidents. When I was in high school, my great love was mathematics, and that was my high-school major. My hopes were to become a physicist. I wrote to Caltech, which was where I wanted to go, and found that I had to spend two years in college before they would take me. So I went to the University of Chicago, and there I got hooked on the philosophy of science. So instead of transferring to Caltech I just stayed on at the university and got a bachelor's degree. After I graduated I went into journalism, and then spent four years in the Navy. After that I went back to Chicago and did some graduate work in philosophy, but didn't get any higher degree. Then I went to New York to become a free-lance writer, and for the first eight years there I was earning my living writing for a magazine for children called *Humpty Dumpty*; I did their puzzles. And then I sold an article on the hexaflexagon to *Scientific American*.

MP: *You just wrote it and sent it off to them?*

Gardner: I had sold them one previous article on logic machines, so I queried them about this second piece, explained what a flexagon was, and they said, "Go ahead." So I made a trip to Princeton and interviewed John Tukey and Bryant Tuckerman; they and Arthur Stone and physicist Richard Feynman were the four who did the pioneer work on hexaflexagons when they were undergraduates at Princeton. After that piece ran in the *Scientific American*, Gerard Piel, the publisher, called me in and asked me if I thought there was enough material in the field of recreational mathematics to justify a monthly feature. At that time I don't think I owned any books on recreational math at all, but I knew there was a big field out there. I rushed around New York and bought all the major references I could find, like Ball's *Mathematical Recreations and Essays*, and started a library. I turned in the column, Gerry Piel liked it, and that was the start of my column. It was Gerry Piel's idea.

I must say this is about the time that Newman had brought out *The World of Mathematics*. Nobody expected that these four volumes would become a best seller, but they did. They went through innumerable printings. So Gerry Piel suddenly realized there was a tremendous market out there. It came as sort of a revelation.

I have enjoyed doing the column ["Mathematical Games"] because I love mathematics, and I have a pretty good head for it in general. The fact that I really don't know too much about mathematics, I think, works to my advantage: If I can't understand what I'm writing about, why my readers can't either.

MP: *Can you explain what it is about mathematics that makes you like it so much?*

Gardner: I can't say anything different from what mathematicians have said before. It's just the patterns, and their order—and their beauty: the way it all fits together so it all comes out right in the end. It exercises your reason. Almost all the great mathematicians said something like that.

MP: *How do you choose your topics?*

Gardner: Well, it's hard to say. I have a lot of things in mind. I have a big file of possible future topics. I try to pick a topic that is as different as possible from the last few so that I get maximum variety from month to month—sort of a surprise element. And then I try to pick topics that are half recreational mathematics but also lead into what I think is significant mathematics.

MP: *How did people react to your April Fool's joke in the April 1975 issue? Were they mostly amused or mostly distressed?*

Gardner: I think that eventually they were mostly amused, but it came as a tremendous surprise to me and also to the editors of the magazine; because when I turned in the column it seemed inconceivable to me that anyone could take it very seriously. I got several thousand letters on it, and the thing that startled me most was that I was getting letters from people who took it all seriously except for one item in the column that was in their speciality. For example, I got about a thousand letters from physicists; and a typical letter would say, "I enjoyed your column very much and you're doing a great service by letting us know about these new breakthroughs, but I think you make a terrible mistake in demolishing relativity." And then there would follow four or five pages of elaborate diagrams showing why my paradox was not a paradox after all. The map-coloring thing drew the most letters. A lot of people colored the map and then send me the colored map. By the way, that hoax is still going on: About a year ago somebody sent me a clipping from an Australian paper announcing the fact that the four-color theorem had been proved by Appel and Haken; and then some reader wrote in and said that this proof couldn't possibly be true because *Scientific American* had published a map which was a counterexample.

MP: *Among your many columns are there one or two that you are especially proud of?*

Gardner: I particularly enjoy writing columns that overlap with philosophical issues. For example, I did a column a few years ago on the marvelous paradox called Newcomb's paradox, in decision theory. It's a very intriguing paradox and I'm not sure that it's even yet resolved. And then every once

Gardner in his study in 1983. Since retiring (?) from *Scientific American*, he has authored *The Whys of a Philosophical Scrivener*; *Wheels, Life, and Other Mathematical Amusements*; *Aha! Gotcha*; *Aha! Insight*; and several articles.

in a while I get a sort of scoop. The last scoop that I got was when I heard about the public-key cryptograph system at MIT. I realized what a big breakthrough this was and based a column on it, and that was the first publication the general public had on it. [The system produces a code that is, in practice, impossible to break.]

MP: *Do you get many scoops?*

Gardner: No, that happens very rarely.

MP: *Besides all the standard books on recreational mathematics you must have developed other more arcane sources, and also, obviously, a network of correspondents and consultants.*

Gardner: That's right. Aside from the books—and I try to buy all the books that come out on recreational math—the second big source is, of course, periodicals. I subscribe to about ten journals that publish recreational problems, including, of course, the *Journal of Recreational Mathematics*. The third major source is just a big correspondence with readers who send me their ideas. It developed very soon once the column became popular and people interested in recreational math started writing to me. Then if there's something that I don't quite understand, I rely on mathematicians that I know, like Stan Ulam and Ron Graham, to give me an opinion.

Pseudoscience

MP: *In "Fads and Fallacies" you said, "It is not at all amusing when people are misled by scientific claptrap." Do you think things have gotten any better since you wrote that in the early fifties, or are people more susceptible to pseudoscience than every before?*

Gardner: Oh, I think things have gotten tremendously worse in the past twenty years as far as science is concerned. I don't discern any great increase in pseudomathematics. And I don't know why that is. There are always people around who are trisecting the angle and things like that; and, of course, I hear from them occasionally. But the interest in pseudoscience on the part of the general public has been on an incredible rise. I have often thought of doing a sequel to *Fads and Fallacies*, but I haven't had enough time. John Wheeler's blast at the AAAS (American Association for the Advancement of Science) is the first time, I think, that a major scientist has taken a strong position on the rise of pseudoscience. Wheeler had been attending a conference in Switzerland. He was speaking on the subject of quantum mechanics and consciousness, and a parapsychologist rushed up to him afterward, embraced him, and said, "I'm so happy, Dr. Wheeler, to learn that you think quantum mechanics provides an explanation for psychic phenomena." Poor Wheeler was taken aback and realized for the first time that the papers he had been writing had been picked up and misinterpreted by the parapsychologists—and that they were quoting him all over the place as justification. So when Wheeler got back and attended the AAAS conference in Houston and found himself on the same panel with Puthoff and Targ of Stanford Research Institute,* who lectured on clairvoyance tests, and Charles Honorton, a parapsychologist from Maimonides, he issued a strong statement which started out by saying that if he had known that they were on the panel he wouldn't have come. In the past ten years the situation has gotten so much worse that funding has been diverted into crazy theories. He pointed out that we now have two thousand professional astronomers whereas there are twenty thousand professional astrologers. So he thought it was time for the AAAS to reconsider whether the parapsychologists should have a special affiliation with the organization [AAAS].

Mathematicians and the Public

MP: *Sherman Stein wants to know if you think mathematicians are doing anything wrong in their public relations and whether other scientists or disciplines do it better.*

Gardner: That's a tough one, because almost all of the really exciting research going on in mathematics is not the sort of thing that the public can understand. It takes a considerable knowledge of mathematics to know what the breakthroughs are whereas big breakthroughs in biology and many other fields are popularized fairly easily. Certainly mathematicians like Sherman Stein have done a marvelous job of doing popular articles on the subject, and books that the layman can read and understand; but I really don't know what could be done that isn't already being done.

---

*Now called SRI International.

MP: *Well, I think that you, Martin, do that. In fact, yesterday Ron Graham gave a marvelous lecture about some esoteric question; and I was wondering during it: Well, why devote so much ingenuity? Then I remembered what, I think, Fourier or Laplace wrote, that mathematics—one reason for its being—is to defend the honor of the human mind. And your column does that, defends the honor—maybe also the sense of humor—of the human brain.*

Gardner: Why, thank you. Well, it would be good if other science magazines would run similar columns on mathematics, and perhaps they will. Right now *Time* magazine is making preparations for a new magazine devoted to science, to be edited by Leon Jaroff. I think he's planning to have some sort of a department devoted to reporting on mathematics. I hope it goes through, because I think Leon Jaroff will do a very good job on it; and it will fill a place that *Omni* should have filled, but I don't think quite does it, because the publisher, Guccione, is pressing for more and more emphasis on the paranormal.

MP: *If perhaps the scientific community made a greater effort to tell the public what they're up to and what's going on, it would make people appreciative of what science is as opposed to what is being ladled out in the media.*

Gardner: Yes, I think so. However, I can understand why scientists are reluctant to take time off from their work in order to appear on a TV program, let's say, or write a popular article attacking pseudoscience. The situation has gotten so bad now that I think you may see more and more of them taking a little time for public relations, but so far only a very small number do that. Carl Sagan and his writings about Velikovsky is an example of an astronomer taking the time to respond to Velikovsky. Most astronomers wouldn't want to waste time even meeting Velikovsky. Sagan has a book coming out, by the way, a collection of essays; and a portion of the book is devoted entirely to pseudoscience—not just Velikovsky, but others as well.

MP: *It seems to me you're saying there's a responsibility to know what the opposition is up to that scientists have perhaps ignored.*

Gardner: Yes, I think so. The other day I was reading some old essays by T. H. Huxley, and I noticed quite a number of them deal with what he regarded as pseudoscience. Huxley, as you know, was one of the great popularizers of science, and not only a good geologist and biologist, but also a very skillful writer. He wrote quite a number of articles that were specific exposés of attacks on the theory of evolution and other kinds of pseudoscience.

MP: *Martin, I wonder if you would mention some books or magazines that you think would be particularly attractive to people who are interested in recreational mathematics.*

Gardner: Certainly the great English classic is Ball's book on recreational math. And then Kasner and Newman did a marvelous job of popularizing modern mathematics in their *Mathematics and the Imagination*. It's not so much recreational math, although they have some in it. It's just a very exciting and well-written survey of certain aspects of modern mathematics. And Honsberger has written two books called *Mathematical Gems* that I think are excellent.

Advice to Teachers

MP: *We as teachers are involved, in a way, in an ongoing public relations effort with our students. What advice can you give to teachers of mathematics to make their subject more interesting?*

Gardner: Well, I can only repeat what I've said before in the introductions to some of my collections, and that is that I've always felt that a teacher can introduce recreational math; and I'm defining it in the very broad sense to include anything that has a spirit of play about it. I don't know of any better way to hook the interests of the students. I think teachers are beginning to discover that it does have pedagogical value. And successful textbooks like Jacobs's *Mathematics, A Human Endeavor* are heavily loaded with recreational math.

MP: *Do you think that this [recreational math] will sell better than the "applications" approach that people are trotting out now?*

Gardner: I think the two should go together. I certainly don't think a teacher should spend all the time on puzzles, of course; then the whole thing becomes trivial. If math can be applied in a way that's useful in the child's experience and things can be introduced so they're challenging and have a play aspect, I think the two sort of go together. And we're entering a revolution now where everybody's going to own a pocket computer. This is another important aspect of the whole thing. Kids love to work things out on a programmable computer. So I think that if the applications of math and the recreational aspects of math are tied in with computer programming then that will do a lot to get to the students.

As you probably know, the big field in games now is no longer the board game, played with counters on a board. All the games that are selling very well now are electronic games that involve some type of computer.

Of course, one of the dangers of having a pocket calculator is that you forget how to do ordinary arithmetic. I had this brought home to me very recently. I had a problem in a column a month or so ago: It was to arrange the first nine positive integers to form a nine-digit number such that the number formed by the first two digits was divisible by two, the number formed by the first three digits was divisible by three, and so on until the entire nine-digit number was divisible by nine. It has a unique solution. I have received now between two and three hundred letters from readers that all say, "I read that it had a unique solution, but I have found two more." And then they all list the same two "solutions," and none of them has the first eight digits divisible by eight. This had me honestly puzzled until suddenly I realized what had happened; and that is that they were all using small pocket calculators with eight-digit read-outs. When they divided the eight it didn't show any remainder. Not one of them bothered to divide by eight by hand. I think in the long run though it's going to turn out that the pocket calculator is a blessing rather than otherwise.

Who is Peter Fromm?

MP: *Peter Renz reminded me before we began this interview that you are principally a writer, and though at the present time you write mainly about recreational mathematics you have ventured apart from that. Your novel, "The Flight of Peter Fromm," deals with a theology student at the University of Chicago in the thirties and forties and prompts an obvious question: Are you Peter Fromm?*

Gardner: Well, you must be one of the thirteen people who read that novel. Yes, basically that's sort of an autobiographical piece. I never went to theological school, and Peter has a different personality from mine, but basically the novel is about the changes that I went through when I was a freshman and sophomore in college. I came from an orthodox Protestant family in Tulsa, the fundamentalist capital of the world. I went through a Protestant fundamentalist stage when I was in high school. I quickly got over it in college, and I drew on that experience when I wrote the novel.

MP: *But there was a dramatic shift for you in moving from high school to college. In high school you were still thinking about doing physics seriously and continuing with the mathematics that would be needed to do physics, and yet when you went to the University of Chicago you did not enroll in a freshman math course.*

Gardner: Well, there really wasn't much opportunity in the first two years. When I was at Chicago they had just instituted what they called the New Plan, under Hutchins, and the idea was for the student to get a general liberal education the first two years. So the first chance really to take a math course would have been the third year, and by that time I had decided to major in philosophy. The other big change, of course, was that I got over my Protestant orthodoxy. I finally got it all off my chest in that novel.

MP: *Was that the purpose of the novel?*

Gardner: Actually, when I began free-lancing I started out writing fiction; and that was what I most wanted to do. When I got out of the Navy and went back to Chicago, the only reason I didn't take back my old job—public relations work at the University of Chicago—was that I sold a short story to *Esquire* magazine. So I was supporting myself for about the first two years by selling fiction to *Esquire*. I had about a dozen stories published in the magazine. It was during that period that I wrote the first draft of this novel, and I found it to be totally unsalable. It was when Bill Kauffman, who had formerly been with W. H. Freeman, came out to see me one day and asked me if I had any manuscripts lying around that I thought of this thing; and I pulled it out and let him see it. He said he would publish it if I revised it; so that's how it got published—an old novel from way back in my youth.

MP: *Do you have another in the works?*

Gardner has been a debunker of pseudoscience and has exposed several "psychics" as frauds.

Gardner: No, the only book that I have in the works that is different from what I've been writing is a book of philosophical essays that will deal with various problems in contemporary philosophy. And again, it's the sort of book that probably will not sell very well. That's the next major project that I would like to do.

The Future of "Mathematical Games"

MP: *I've been hearing things suggesting you might stop doing the column for* "Scientific American."

Gardner: Well, that's true. I'll be sixty-five in October and eligible for retirement benefits. Even though I contribute on a free-lance basis, I'm actually considered to be on the staff of the magazine, so I'm eligible for their health benefits, retirement benefits, and so on. I have an option to resign from the column in October, and I also have an option to continue it—in other words, it's not a forced retirement. I just haven't made up my mind yet. And I don't know quite how to work it out, because if I could stop doing the column I would have time to do books I would like to do—for example, the book of philosophical essays. I seem to get to work on it about two days a month, and at that rate I'll never finish it.

MP: *I believe that* "Scientific American" *might have some qualms about letting the column evaporate*.

Gardner: They've asked me to continue, and I don't know quite how to work it out. I think that my only suggestion was that it might be good to have it replaced by a column that would emphasize computer recreation, because that's really a growing field.

There might be a compromise in which maybe I would alternate columns with somebody who would do columns more on the computer side. Up until now I don't think it [computer recreations] would have had a big readership, but I'm thinking that it would in the next four or five years. I don't have any regrets about doing the column, because I've enjoyed every column that I've written. The success of the column has made it possible for me to sell books in other fields that I don't think I could have sold otherwise. It's just a question of whether I want to go on for the next ten years writing the same kind of thing, or whether I should at this point stop the column and get into the things that I want to write about while I still have my wits about me.

Alice, the Snark, and Casey

MP: *That's almost an hour about mathematics. We could spend another hour dealing with the marvelous things you've done in physics, relativity, articles and books—and also "The Annotated Alice." That's entirely different. You haven't mentioned it yourself.*

Gardner: No, but I would like to do more books of that sort too. It's just very, very difficult to do them and keep the column going. Each year the column gets more difficult to do because my correspondence on it increases.

The Annotated Alice, of course, does tie in with math, because Lewis Carroll was, as you know, a professional mathematician. So it wasn't really too far afield from recreational math, because the two books are filled with all kinds of mathematical jokes. I was lucky there in that I really didn't have anything new to say in *The Annotated Alice* because I just looked over the literature and pulled together everything in the form of footnotes. But it was a lucky idea because that's been the best seller of all my books.

MP: *And you did "The Annotated Snark"?*

Gardner: I later did *The Snark*, but that didn't do too well.

MP: *And "The Annotated Casey at the Bat," what prompted that?*

Gardner: Oh, I have always been interested in the fact that there are poems that are not great poetry but seem to outlast the entire poetic output of poets who were very famous in their day. I guess the best term for them is popular verse. They don't pretend to be great poems and yet a single poem written by an individual like Thayer, who wrote *Casey at the Bat*, can go on and on and on, and everybody knows about it. It will probably be remembered after everybody's forgotten every poem ever written by, say, Ezra Pound. This has always struck me as a very curious phenomenon. So I did an article on the history of *Casey at the Bat* that I sold to *Sports Illustrated*. That was how it started. After it appeared, it occurred to me that I might put together an anthology of sequels to *Casey*. That's what the book is, a collection of the original poem with sequels by various other people. Then I annotated all the poems with sort of fake annotations to tie them all together in a connected story. The book is done as kind of a joke. It didn't sell too well either.

MP: *How about things in physics like your "Ambidextrous Universe"?*

Gardner: *The Ambidextrous Universe* was great fun to write, and I'm happy to say that Scribner's is bringing it out in a new edition this month with four new chapters. When the book came out, all the work on time-reversal was too late for me to catch, so the original book doesn't have anything in it about the recent work that's been done on time asymmetry. This gave me a chance to write four new chapters that deal with speculation about time-reversed galaxies and that sort of thing.

Magicians and Debunkers

MP: *How about magic?*

Gardner: It's just been a minor hobby of mine ever since I was a boy. I've alway been interested in the field. I've written some books that sell only in magic stores.

MP: *You are affiliated with a group of magicians and scientists who are debunkers, right?*

Gardner: Yes, we have an organization called the Committee for the Scientific Investigation of Claims of the Paranormal, and we publish a magazine called the *Skeptical Inquirer*. It's now a quarterly, but it first came out twice a year. The idea was to put out a magazine that would try to tell the other side of the story, so the editors and TV management could turn to it if they wanted to find out how the scientific community felt about something.

MP: *Is the initial response encouraging?*

Gardner: Yes, I think so. I think the membership of the committee has been growing and the circulation of the magazine has been increasing. It's been very well edited by Ken Frazier, who used to be the editor of *Science News* magazine. It's doing very well considering the fact that it's only about a year old and we haven't had too much publicity about it.

MP: *What do magicians do in the organization? Why would they be interested?*

Gardner: Well, because right now so many parapsychologists are being taken in by "psychics" who are just simply magicians in disguise. The outstanding example, of course, is Uri Geller, who—in my opinion, and the opinion of almost every magician who has studied Geller—is nothing more than a magician who is pretending to be a psychic. A lot of the physicists who are into parapsychology, like Puthoff and Targ at SRI, for example, took Geller very seriously. There's really a rather surprising number of top physicists who are convinced that the kind of psychic phenomena that Uri produces is really genuine. There's Brian Josephson, who won a Nobel Prize for his work on the tunneling effect, and Gerald Feinberg, at Columbia. They thought Geller was a genuine psychic. What our committee is really saying is that before you go overboard and write articles and books about this kind of psychic phenomenon, at least have the sense to consult a magician and get him in as an observer. Of course, the outstanding example of a man who did make a real ass out of himself is John Taylor, the mathematical physicist from England, who fell for Geller and wrote this gigantic book called *Superminds*. He was finally convinced that he was taken in and now has retracted his whole position in articles in *Nature*.

Martin Gardner with the Alice Statue in Central Park, New York City.

MP: *What about the mathematical side of this parapsychological research?*

Gardner: I'm going to do a column that will discuss this whole aspect of contemporary parapsychology, and the need for a more sophisticated understanding of some of the statistics involved.

MP: *Just to cheer you up, I noticed that some national poll indicated that something like 70 percent of all high-school students interviewed believed that ESP was a proven fact.*

Gardner: Absolutely, there's no question that most people believe that. If you try to tell them that 99 percent of the professional psychologists around the country take an opposite position, they can't believe it. They see these pseudodocumentaries on TV in which these things are treated as though they're genuine scientific break-throughs. I think it's damaging to American education and damaging to American science. There was one very good documentary on the Bermuda Triangle from the skeptical point of view that I think "Nova" did, but it's very hard to get the major networks to back such programs because the public is interested in the other side of the story. One program that I hope will get off the ground in a year or two is the Children's Workshop Theater. They're the ones who produce "Sesame Street," and they have finally gotten funding for a program dealing with science for children. I think that the people involved are going to be very careful to keep pseudoscience out, although one of the advisors is dean of the engineering school at Princeton University—a man named Robert Jahn. To everybody's surprise he has suddenly become a new convert to parapsychology. And the latest news is that Charles Honorton is leaving Maimonides and teaming up with Dr. Jahn at Princeton. They're going to have a parapsychology laboratory in the engineering school at Princeton. So that's a little bit discouraging. I don't know how that's going to work out.

BIBLIOGRAPHY

Martin Gardner

Articles in Scientific American:

Flexagons (December 1956).
Free will revisited, with a mind-bending prediction paradox by William Newcomb (July 1973).
Reflections on Newcomb's problem: A prediction and free-will dilemma (guest column by R. Nozick) (March 1974).
Six sensational discoveries that somehow or another have escaped public attention (April Fool's hoax) (April 1975).
Concerning an effort to demonstrate extrasensory perception by machine (Puthoff and Targ) (October 1975).
A new kind of cipher that would take millions of years to break (August 1977).

Books:

The Ambidextrous Universe, Scribner's, 1978.
The Annotated Alice, Meridian Books, The New American Library, 1960.
The Annotated Casey at the Bat, C. N. Potter, 1967.
The Annotated Snark, Simon & Schuster, 1962.
Fads and Fallacies in the Name of Science, Dover, 1957.
The Flight of Peter Fromm, William Kauffman, Inc., 1973.

W. W. Rouse Ball and H. S. M. Coxeter, Mathematical Recreations and Essays, 12th ed., University of Toronto Press, 1974.
Ross Honsberger, Mathematical Gems (1973) and Mathematical Gems II (1976), Mathematical Association of America.
Harold Jacobs, Mathematics, A Human Endeavor, W. H. Freeman, 1970.
Edward Kasner and James R. Newman, Mathematics and the Imagination, Simon & Schuster, 1940.
James R. Newman, The World of Mathematics, Simon & Schuster, 1956.
Sherman K. Stein, Mathematics: The Man-made Universe, W. H. Freeman, 1976.
John Taylor, Superminds: An Enquiry into the Paranormal, Macmillan, London and Basingstoke, 1975.
John Taylor and E. Balanovski, Can electromagnetism account for extrasensory phenomena? Nature, 276, 64(November 1978).
John Archibald Wheeler, Not Consciousness, but the Distinction Between the Probe and the Probed, as Central to the Elemental Quantum Act of Observation; Appendix A: Drive the Pseudos Out of the Workshop of Science; Appendix B: Where There's Smoke There's Smoke, American Association for the Advancement of Science Annual Meeting, Houston, January 8, 1979. [See also Psi-Fi in Science and the Citizen, Scientific American, April 1979.]

RONALD L. GRAHAM

RONALD L. GRAHAM

by Bruce Schecter

The artistically landscaped buildings contain some of the most advanced scientific apparatus in the world. But in one small room at the Bell Laboratories complex in Murray Hill, New Jersey, the only equipment in sight seems more appropriate to a novelty shop. The room is Ronald Graham's office, and it is cluttered with Rubik's Cubes, geometric puzzles, Chinese illusions, juggling balls, and other curiosities. As Graham chats with a visitor, his hand moves across his desk toward a blue plastic object shaped like a cigar flattened on one side. Catching the visitor's eye, Graham gives the object an offhand flick that sets it spinning clockwise.

After a moment, something strange happens: the spinning object slows, stops, and then—reverses! Graham does an exaggerated double take, his eyebrows shoot up, he shrugs, and he asks with mock surprise, "How did *that* happen?" The answer, he subsequently explains, has something to do with the cigar's subtly asymmetrical shape and some complicated physics and mathematics.

Graham, 46, is the head of Bell Labs' Mathematical Studies Center, and he also delights in mysteries, mystification, and puzzles. One of the world's leading combinatorial mathematicians, he has solved professional problems that, to the initiated, are as puzzling and wonderful as the reversing spin of the blue cigar. For the past 20 years he has confronted the formidable challenges that arise from the need to route hundreds of millions of telephone calls through the intricate communications web of cables, microwaves, and satellites that embraces the earth. The mathematical techniques and theorems he has developed in the process can be applied not only to the routing of information with a computer, but also to the efficient scheduling of an astronaut's day, or even to the allocating of an entire nation's resources. He has the rare ability to translate real-world problems into mathematics, and as colleague Persi Diaconis, a mathematician at Stanford University, puts it, "Ron, as much as anybody, is responsible for bringing high powered math to bear on computer science."

Graham's life is itself a scheduling problem that would try the capacity of any computer. He is a remarkably prolific mathematician, publishing more than a dozen papers a year. He sits on the editorial boards of some 20 mathematics journals, travels extensively, and lectures frequently. He is also a talented and dedicated juggler, and has been honored for his skills by being elected president of the International Jugglers Assocation. He constantly works at improving his juggling technique; a net hangs from his office ceiling to snare the occasional ball that escapes him. In his younger years he earned money as a trampoline acrobat, and he still stays in shape by bouncing and flipping on his home trampoline.

Even now he delights in finding new skills to master—and he masters them better than most. "You should never be afraid to be a beginner," he says. "It keeps your mind open and flexible." Since reaching adulthood, Graham has learned to bowl (he has rolled a couple of 300 games), throw a boomerang, play Ping-Pong (he was the Bell Labs champion), parachute jump, speak Chinese, and play the piano. He has run in a marathon, and is now learning to play tennis. How does he do it? "Well," he says in a slow, quiet voice, "there are a hundred and sixty-eight hours in a week."

Many of those hours are consumed by his job. Besides administering a large department, directing its research, and solving mathematical problems, Graham thinks nothing of dropping everything to help a graduate student untangle a messy proof, or to locate an obscure reference for a colleague down the hall. "I think all of us wish we had more time with him," says Diaconis, "but when you're really in trouble with a mathematical proof or something, he'll never leave you hanging. He's a nice genius."

He also thinks very fast. Recently, Diaconis asked him for advice on a problem involved in producing images from data generated by the brain-scanning apparatus. Three days later Graham returned with 60 pages of calculations that solved the difficulty. Another time, Diaconis was giving a talk describing research he and Graham were conducting, and concluded with "This case is unsolved." At that point, Graham rose from his seat in the audience and outlined a solution he had just thought out. The audience was so impressed it applauded—a rare show of emotion at a meeting of mathematicians.

One reason for Graham's prodigious productivity, he says, is that he is often on the move and difficult to reach. That gives him time for uninterrupted thought. As he explains, "So far there are no telephones on airplanes," and indeed many of his best mathematical ideas emerge while he is belted into an airplane seat. Furthermore, traveling keeps him in touch with almost everything that is happening in his field of mathematics. If a mathematician at Stanford has a problem, Graham is likely to know of some mathematician on the other side of the country who has a solution. He serves as a mathematical matchmaker, bringing together people and problems.

In the jargon of mathematicians, the problems that Graham specializes in are known as "hard"— because of the mind-boggling complexity they assume in successive steps. An example is the infamous traveling salesman problem: a salesman must pick the shortest possible route for his visits to various cities. If there are only a few cities on his route, the problem is almost trivial; he has merely to look at a map, calculate the lengths of possible routes, and choose the shortest one. But as the number of cities increases, the problem quickly grows beyond the capacity of even the fastest computers.

Graham helping a student prove a tricky theorem.

No general and perfect solution has ever been found for the traveling salesman problem, but mathematicians like Graham keep working to improve the partial solutions they have discovered. The results of their labors have practical applications in scheduling, economics, cryptography, and computer science, all of which involve "hard problems," technically described as "nondeterministic polynomial complete" problems. But the hard problem of greatest interest to Graham's employer is the efficient routing of telephone messages between tens of thousands of communities—a problem for which a bad solution could cost untold millions of dollars. By finding better approaches to the traveling salesman and other, related problems, Graham and his colleagues continue to improve long-line telephone efficiency, thereby earning their salaries many times over.

At times Graham's special gifts have helped other organizations, including NASA. During the Apollo moon program, the space agency needed to evaluate mission schedules so that the three astronauts aboard a spacecraft could find the time to perform all necessary tasks—experiments, eating and sleeping, and managing their vehicle. The number of ways to allot these tasks was astronomical—too vast even for a computer to sort out—and NASA officals were concerned about wasting valuable time through inefficient scheduling. NASA had based its schedules on mathematical techniques that produced good but not perfect solutions. The agency wanted to know just how badly it might be erring, so it turned to Graham, who had pioneered a field of mathematics known as worst case analysis. Graham provided a reassuring answer; he proved that NASA's methods never produced a schedule that was more than a few percentage points worse than the theoretical optimum.

But it is in pure mathematics—mathematics with no obvious application to telephone lines or spaceships—that Graham has earned his distinguished reputation. His contributions have been mainly in a field known as Ramsey theory, named for Frank Ramsey, a brilliant mathematician who, before his death in 1930 at the age of 26, published a paper that was to influence a generation of mathematicians. As Graham explains it, "Ramsey theory says that complete disorder is impossible. There is always structure somewhere." Since, according to Graham, mathematics is the study of order, Ramsey theory is particularly appealing to him—so much so that the license plate on his car reads RAMSEY.

A simple prediction of Ramsey theory can be stated in human terms. If two people are selected at random from any group, they will either know each other or be strangers. But Ramsey theory shows that in any group of six or more people, three will either all know one another or all be strangers. This fact is fairly easy for mathematicians to prove, but as the numbers grow larger Ramsey theory becomes more difficult to apply. For instance, it can be shown that in a group of 18 people there must be four who all know each other or are all strangers. But how big a group is required so that it will always include five mutual friends or five mutual strangers? Nobody·yet knows.

Graham's research has increased the scope of Ramsey theory well beyond this type of problem, and it was for this work that in 1972 he shared, along with Bruce Rothschild and Klaus Leeb, the prestigious Polya Prize, awarded by the Society for Industrial and Applied Mathematics. His work in Ramsey theory has also been recognized by no less an authority than the *Guinness Book of World Records*, which acknowledges that Graham holds the record for identifying the largest number ever used in a mathematical proof. The number is so big that it has to be written using a special notation, developed by Stanford computer scientist Donald Knuth, in which exponentials pile upon exponentials in dizzying, astronomically large towers of powers.

Diaconis believes that Graham's work on Ramsey theory "will take a hundred years to be applied, if it ever is." But Graham is convinced that all mathematics is related to the real world, that math really "exists" in some sense. "Is mathematics a creation of the human mind," he asks, "or is it really out there? I happen to believe that it is there, it was always there, and would be there without us. Our job is to discover, not to create. But the act of discovery is a creative process." A field of mathematics known as Riemannian geometry, the geometry of curved space, was investigated, as Graham points out, by 19th century mathematicians who had no notion of applying it to anything "real." But years later Einstein used Riemannian geometry as one of the cornerstones of his theory of general relativity. Concludes Graham, "From my point of view, in some sense, the basic reality is mathematics."

Graham's approach to solving mathematical problems is similar, he says, to his approach to juggling or learning any other new skill: systematic, determined, and intuitive. As he explains, "I may have abstracted this essential feature more than a lot of other people have: trying to break a large problem into smaller problems that are a little easier to deal with." It is like learning Chinese, he says, first mastering the individual sounds, and only then moving on to words and sentences.

Graham also tries to instill these principles when he teaches, and he has taught many subjects, including all levels of mathematics, gymnastics, and juggling. "In order to teach," he says, "you have to be able to focus on exactly where the difficulty is, and that means splitting the problem apart until you can really locate exactly what it is that is difficult." Still, the basic unpredictability of mathematics, Graham concedes, makes it harder to teach than juggling.

It is this unpredictability, he says, that often causes the systematic, analytic approach to break down. "There's really more insight than mechanics in mathematics, more leaps of intuition, more making of connections," and, he says, many of the best mathematical proofs seem to start out with a

**Mathlete Graham limbers up in the hall hoping perhaps
to gain a new perspective on a mathematical problem.**

nonsequitur, in which the mind jumps from the main stream of thought to something seemingly unrelated that ultimately provides the key to the problem. "I feel strongly that your brain is working on many channels at the same time. In a way, it's a kind of mental juggling. A lot of things going on at once. Even when I'm talking, when I'm sleeping, eating, or on the trampoline, part of my brain is dedicated to trying and experimenting, piecing things together." This tangential type of thinking is reflected in Graham's style of speaking; the words come out like accumulating pieces of a puzzle that grows in what sometimes appears to be haphazard patterns.

Graham acknowledges that some problems defy his preferred approach and call for outright wrath. "You can start to get mad at a problem," he says. "Did you ever see a Daffy Duck cartoon? When a few bad thing happend to Daffy—Elmer Fudd or somebody blows him up in a number of different ways—he says, 'Of course, you know this means war!' And you often get a very appealing, attractive problem that gets under your skin. You're living the problem, it's part of you, it's always in the background, running. In some sense this means war. It's kind of life and death."

Graham's wars rage on for a week or so at a time, but then he must back off. As he explains, using words drawn from the vocabulary of mathematics, "It isn't always optimal for me to stay in this mode for weeks and weeks." When a problem *really* gets under his skin, Graham will occasionally resort to putting out a contract on the problem by offering a cash prize for its solution. "It's putting your money where your mouth is," he says. "You say, 'Look, I'd be happy to pay a hundred dollars to have somebody solve this thing and get it out of my hair and relieve my misery. Either prove it or show me that I'm wrong, but let me know!'"

Neither of Graham's children, Cheryl and Marc, is interested in becoming a mathematician, but both are fine trampolinists and jugglers.

Failures sometimes lead Graham to doubt mankind's fitness to do mathematics. "Somehow I feel that human beings and the human brain are relatively recent developments, not designed to prove the Riemann hypothesis or study the space-time structure of the universe," he muses. "Our brains were designed to keep us out of the rain, pick berries, and keep us from being killed. So the brain did that, but now it's got a whole new set of challenges—and we're getting better, but we're still a long way from being good at them."

Graham began to excell in mathematics almost from the start. Born in Taft, California, a small town northwest of Los Angeles, he spent most of his youth moving back and forth between Georgia and California as his father kept changing jobs, moving from oil fields to shipyards. Graham never went to the same school for more than a year and a half. He managed to acquire and shed a Southern accent alternately as he repeatedly crossed the country. "Because I was always kind of a new kid," he recalls, "I was never really accepted socially into the in-groups." Also, because he skipped a few grades, he was always younger and smaller than his classmates. "I was of negative value to the team, so I didn't get involved."

Instead, Graham became interested in mathematics and astronomy. The stars have always fascinated him, but, largely because of a few good teachers and the fact that "math is portable," he fixed on mathematics. In the fifth grade a teacher taught him how to extract square roots of numbers, and with a mathematician's gift for generalization, he wondered whether he could extend the technique to finding cube roots. It was not until much later than he worked out a way to do it, but thinking about the problem led him to higher mathematics.

Eventually Graham's parents were divorced, and he moved to Florida with his mother. Alert to scholarship possibilities, she signed him up to try for a Ford Foundation scholarship. He did well on the examination, got the scholarship, and at fifteen, without graduating from high school, he enrolled at the University of Chicago.

Finally among other bright young students from all over the country, Graham flourished. Still small for his age—though he would eventually reach 6'2"—he began to learn gymnastics, a sport in which the short are not at a disadvantage. He went on to master juggling and the trampoline, and for the first time found himself admired for talents other than his obvious intelligence. Juggling also gave Graham a new skill that he could carry anywhere and that would give him instant acceptance—at least by some people. "If you see another juggler juggle five balls, you already have a certain tie to this person," he says. "You know that he took the time, the energy. You know that what he's going through you went through yourself. There's a certain feeling of comradeship, and it's just as true in mathematics."

Graham became so fascinated with wonderful mathematical objects called Penrose tiles that he and David Hagelbarger mass-produced 100,000 of them at Bell Labs.

When his scholarship ran out, he transferred to Berkeley for a while, and then, facing the draft, enlisted in the Air Force, where he thought he could work hard and get a choice assignment. He did work hard, finishing at the top of his class in communications, but he found himself posted to Fairbanks, Alaska, which he did not regard as choice. Undeterred, he signed up at the University of Alaska by day and worked at his Air Force unit at night. Although he majored in mathematics, the university was not accredited in math, and he was awarded a degree in physics instead.

Mathematical People

Serving out his enlistment, he returned to Berkeley to earn a Ph.D. While he was there, he and two other students formed a professional trampoline troupe, earning money by performing at schools, supermarket openings, and even the circus. He also met and married Nancy Young, a brilliant fellow math major. In 1962, with a fresh Ph.D., he joined the staff of Bell Labs.

Graham moved to New Jersey with Nancy, where they had two children—Cheryl, now twenty and a journalism major at Northwestern University, and Marc, fifteen, a tenth-grader. Neither Marc nor Cheryl wants to follow in Graham's mathematical footsteps, but both are expert trampolinists and competent jugglers. Their father has offered them a prize of $25 for juggling three balls, $100 for juggling four balls, and a whopping $1,000 for juggling five balls. So far both have collected $25. Unlike the mathematical prizes that Graham offers, the juggling prizes are good only within his family.

Divorced four years ago, Graham has since lived alone in a small house about two miles from his office. There, the walls are decorated with interlocking patterns of Penrose tiles (named for their British inventor, the mathematical physicist Roger Penrose)—small plastic panels, shaped like kites and arrowheads, that can fit together in an infinite number of ways. Graham became so fascinated with Penrose tiles that he and a colleague made a plastic mold to mass-produce them. Of the 100,000 he has made, many have been sent to mathematicians around the world for use as both playthings and devices to help prove interesting new theorems.

Gadgets are everywhere in Graham's house. He owns a bewildering array of calculators, as well as several shortwave receivers and two dozen Rubik's Cubes (he is one of the world's leading experts on the cube and a consultant to the manufacturer on copyright-infringement cases). Other belongings include a computer chess game, a juggling apparatus, a video tape recorder, and many more electronic gadgets.

When he needs reassurance about his life and his profession, Graham thinks about Gödel's theorem, which states, roughly, that there is no end to mathematics, that the adventure of mathematical discovery will continue forever. "Mathematics is to me, and to a lot of mathematicians, a very exciting thing," Graham says. "It's an open-ended challenge. No one's good enough to do even a small fraction of what there is to be done. The problems are more than adequate to challenge anyone, and as far as I can tell, that's always going to be the case. It's like juggling. When have you become the absolute juggler? When you can do all the tricks? Well, there's always one more ball."

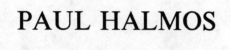

PAUL HALMOS

PAUL HALMOS

Interviewed by Donald J. Albers

Paul R. Halmos is Distinguished Professor of Mathematics at Indiana University and Editor of the *American Mathematical Monthly*. He received his Ph.D. from the University of Illinois and has held positions at Illinois, Syracuse, Chicago, Michigan, Hawaii, and Santa Barbara. He has authored ten books and 100 articles. He is a member of the Royal Society of Edinburgh and of the Hungarian Academy of Science.

The writings of Halmos have had a large impact on both research in mathematics and the teaching of mathematics. He has won several awards for his mathematical exposition, including the Chauvenet Prize, and has twice won the Lester R. Ford Award.

In August of 1981, I interviewed Paul Halmos in Pittsburgh at the combined annual summer meetings of the Mathematical Association of America and the American Mathematical Society. During the course of the interview, Halmos confessed to being a *maverick mathologist*. A *mathologist* is a pure mathematician and is to be distinguished from a *mathophysicist*, who is an applied mathematician. (Both terms were coined by Halmos.) A few of his statements from the interview help to underscore his maverick nature:

> "I don't think mathematics needs to be supported."
>
> "If the NSF had never existed, if the government had never funded American mathematics, we would have half as many mathematicians as we now have, and I don't see anything wrong with that."
>
> "The computer is important, but not to mathematics."

In the pages that follow, Halmos in his inimitable style talks about teaching mathematics, writing mathematics, and doing mathematics. After a short time with him, I was convinced that he is a *maverick* and a *mathologist*.

A Downward-Bound Philosopher

MP: *You have described yourself as a downward-bound philosopher. What does that mean?*

Halmos: Most mathematicians think of a hierarchy in which mathematics is above physics, and physics is higher than engineering. If they do that, then they are honor-bound to admit that philosophy is higher than mathematics. I started graduate school with the idea of studying philosophy. I had studied enough mathematics and philosophy for a major in either one. My first choice was philosophy, but I kept a parallel course with mathematics until I flunked my master's exams in philosophy. I couldn't answer all the questions on the history of philosophy that they asked, so I said the hell with it—I'm going into mathematics. I made philosophy my minor, but even that didn't help; I flunked the minor exams too.

MP: *So mathematics was not your original calling, if you like?*

Halmos: As a philosophy student, I played around with symbolic logic and was fascinated with all the symbols in *Principia Mathematica*. Even as a philosopher, I tended toward math.

MP: *You are the third Hungarian we have interviewed.*

Halmos: I reject the appellation.

MP: *We know that you were born in Hungary and that you lived there until the age of thirteen, but you still reject the appellation.*

Halmos: I don't feel Hungarian. I speak Hungarian, but by culture, education, world view, and everything else I can think of—I feel American. When I go to Hungary, I feel like an American tourist, a stranger. I speak English with an accent, but I speak it infinitely better than I speak Hungarian. I can control it, and I cannot do that in Hungarian. In every respect, except accent, I am an American.

MP: *You may not claim Hungary, but I wouldn't be surprised if Hungary claims you. In fact, you are a member of the Hungarian Academy of Science.*

Halmos: I was elected a member of the Hungarian Academy of Science only a couple of years ago, in recognition of my work, I hope, but I am sure that my having been born in Hungary helped. In theory, it needn't help, as there are a certain number of foreign members elected each year. But if they are in some sense ex-Hungarians or have Hungarian roots, that doesn't hurt. I am not ashamed of my Hungarian connection, but just as a matter of fact I try to straighten out my friends and tell them that they shouldn't attribute to my country of origin whatever properties they ascribe to me.

MP: *How did you come to leave Hungary?*

Halmos: I give full credit to my father. In 1924, when he was in his early forties, he left Hungary, where he had been a practicing physician with a flourishing practice. The country was at peace and in good shape, but he thought it was a sinking ship. He arranged for his practice to be taken over by another physician, who was also foster father to his three boys, of whom I am the youngest. (My mother died when I was six months old, and I never knew her.) He came to this country with the feeble English that he had learned. After working as an intern at an Omaha hospital for a year in order to prepare for and pass the state and national boards, he started a practice in Chicago. Five years later he became a citizen and imported his sons. Coming to America wasn't a decision on *my* part; it was a decision on *his* part. It turned out to be a very smart move.

MP: *Did you have any glimmerings of strong mathematical interests as a child? We know the Hungarians do a remarkably good job of producing superior mathematics students.*

Halmos: Yes and no. I cannot give credit to the Hungarian system, which I admire and about which I am somewhat puzzled (as are most Americans), as to how they produce Erdös's, Pólya's, and Szegö's, and dozens more that most of us can rattle off. I know the rumor that they look for them in high school and encourage them and conduct special examinations to find them. Nothing like that had a chance to happen to me. By the age of thirteen I was exposed to a lot more mathematics than American students are exposed to nowadays, but not more than American students were exposed to in those days. I was exposed to parentheses and quadratic equations, two linear equations in two unknowns, a few applied ideas, and the basic things in physics. I remember that I enjoyed drawing the design of a water pump and other things like that. I was good at it, the way good students in calculus are good at calculus in our classrooms, but not a genius. I just enjoyed it and fooled around with it. In mathematics classes, I usually was above average. I was bored when class was going on, and I did things like take logarithms of very large numbers for fun.

The American system in those days was eight years of elementary school and four years of high school. In Europe it was the other way around—four years of elementary school followed by eight years of secondary school, adding up to the same thing. I left Hungary when I was in the third year of secondary school, which would have been the equivalent of the seventh grade in this country.

MP: *So the thirteen-year-old Halmos came to the U.S. and entered a high school in the Chicago area. You spoke Hungarian and German and knew a little Latin, and yet instruction was in English. That must have posed a few problems.*

Halmos: For the first six months it was a hell of a problem. On my first day, somebody showed me to a classroom in which, I still remember, a very nice man was talking about physics. I listened dutifully for the first hour and didn't understand a single word of what was being said. At the end of the hour, everyone got up and went to some other room, but I didn't know where to go, so I just sat there. The instructor, Mr. Payne, came over to my seat and asked me something and I shrugged my shoulders helplessly. We tried various languages. I didn't know much English, and he didn't know

German. We both knew a few Latin words and a few French words, and he finally succeeded in telling me that I had to go to Room 252. I went to Room 252, and that was my first day in an American high school. Six months later I spoke rapid, incorrect, ungrammatical, colloquial English.

MP: *Were there any special events in high school that stand out in your memory?*

Halmos: Well, there was a little chicanery surrounding my admission. I explained this business of eight years followed by four in this country, and four years followed by eight in Europe. There was some confusion about that. I *hinted* to the school authorities that I had completed three years of *secondary* school, and I was *believed*. There was, to be sure, a perfunctory examination of my record, and, after being translated by an official in the Hungarian consulate, it said three years of secondary school. That means in effect that I skipped four grades at once, and I went from what was the equivalent of the seventh grade to the eleventh grade; and a year and a half later, at the age of fifteen, I graduated from high school.

A College Freshman at Age Fifteen

MP: *So you were a very young high-school graduate.*

Halmos: Yes. I entered the University of Illinois at the age of fifteen.

MP: *That's very young to be entering college. Did that produce any difficulty?*

Halmos: There were no problems. I was tall for my age and cocky. I pretended to be older and got along fine.

MP: *When did you become interested in mathematics and philosophy?*

Halmos: I started out in chemical engineering, and at the end of one year decided that it was for the birds; I got my hands too dirty. That's how mathematics and philosophy came into the act.

MP: *Can you remember what attracted you to mathematics and philosophy? Can you separate them?*

Halmos: It is difficult. I remember calculus was not easy for me. I was a routine calculus student—I think I got B's. I didn't understand about limits. I doubt that they taught it. At that time, they probably wouldn't have dared. But I was good at integrating and differentiating things in a mechanical sense. Somehow I liked it. I kept fooling around with it. In philosophy, it was symbolic logic that interested me. What attracted me is hard to say, just as it is hard for any of us to say what attracts us to a subject. There was something about abstraction. I liked the cleanness, the security of the ideas. When I learned something about history, I was at the least very suspicious; and strange as it may sound, when I learned something about physics and chemistry, I was most suspicious: I was practically doubtful, and I thought it might not even be true. In mathematics and in that kind of philosophy (logic), I knew exactly what was going on.

"Suddenly I Understood Epsilons!"

MP: *Was there some point when you decided that you were going to be a mathematician?*

Halmos: There was *no* point when I decided that I was going to be an *academic*. That somehow was just taken for granted, not by anybody else, but by me. I just wanted to take courses and see what happened. I was studying for a master's and flunking the master's exam in philosophy, but nothing would stop me. I continued taking courses. I finished my bachelor's quickly, in three years instead of four. As a first-year graduate student, I took a course from Pierce Ketchum in complex function theory. I had absolutely no idea of what was going on. I didn't know what epsilons were, and when he said take the unit circle, and some other guy in class said "open or closed," I thought that silly guy was hair-splitting, and what was he fussing about. What difference did it make? I really didn't understand it.

Then one afternoon something happened. I remember standing at the blackboard in Room 213 of the mathematics building talking with Warren Ambrose and suddenly I understood epsilons. I understood what limits were, and all of the stuff that people had been drilling into me became clear. I sat down that afternoon with the calculus textbook by Granville, Smith, and Longley. All of that stuff that previously had not made any sense became obvious; I could prove the theorems. That afternoon I became a mathematician.

According to Halmos, "Mathematics is Security. Certainty. Truth. Beauty. Insight. Structure. Architecture. I see mathematics as one great, glorious thing."

MP: *So there **was** a critical point. You even remember the room number.*

Halmos: I *think* I remember the room number.

MP: *After earning your Ph.D., you became a fellow at the Institute for Advanced Study, where you served as an assistant to Johnny von Neumann. How did you come to be his assistant? What was it like being an assistant to someone with that kind of power?*

Halmos: Let me back up a little. I got my Ph.D. in 1938, and preparatory to graduation, I applied for jobs. Xerox was not known in those days, and secretarial service was not available to starving graduate students. I typed 120 letters of application, mailed them out, and got two answers, both *no*. I got no job. The University of Illinois took pity on me and kept me on for one year as an instructor. So in '38–'39 I had a job, but I kept applying. I did get a job around February or March at a state university. I accepted it without an interview. It was accomplished with correspondence and some letters of recommendation. Two months later my very good friend Warren Ambrose, who was one year behind me, got his degree. He had been an alternate for a fellowship at the Institute for Advanced Study; and when the first choice declined, he got the scholarship, and that made me mad. I wanted to go, too! I resigned my job, making the department head, whom I had never met, very unhappy, of course. In April, I resigned my job, and went to my father and asked to borrow a thousand dollars, which in those days was a lot of money. The average annual salary of a young Ph.D. was then $1,800. I wrote Veblen and asked if I could be a member of the Institute for Advanced Study even though I had no fellowship. It took him three months to answer. He answered during summer vacation, and said, "Dear Halmos, I just found your letter, and I guess you mean for me to answer. Yes, of course, you are welcome." That's all it took; I moved to Princeton.

But, of course, Veblen wasn't giving me anything except a seat in the library. Six months after I got there, the Institute took pity on me and gave me a fellowship. During the first year, I attended Johnny von Neumann's lectures, and in my second year I became his assistant. I followed his lectures and took careful notes. The system of the Institute was that each professor had an assistant assigned to him. The duties of the assistant depended upon the professor. Einstein's assistant's duties were to walk him home every day and talk German to him. Morse's assistant's duties were to do research with him —eight hours a day sitting with Morse and listening to him talk and talk. Von Neumann's assistant had very little to do—just go to the lectures and take notes; and sometimes those notes were typed up and duplicated. Von Neumann's assistant that year was Hugh Dowker, who is a mathematician *par excellence*, but not in the least interested in matrices and operator theory and all those things that von Neumann lectured on. On the other hand, I was fascinated by them; that was my subject. So, I took careful notes and Dowker used them and took them to Johnny. There was no duplicity about it. He told Johnny what he was doing. When his job was up, I became Johnny's assistant.

How was it? Scary. The most spectacular thing about Johnny was not his power as a mathematician, which was great, or his insight and his clarity, but his rapidity; he was very, very fast. And like the modern computer, which no longer bothers to retrieve the logarithm of 11 from its memory (but, instead, computes the logarithm of 11 each time it is needed), Johnny didn't bother to remember things. He computed them. You asked him a question, and if he didn't know the answer, he thought for three seconds and would produce the answer.

Inspirations

MP: *You have described an inspirational day with Warren Ambrose when you decided to become a mathematician. Are there other individuals who have been inspirations for you?*

Halmos: I'm not prepared for this question. Therefore, my answer is bound to be more honest than for any other question. The first two names that occur to me are two obvious ones. The first is my supervisor, Joe Doob, who is only six years older than I. I was twenty-two when I finished my Ph.D., and he was twenty-eight, both young boys from my present point of view. He arrived at the University of Illinois when he himself was about twenty-five. I was already at the stage where I was signed up to do a Ph.D. thesis with another professor. I remember having lunch with Joe one day at a drugstore and hearing him talk about mathematics. My eyes were opened. I was inspired. He showed me a kind of mathematics, a way to talk mathematics, a way to think about mathematics that wasn't visible to me before. With great trepidation, I approached my Ph.D. supervisor and asked to switch to Joe Doob, and I was off and running.

The other was Johnny von Neumann. The first day that I met him he asked if it would be more comfortable for me to speak Hungarian, which was his best language, and I said it would not. So we spoke English all the time. And as I said before, his speed, plus depth, plus insight, plus inspiration turned me on. They—Doob and von Neumann—were my two greatest inspirations.

MP: *In* 1942, *you produced a monograph called "Finite-Dimensional Vector Spaces." Was it a result in part of notes that you had taken?*

Halmos: Yes. Von Neumann planned a sequence of courses that was going to take him four years. He began at the beginning with the theory of linear algebra—finite-dimensional vector spaces from the advanced point of view. And just as van der Waerden's book was based on Artin's lectures, my book was based on von Neumann's lectures and inspired completely by him. That's what got me started writing books.

MP: *Most people who read that book remark that it is written in an unusual way; the Halmos style is quite distinctive. I studied from your book, and I still remember that it gave me fits because your problems were not of the classical type. You didn't set* **prove** *or* **show** *exercises; more often than not you gave statements that the student was to prove if true or disprove if false. I am sure that it was deliberate, and it seems to underscore a philosophy of teaching that you have spoken about in a recent article in the Monthly, "The Heart of Mathematics." In that article, you said that it is better to do substantial problems on a lesser number of topics than to do oodles of lesser exercises on a larger number of topics. Had you thought a great deal about that before writing problems for "Finite-Dimensional Vector Spaces"?*

Halmos: No. That wasn't a result of thought; it was just instinctive somehow. I felt it was the right way to go, and thirty years later I summarized in expository articles what I have been doing all along. You said it very well. I strongly believe that the way to learn things is to do things—the easiest way to learn to swim is to swim—you can't learn it from lectures about swimming. I also strongly believe that the secret of mathematical exposition, be it just a single lecture, be it a whole course, be it a book, or be it a paper, is not the beautifully written sentence, or even the well-thought-out paragraph, but the architecture of the whole thing. You must have in mind what the lecture or the whole course is going to be. You should get across *one* thing. Determine that thing and then design the whole approach to get at it. Instinctively in that book, and I must repeat it was inspired by von Neumann, I was driving at one thing—that matrix theory is operator theory in the most important and the most translucent special case. Every single step, and in particular every exercise (they were not different from any other step), was designed to shed light on that end.

MP: *In the article, "The Heart of Mathematics," you discussed courses that went down as low as calculus. What do you think about that approach for precalculus or for high-school algebra? Would you also advocate that approach for such courses?*

Halmos: Yes and no. I think, and I repeat, the only way to learn anything is do that thing. The only way to learn to bicycle is ride a bicycle. The only way to teach bicycling is to put challenges in front of the prospective bicyclist and make him conquer them. So, yes, I believe in it. I have tried it, not only in calculus, but in as low-level college courses as precalculus and high-school trigonometry with a great deal of joy and enthusiasm many times. To the extent possible, I have tried to follow that kind of system. But let's be honest. The so-called Moore method, which is a way to describe the Socratic question-asking, problem-challenging approach to teaching, doesn't work well when you have forty people in the class, let alone when you have one hundred and forty. It is beautiful if you have two people sitting at two ends of a log, or ten or eleven sitting in a classroom facing you. Obviously, there are practical problems that you have to solve, but they can be solved. Moore, for instance, did teach first-year calculus that way. So a one-word answer to your question is, *yes*, I do advocate it in all teaching; but *no*, one has to be careful. One has to be wise. One has to face realities and adapt to economic circumstances.

Mathophysics

MP: *A few have said that you have been a strong exponent of what is called the New Mathematics.*

Halmos: Absolutely not! I was a reactionary all the time. The old mathematics was just fine. I think high-school students should be taught high-school geometry *à la* Euclid. You should teach them step one, reason; step two, reason; and all that stuff. I thought that was wonderful. I got my training that way. Morris Kline and I hardly know each other, but we seem to disagree on everything; and he is (a) strongly against the New Math, and (b) strongly against many things I advocate. It's quite possible that people who agree with him identify me as a champion of the New Math because we disagree on most things.

MP: *You say you think that you and Kline disagree on just about everything. He stood in strong opposition to the New Math, and you just said that you were absolutely not an exponent of the New Math. Now there is some agreement there.*

Halmos: I hate to admit it. He may be against the right thing, but he certainly is against it for the wrong reason.

MP: *It seems to me that you and Kline have another strong point of agreement, which surprised me a bit. In your article, "Mathematics as a Creative Art," that appeared in American Scientist back in the late sixties, you surprised me by saying that virtually all of mathematics is rooted in the physical world. Kline, as you know, wrote a book entitled "Mathematics and the Physical World."*

Halmos: I think we understand different things by it. I get the feeling that Kline either thinks, or would love to think, that all mathematics is not only rooted in the physical world but must aim toward it, must be applicable to it, and must touch that base periodically.

MP: *So he is what you would call a mathophysicist?*

Halmos: And how! But it's another thing to say, almost a shallow, meaningless thing to say, that we are human beings with eyes, and we can see things that we think are outside of us. Our mathematics— our instinctive, unformulated, undefined terms—come from our sense impressions; and in that sense at least, a trivial sense, mathematics has its basis in the physical world. But that is an uninformative, unhelpful, shallow statement.

$$\begin{pmatrix} A & X \\ 0 & Y \end{pmatrix} = B \quad (normal)$$

$$A^*A \geq AA^* \quad hyponormal$$

$$hyponormal \Rightarrow subnormal$$

"I am proud to be a teacher.—Teaching is an ephemeral subject. It is like playing the violin. The piece is over, and it's gone. The student is taught, and the teaching is gone."

What Is Mathematics?

MP: *This prompts the next question for which I can't expect you to give a complete answer in such a short time, but I will ask it anyhow. What is mathematics to you?*

Halmos: It is security. Certainty. Truth. Beauty. Insight. Structure. Architecture. I see mathematics, the part of human knowledge that I call mathematics, as one thing—one great, glorious thing. Whether it is differential topology, or functional analysis, or homological algebra, it is all one thing. They all have to do with each other, and even though a differential topologist may not know any functional analysis, every little bit he hears, every rumor that comes to him about that other subject, sounds like something else that he does know. They are intimately interconnected, and they are all facets of the same thing. That interconnection, that architecture, is secure truth and is beauty. That's what mathematics is to me.

Federal Support of Mathematics?

MP: *In "Mathematics as a Creative Art," you were addressing lay readers when you said: "I don't want to teach you what mathematics is, but that it is." This reflects a concern that you had at the time about mathematics in the mind of the layman. (You said, "A layman is anyone who is not a mathematician.") Is your concern still there—that a great body of intelligent, well-educated people don't know perhaps that your subject is? Is that concern stronger or weaker than it was in '68, when you wrote the article?*

Halmos: The same, I would say. Let me first of all explain that I am a maverick among mathematicians. I don't think it is vital and important to explain to members of Congress and administrators in the National Science Foundation what mathematics is and how important it is and how much money it must be given. I think we have been given too much money. I don't think mathematics needs to be supported. I think the phrase is almost offensive. Mathematics gets along fine, thank you, without money, and I look back with nostalgia to the good old days, three or four hundred years ago, when only those did mathematics who were willing to do it on their own time.

In the fifties and sixties, a lot of people went into mathematics for the wrong reasons, namely that it was glamorous, socially respected, and well-paying. The Russians fired off Sputnik, the country

Paul Halmos

became hysterical, and then NSF came along with professional, national policies. Anything and everything was tried; nothing was too much. We had to bribe people to come to mathematics classes to make it appear respectable, glamorous, and well-paying. So we did. One way we did it, for instance, was to use a completely dishonest pretense—the mission attitude towards mathematics. The way it worked was that I would propose a certain piece of research, and then if it was judged to be a good piece of research to do, I would get some money. That's so dishonest it sickens me. None of it was true! We got paid for doing research because the country wanted to spend money training mathematicians to help fight the Russians.

Many young people of that period were brought up with this Golden Goose attitude and now regard an NSF grant as their perfect right. Consequently, more and more there tends to be control by the government of mathematical research. There isn't strong control yet, and perhaps I'm just building a straw man to knock down. But time and effort reporting is a big, bad symptom, and other symptoms are coming I am sure. Thus, I say that it was on balance a bad thing. If the NSF had never existed, if the government had never funded American mathematics, we would have half as many mathematicians as we now have, and I don't see anything wrong with that. Mathematics departments would not have as many as eighty-five and one hundred people in some places. They might have fifteen or twenty people in them, and I don't see anything wrong with that. Mathematics got along fine for many thousands of years without special funding.

MP: *But we certainly have seen a great increase in demand for mathematical skills which means a need for people who are able to teach mathematics. You certainly need someone to deliver the mathematics.*

Halmos: That's a different subject. The demand for teaching mathematics that seems to be growing is again because of a perceived threat by the Russians and Chinese. In other words, we want people in computer science; we want people in statistics; and we want people in various industrial and other applications of mathematics. We have to teach them trigonometry and other subjects so that they can do those things. That's not mathematics; that's a trade. It's doing mathematicians good only insofar as it enables them to buy an extra color TV set or more diapers for the baby.

MP: *Can we return to your concern about letting laymen know that mathematics is?*

Halmos: My interest was more on the intellectual level. I have absolutely no idea of what paleontology is; and if somebody would spend an hour with me, or an hour a day for a week, or an hour a day for a year, teaching it to me, my soul would be richer. In that sense, I was doing the same thing for my colleague, the paleontologist: I was telling him what mathematics is. That, I think, is important. All educable human beings should know what mathematics *is* because their souls would grow by that. They would enjoy life more, they would understand life more, they would have greater insight. They should, in that sense, understand all human activity such as paleontology and mathematics.

Why Write about Mathematics?

MP: *So you were performing a service to paleontologists perhaps by explaining to them that your subject is rather than what it is. How do you explain the motivation for your other writing activities? Writing is hard work. In fact, when I reread your "How to Write Mathematics" last night, I was more convinced than ever that you must work very hard when you write. Why do you do it? Now you aren't talking to paleontologists; you are talking to mathematicians.*

Halmos: It is the same thing. Why do I do it? It is a many-faceted question with many answers. Yes, writing is very hard work, and so is playing the piano for Rubinstein and Horowitz, but I am sure they love it. So is playing the piano for a first-year student at the age of ten, but many of them love it. Writing is very hard work for me, but I love it. And why do I do it? For the same reason that I explain mathematics to the paleontologist. The answer is the same—it is all communication. That's important to me. I want to make things clear. I enjoy making things clear. I find it very difficult to make things clear, but I enjoy trying, and I enjoy it even more on the rare occasions when I succeed. Whether it is making clear to a medical doctor or to a paleontologist how to solve a problem in the summation of geometric series, or explaining to a graduate student who has had a course in measure theory why L_2 is an example of a Hilbert space, I regard them as identical problems. They are problems in communication, explanation, organization, architecture, and structure.

MP: *So you enjoy doing it. It makes you feel good, and it makes you feel perhaps even better if you can sense that the receiver understands. That sounds like a teacher—the classical reasons that people give for teaching—the joy of seeing the look of understanding.*

Halmos: That's very good. Yes, I accept the word. I am proud to be a teacher and get paid for being a teacher, as do many of us who make a living out of mathematics. But it is also something else. Mathematics is, as I once maintained, a creative art and so is the exposition of mathematics. Teaching is an ephemeral subject. It is like playing the violin. The piece is over, and it's gone. The student is taught, and the teaching is gone. The student remains for a while, but after a while he too is gone. But writing is permanent. The book, the paper, the symbols on the sheets of papyrus are always there, and that creation is also the creation of the rounded whole.

MP: *You've also written about talking mathematics. Based on what you have just said, my strong suspicion is that you get a greater joy out of writing than talking, although you also seem to have a lot of joy when you talk about mathematics.*

"Writing is permanent. The book, the paper, the symbols on the sheets of papyrus are always there.—Writing is very hard work for me, but I love it."

Halmos: They're nearly the same thing, but writing is more precise. By more precise, I mean the creator has more control over it. I myself feel that I am a pretty good writer, A − or B +; and a good, but less good speaker, B or B −; and therefore, I enjoy writing more. But they are similar and are part of the art of communication.

MP: *A short time ago, someone talked with me about your book "Naive Set Theory". She said that it has a smooth, conversational style, and is in some ways like a bedtime story. What motivated you to write it?*

Halmos: *Naive Set Theory* was the fastest book I ever wrote. *Measure Theory* took eighteen months of practically full-time work and *Naive Set Theory* took six months. Bedtime story is an apt description, for most of it was written while perched on the edge of a bed in a rented house in Seattle, Washington. It was being written because I had just recently learned about axiomatic set theory, which was a tremendous inspiration to me. It was a novelty to me. I didn't realize that it existed and what it meant, and at once I wanted to go out and tackle it. So I wrote it down. It wrote itself. It seemed 100% clear to me that you have to start here and you have to take the next step there, and the third step suggested itself after the first two. I had almost no choice, and, as I keep emphasizing, that's the biggest problem of writing, of communication—the organization of the whole thing. Individual words that you choose you can change around; you can change the sentence around. The structure of the whole thing you cannot change around—that's what was created while I perched on the edge of the bed the first day, and from then on the book wrote itself.

MP: *Was that a unique writing experience for you?*

Halmos: In that respect, yes, because it had a much better defined subject than usual. When I wrote on Hilbert space theory, I had, subjectively speaking, an infinite area from which to carve out a small chunk. Here was an absolute, definite thing. There is much more to axiomatic set theory than I exposed in *Naive Set Theory*, but it was a clearly defined part of it that I wanted to expose, and I did.

Is Applied Mathematics Bad?

MP: *You have recently written an article with another intriguing title, "Applied Mathematics is Bad Mathematics."*

Halmos: I have been sitting here for the last 55 minutes dreading when this question was going to be asked.

MP: *What do you mean when you say applied is bad?*

Halmos: First, it isn't. Second, it is. I chose the title to be provocative. Many mathematicians, whom everybody else respects and whom I respect, agree with the following attitude: There is something called mathematics—put the adjective "pure" in front of it if you prefer. It all hangs together. Be it topology, or algebra, or functional analysis, or combinatorics, it is the same subject with the same facets of the same diamond; it's beautiful and it's a work of art. In all parts of the subject the language is the same; the attitude is the same; the way the researcher feels when he sits down at his desk is the same; the way he feels when he starts a problem is the same. The subject is closely related to two others. One of them is usually called applied mathematics, and its adherents frequently deny that it exists. They say there is no such thing as applied mathematics and that there isn't any difference between applied mathematics and pure mathematics. But, nevertheless, there is a difference in language and attitude. I am about to say a bad word about applied mathematicians, but, believe me, I mean it in a genuinely humble way. They are sloppy. They are sloppy in perhaps the same way that you and I are sloppy, as ordinary mathematicians are sloppy compared with the requirements of a formal logician. And a formal logician would probably be called sloppy by a computing machine.

There are at least three different kinds of language, which can roughly be arranged in a hierarchy: *formal logic* (that's equated nowadays with computer science), *mathematics*, and *applied mathematics*. They have different objectives; they are different facets of beauty; they have different reasons for existence and have different manners of expression and communication. Since communication is so important to me, that is the first thing that jumps to my eye. A logician just cannot talk the way a topologist talks. And an algebraist couldn't make like an applied mathematician to save his life. Some geniuses like Abraham Robinson can be both. But they are different people being those different things. So, in that sense, what I wanted to say in that article is that there are at least two subjects. Now I am saying there are three or more, and I wanted to call attention to what I think the differences are.

There is a sense in which applied mathematics is like topology, or algebra, or analysis, but (and shoot if you must this old grey head) there is also a sense in which applied mathematics is just bad mathematics. It's a good contribution. It serves humanity. It solves problems about waterways, sloping beaches, airplane flights, atomic bombs, and refrigerators. But just the same, much too often it is bad, ugly, badly arranged, sloppy, untrue, undigested, unorganized, and unarchitected mathematics.

MP: *Computers are still relatively new objects within our lifetimes and intimately linked to what many call applied mathematics. What do you think of them? Are they important to you?*

Halmos: Who am I? A citizen or mathematician? As a mathematician, no, not in the least.

MP: *Let's take something specific, the work of Appel and Haken and the computer.*

"When I was forty, I had every disease in the book, I *thought* hypochondriacally. I went to the doctor with a brain tumor, with heart disease, with cancer, and everything else, I *thought*. He examined me and said, 'Halmos, there isn't anything wrong with you. Go take a long walk.'"

Halmos: On the basis of what I read and pick up as hearsay, I am much less likely now, after their work, to go looking for a counterexample to the four color conjecture than I was before. To that extent, what has happened convinced me that the four color theorem is true. I have a religious belief that some day soon, maybe six months from now, maybe sixty years from now, somebody will write a proof of the four color theorem that will take up sixty pages in the *Pacific Journal of Mathematics*. Soon after that, perhaps six months or sixty years later, somebody will write a four-page proof, based on concepts that in the meantime we will have developed and studied and understood. The result will belong to the grand, glorious, architectural structure of mathematics (assuming, that is, that Haken and Appel and the computer haven't made a mistake).

Paul Halmos

I admit that for a number of my friends, mostly number theorists and topologists, who fool around with small numbers and low-dimensional spaces, the computer is a tremendous scratch pad. But those same friends, perhaps in other bodies, got along just fine twenty-five years ago, before the computer became a scratch pad, using a different scratch pad. Maybe they weren't as efficient, but mathematics isn't in a hurry. Efficiency is meaningless. Understanding is what counts. So, is the computer important to mathematics? My answer is *no*. It is important, but not to mathematics.

MP: *Do you sense the same attitude about computers among most of your colleagues?*

Halmos: I think the ones who share my attitude are perhaps in the minority.

MP: *There are now mathematicians who seem to have a hybrid nature. Let's take someone like Don Knuth, who earned his Ph.D. in mathematics, and along the way discovered the art of computing.*

Halmos: It's not fair arguing by citing examples of great men. How could I possibly disagree? Don Knuth is a great man. Computer science is a great science. What else is there to say? In many respects, that science touches mathematics and uses mathematical ideas. The extent to which the big architecture of mathematics uses the ideas rather than the scratch pad aspect of that science is, however, vanishingly small.

Nevertheless, the connection between computer science and the big body of pure mathematics is sufficiently close that it cannot be ignored, and I advise all of my students to learn computer science for two reasons. First, even though efficiency is not important to mathematics, it may be important to them; if they can't get jobs as pure mathematicians, they need to have something else to do. Second, to the layman, the difference between this part and that part and the third part of something, all of which looks like mathematics to him, looks like hair-splitting. My students and all of us should represent this science in the outside world.

MP: *It is rumored that you're one of the world's great walkers.*

Halmos: That's certainly false. I enjoy walking very much. It is the only exercise I take. I do it very hard. I walk four miles every day at a minimum. I just came from a ten-day holiday, most days of which I walked fast for 10, 12, 15 miles, and I got hot and sweaty. I love it because I am alone; I can think and daydream; and because I feel my body is working up to a healthy state. To call me one of the world's great walkers is an exaggeration, I'm sure.

MP: *Have you been a walker, in a strong sense, for many years?*

Halmos: Twenty-five years. When I was forty, I had every disease in the book, *I thought* hypochondriacally. I went to the doctor with a brain tumor, with heart disease, with cancer, and everything else, *I thought*. He examined me and said, "Halmos, there isn't anything wrong with you. Go take a long walk." So I took a five-minute walk. And then the next week, I increased it to six minutes, and seven, and eight, and nine, until I got to sixty, and I stopped. On weekends, I walk greater distances. When I was young, I drank like a fish, smoked heavily, and had every other vice that you can imagine. Then when I started worrying about such things, I really started worrying about such things. How old are you?

MP: *I just turned forty.*

Halmos: Then start worrying!

MP: *I fully expect you soon to write another article that would pick up on two previous articles. You have done "How to Write Mathematics" and "How to Talk Mathematics." May I soon expect to see "How to Dream Mathematics"?*

Halmos: I'm ahead of you, but I haven't written the article. I have half-planned an entire book. If I live long enough and really have the guts to stand up in public to do it, I might write a book on how to be a mathematician. I have outlined it on paper.

It will include all aspects of the profession, except how to do research. I won't pretend to tell anyone how to do that. What I think I can do is describe the mechanical steps that people go through (and apparently we all have to go through) to do research, to be a referee, to be an author, to write papers, to teach classes, to deal with students—in short, to be a member of the profession—the most glorious profession of all.

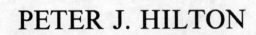

PETER J. HILTON

PETER J. HILTON

Interviewed by Lynn A. Steen and
G. L. Alexanderson

Peter J. Hilton, Distinguished Professor of Mathematics at the State University of New York at Binghamton, has held, in addition to lecturerships at Cambridge and Manchester, professorships at Cornell and the University of Washington, the Mason Professorship of Pure Mathematics at Birmingham and the Louis D. Beaumont University Professorship at Case Western Reserve. He is among the most peripatetic of modern mathematicians, lecturing around the world on mathematics and mathematical education. In the past few years he has lectured in South Africa, Brazil, Australia and throughout Southeast Asia, as well as in a number of countries in Europe and North America.

His research has been in algebraic topology, categorical algebra, homological algebra, and in mathematical education, and he has published almost 300 papers in these fields. He holds numerous editorships and offices in professional organizations and has chaired a number of prestigious committees, one of the most notable being the National Research Council Committee on Applied Mathematics Training. His books include *An Introduction to Homotopy Theory* (Cambridge University Press, 1953), *Differential Calculus* (Routledge and Kegan Paul, 1958), *Partial Derivatives* (Routledge and Kegan Paul, 1960), *Homology* (coauthored with S. Wylie; Cambridge University Press, 1960), *Homotopy Theory and Duality* (Gordon & Breach, 1965), *Classical Mathematics* (coauthored with H. B. Griffiths; Van Nostrand Reinhold, 1970; reissued by Springer, 1980), *Algebraic Topology* (Courant Institute of the Mathematical Sciences, 1969), *General Cohomology Theory and K-Theory* (London Mathematical Society Lecture Note Series 1, Cambridge University Press, 1971), *Lectures on Homological Algebra* (American Mathematical Society, 1971), *A Course in Homological Algebra* (coauthored with U. Stammbach; Springer, 1971), *A Course in Modern Algebra* (coauthored with Y.-C. Wu; John Wiley, 1974), *Localization of Nilpotent Groups and Spaces* (coauthored with G. Mislin and J. Roitberg; North Holland, 1975), and *Fear No More* (coauthored with J. Pedersen; Addison-Wesley, 1983).

MP: *In the early years you were exclusively a topologist, influenced by Whitehead and Hopf. How did you get into topology? Was it a certain mathematician who led you to topology, or did you just get interested as a student and seek out a topologist to study with?*

Hilton: It was very much the first way. During World War II, I was working with British military intelligence and two of my colleagues there were Henry Whitehead and Max Newman. Due to the peculiar circumstances of the war, I was on an equal footing with them. In fact, I was simply a young man who had taken a war-time degree and they were eminent mathematicians. I became, in particular, very, very friendly with Henry Whitehead, on first-name, beer-drinking terms.

After the war, Henry Whitehead invited me to come back to Oxford and be his first post-war research student. I said, "But I don't know anything about topology." And he said, "Oh, don't worry, Peter, you'll like it." So in fact, I didn't even know what the subject was.

I went back to Oxford, and I studied topology. Whatever Henry Whitehead had specialized in, I would have studied. His personality was so attractive, that it was clear that it was going to be great fun to work with him. It turned out to be not only fun but extremely exacting and demanding—it was a marvelous experience. I really took up topology because it was Whitehead's field.

MP: *Was personal contact the main influence that your war-time work at Bletchley on the Enigma Project* had on your subsequent career, or were there mathematical streams also that came out of that work?*

Hilton: No, there weren't many mathematical streams that came out of that work so far as my research was concerned. There were certain attitudes that I developed there, for example, towards the key question of the roles of pure and applied mathematics in an overall mathematical education. But the principal effect on me was getting to know Henry Whitehead, so I could simultaneously be his friend and his student. At that time, particularly in England, it was very unusual to be on such friendly terms with a great man, when one was simply a student. Perhaps it is a little commoner now, but it is still not common. Then, the gulf would have been enormous. Because I had even trained Henry in some of the work he had done in the war, we had this relationship of apparent equality.

It was not only with Henry that I had this sense of being a colleague of mathematicians. There were many, many other mathematicians at Bletchley, so I was on good terms with a number of mathematicians. Another, for example, was Shaun Wylie. It was with him that I wrote a book on homology theory. Bletchley provided an entry into the world of mathematics when I was really very, very green, barely ready to become an apprentice. So that had an enormous effect on my sense of the possibility of becoming a mathematician.

Henry said to me that it wasn't necessary for me to complete all the courses that would normally have gone towards the bachelor's degree. I had only had time to take what was called a war-time degree before going to Bletchley. He said, "Don't worry about that, Peter. You can start research." So therefore I was able to begin research straight after the war. And I might not have had the patience to go back and fill in all the gaps in my undergraduate preparation, so that, for example, I might have been tempted to stay in the scientific civil service, which I did have a chance of doing. But because

Hilton on a cricket tour with the Oxford University Crocodiles in 1948.

---

*The Enigma Project was a special effort of the British to crack the code used by the German cipher machine, Enigma.

Henry could arrange for me to start doing research immediately, I decided to do that. I hadn't had the feeling at all till I went to Bletchley that I was going to become a professional mathematician. There wasn't a feeling that I wasn't—I just hadn't thought in those terms. It was during the war that the realization came to me that it was a possibility.

MP: *I wondered how you came by your mathematical talent. Is there mathematics in the genes? Were your parents mathematically inclined?*

Hilton: No, they were not. There was an uncle of mine who certainly loved mathematics. He didn't follow it up. He became a doctor. But he did love mathematics so I had his encouragement. He also lived quite close to us and always claimed that he would have done mathematics, but it was impossible to earn a living at it. He seemed to me not really the right type to be a doctor and I don't think he was ever very successful at it.

Struck by a Rolls Royce

MP: *I heard that an automobile accident played some role in your development as a mathematician. Is there a story behind that?*

Hilton: There is a story behind that, yes. At the age of 10, I was run over by a Rolls Royce, no less. It was an extraordinary incident. A boy in school snatched the cap off my head and ran across the road. I was angry and ran after him. I didn't notice the on-coming Rolls Royce. The boy in question was Derek Godfrey, who became an extremely distinguished Shakespearean actor. Maybe my own attitude toward the theatre was somehow affected—I have a very warm feeling towards the theatre and my wife is a professional actress. But what happened was that I had a long period of recuperation, much of which was spent in a hospital bed with plaster of Paris on my left leg, all the way up, in fact, to my navel. So I had this sort of white board permanently available to me sitting on my stomach.

It simply turned out that I spent a lot of my leisure time doing mathematical problems, writing them on the plaster of Paris, and erasing them each morning. It gave me the opportunity to realize that I had this intensive love of mathematics. I realized earlier that I had a certain proficiency, but I hadn't realized before that it was the sort of thing I would do when I had the choice of doing other things. So then it came to me that I really loved mathematics and thoroughly enjoyed doing it. I recall even having unkind thoughts about visitors who came to see if I was all right, as they would interrupt me when I was really enjoying what I was doing.

MP: *Now let us move ahead a few years. We know that you are married to a professional actress. How did you meet your wife—was this in a theatrical context or in a mathematical one or neither?*

Hilton: I met her in neither context. I met her in a beautifully ironic situation. Her parents wanted to take her to a Christmas dance she didn't want to go to. Her parents asked their closest friends if they knew of a young man who might come along to accompany their daughter. Of their closest friends, the husband had been a colleague of mine during my work in British Military Intelligence with whom I had kept in touch. I was then at Manchester University. This man, Norman Barnes, said, "Look, Peter, I know this is not the sort of thing you enjoy, but would you mind helping us out?" So I went reluctantly, and Meg went reluctantly and we thoroughly enjoyed ourselves. My next contact with her was to see her acting. I had always had a great love of the theatre and had done a certain amount of amateur acting myself. She was superbly good in the part she played. I remember it was a play called "The Wishing Well." So I moved towards her special field at a far more rapid rate than she moved towards mine! I think she would agree, although I would also add that over the years she has developed a very considerable feel for what we mathematicians are trying to do. She does not have the technical competence, but I think she does understand the source of our inspiration in a very profound way.

Theatrical Aspirations

MP: *You had something of a career in the theatre yourself, though it was short-lived.*

Hilton: I love the theatre and, of course, being married to Meg, I did more amateur work in the early years of my academic profession. I had the extraordinary experience of going from the peak of my career to its nadir in a remarkably short space of time. I first played the part of Hermann in a dramatization of Pushkin's "The Queen of Spades" and I had the wonderful and unexpected joy of being written up in the newspapers as having given the finest amateur performance that had been given that season. I believe, though I cannot be sure at this distance, that I had quite serious delusions that I was on the threshold of something really very remarkable. And perhaps I was. My next attempt was in a play in which I played the role of a priest. I am not a natural priest. But I interpreted my situation in this play as a priest who had given up all hope of saving the soul of a condemned man. And this was also the interpretation of the director. The only way I knew how to play the part of a priest who had really abandoned hope of saving the soul of a condemned man was to play a thoroughly depressed character. This was in a festival, so called, and the adjudicator said of our particular company, that it would have won an award had it not been for the performance of the priest. So at that stage I realized that it was at least not a certainty that I was destined for great things on the stage. I still have hopes and I annoy Meg from time to time deliberately by saying that, if and when I retire as a mathematician, I look forward to having a second career on the stage. I love the theatre and I believe that I have it in me to become one of the best hams the theatre has ever seen.

MP: *You have also had a career in television.*

Hilton: Yes, in England I did a lot of work in television. I first of all did quite a lot of work on radio for the BBC and, in particular, in an organization called the Fifty-One Society which was a group of people—the number fifty-one testified to the fact that it was formed in 1951 and also to the fact that it had fifty-one members. Each week a celebrity of some kind would be invited and he or she would initiate a discussion and we would ask questions of the celebrity and make short contributions ourselves. It was decided that that did not transfer naturally to television. It was too static. I had in that way an entree into that world and when commercial television was beginning in England, they were looking for people they could use in discussion programs, and there was a program that Granada Television introduced called, "Youth Wants to Know"—in fact my role in that program was due to the fact that youth doesn't really want to know. That is to say, the idea of the program was that the avid youth would be asking keen and interesting questions of some personality. Since the youth had no idea who the personality was or what sort of questions to ask, I would first of all meet with the youth who would be children, chosen from a school or some youth group. I would meet them before the actual transmission of the program and explain to them who the celebrity was, what sort of questions to ask, and I would be ready myself with further questions in case the exchange between the youth and the celebrity faltered. I think I was rather successful in this and throughly enjoyed it myself and it also meant that I could dine with this celebrity. I had the opportunity of dining with Randolph Churchill, with Sybil Thorndyke—an absolutely wonderful opportunity—and with Harold Wilson, less wonderful but worthwhile. And then I could, as I say, ask my own questions if there was a lull in the questioning from the young people. This opportunity continued, as I recall, for a considerable time until I was overcome by the general policy of employing professionals rather than people from the universities. I was replaced by a young lady of much more interesting vital statistics and I am very happy to tell you —and this is sheer *Schadenfreude*—the program collapsed in six months.

MP: *What was so wonderful about Sybil Thorndyke?*

Hilton: Her liveliness, her intelligence, her brightness and her interests in things outside the theatre. She was a very political lady. Her politics and mine were very close. She was wonderfully open. She was in her late eighties then, but her conversation had all the liveliness of a young person. Her wonderfully wicked anecdotes were beautiful. She described making a film, *The Prince and the Showgirl*, with Laurence Olivier and Marilyn Monroe. She said that at first she thought, "This woman has no idea of how to act at all. She doesn't understand anything." She said she later came to realize that film acting was an entirely different technique from stage acting and Marilyn Monroe did it to perfection. She said that Laurence Olivier, after one of his scenes with Marilyn Monroe, asked her, "Sybil, do you think I am playing this scene well?" She said she replied, "Larry, it doesn't matter. As long as Marilyn is there nobody is going to be looking at you."

MP: *What was your experience with Randolph Churchill?*

Hilton: Yes, Randolph Churchill was another one interviewed. Although, on that particular evening I don't say that there was any evidence that he was the worse for drink, I think that, in general, his ability to offer worthwhile opinions on matters of general interest was seriously impaired.

MP: *You thought that Sybil Thorndyke was somewhat more scintillating than Harold Wilson? How can that be?*

Hilton with his son, Tim, in Cambridge in 1955.

Hilton: Harold Wilson was very smug. He was very complacent and very sure of himself, absolutely convinced of his rectitude. Sybil Thorndyke was nothing like that at all. She obviously did not take herself that seriously. She just enjoyed life and even succeeded in giving me the impression that she was delighted to meet me. I must say that was extraordinarily flattering to a young man.

Adopted Children in Africa

MP: *I understand that you have some adopted children in Africa?*

Hilton: Meg and I and our youngest son, Tim, went to Mombasa in the summers—that is, our summers— of 1967 and 1968, for the Africa Mathematics Program. While we were there, we made contact with the deputy headmaster of the local school. He was a Goan and was, of course, therefore a Christian. He took us to a little home that a Catholic community ran for orphans and we felt that we would like to do something for them. So we decided in our first year that we would adopt one of them and we announced this intention and chose one of them, Christopher Mwachagga. We were told he was the one most interested in mathematics. But when we were treated so nicely and so warmly by all the boys, Meg and I said, "Shouldn't we make this a major effort? How can we really pick one out?" So in a way we adopted them all. By that I mean we said we would be responsible for all expenses connected with their education as long as they were receiving full-time education. That included cost of clothing as well as some outings each year in addition to the direct costs of their education. We maintained that undertaking and it was a very nice connection for us, but of course, gradually, one by one, they left school, went into employment, and they were no longer our responsibility. We were then no longer in touch with them, but we adopted the new boys in the home in the same way. But I must say that does not mean as much to us since these are boys we have never met. Of course we keep up contact with Christopher Mwachagga, but we don't have the same contact with the boys in the home.

MP: *How long has it been since you saw any of them?*

Hilton: Since 1967. We would like to go back but we just haven't been able to.

MP: *You are obviously in pure mathematics and yet you have spent a number of years at Battelle and surely Battelle is interested in applied mathematics. Have you found it easy to span the difference between pure and applied mathematics?*

Hilton: Actually my first experience with Battelle was through the Advanced Study Center in Geneva. There they were pursuing research in pure mathematics and I was a consultant to them. That goes back to the 1960's. When I became associated with Battelle in Seattle, I first of all became an organizer of specialized conferences in mathematics. When I became associated formally with them as a fellow of Battelle-Seattle Research Center, I was still largely free to do my own research but I was available for consultation on a general basis. The extent to which I participated in their applied mathematics program principally was to put them in touch with mathematicians who could help them. You might say I was a diagnostician for their problems. They would say, "We are interested in these areas. Whom should we talk to?" I was able to tell them whom they should talk to and whom they should bring in. I did not contribute too much myself to Battelle research in applied mathematics but I did discuss with them, to a considerable extent, the kind of mathematics I thought they should know. I was able to anticipate, to some extent, the sort of revolution that has taken place today—I could say to them that they must anticipate the computer and the sort of mathematics that the computer has brought into prominence. I foresaw they would have to understand some graph theory and combinatorics because this is the sort of mathematics that is being applied. I would organize a series of lectures for them on modern areas of applications of mathematics in which I myself and university mathematicians, as well as mathematicians working at Battelle, would talk about contemporary areas of research in applied mathematics. One of the nice by-products of this was that the people at Battelle came to know what others were doing. They came to find that many times their interests were very close, although they had worked in fairly hermetically-sealed compartments of the Battelle structure. So I never really presented myself as a researcher in applied mathematics, which I'm not, but I was able, through my concern for the kind of mathematics that is now being applied, to explain what sort of mathematics the Battelle scientists should know and what sort of mathematicians they would find the most useful to contact as consultants.

MP: *Do you still maintain your Battelle connection?*

Hilton: Yes, I am happy to say. It is now on a very loosely structured basis. I can go there whenever I like and they are very, very nice to me. They make an office available, copying machine, secretarial service, telephones and so forth and, from time to time, they consult me on something. If they consult me about something, then that arrangement is put on a more formal basis. But I can go there and do my research and, if I publish, I give the Battelle address. They say they are very happy with this arrangement. So, I continue in my publications to express my appreciation for it.

MP: *How long have you been associated with the Seattle branch of Battelle?*

Hilton: That has been continuous since 1970, but as I said, the Battelle connection in some form dates from the early 1960's. In those days, and up to the middle 1970's, within Battelle there was what was known as the Battelle Institute. That was the umbrella organization for pure research that was funded entirely from the Battelle portfolio, and I was, in some sense, both the beneficiary of the program and one of its designers, through an advisory committee that Battelle set up in order to assure comparability of standards between what was done at Battelle and what was done in our best universities. Unfortunately, due to straitened circumstances, Battelle is no longer able to support pure research in science.

The Advantages of Collaboration

MP: *In the writing and research that you have done, starting with Hilton and Wylie, and now in your most recent book, Hilton and Pedersen, you have often collaborated with other people. Is that a conscious decision? Do you find that it is a better way to get a good product—or is it just circumstance?*

Hilton: First I must say that I do enjoy it. I very much enjoy collaborating with friends. Second, I think it is an efficient thing to do because it may very well happen if you are just working on your own that you run out of steam, that the project loses some of its appeal for you. But with two of you, what tends to happen is that when one person begins to feel a flagging interest, the other one then provides the stimulus. In that sense I think it is an efficient procedure.

The third thing is, if you choose people to collaborate with who somewhat complement rather than duplicate the contribution that you are able to make, probably a better product results. For example, Shaun Wylie—since you mentioned that book—had a much vaster knowledge at that time of what I might call classical algebraic topology. He was also, frankly, much more meticulous than I was. I, on the other hand, had I think a somewhat better sense at that time of what areas to emphasize for people who really wanted to work in that field rather than just to learn. I had the conviction that the book would be read principally by people who had in mind the possibility of doing original work. So I think that we did complement each other. Fortunately, I was somewhat more energetic than he, so in that case, I think I kept him on the mark where he might have felt, perhaps, that he would prefer to put the work aside and start again in six months' time. So very often I do think you get a better product with this type of collaboration, if there is this kind of complementarity between the collaborators. But I put first what I said before: it is enjoyable too. It is very nice to develop that sort of relationship with a colleague. It's a very close relationship writing a book together.

There is also an important feature of much of my collaboration in research, namely, collaboration with a younger person. This has many, and I would like to say mutual, advantages. My particular research collaborators—Beno Eckmann, Urs Stammbach, Guido Mislin, Joe Roitberg and Aristide Deleanu—are all my very close friends and, with the exception of the first, very much younger than myself.

MP: *Collaboration seems to be the hallmark of what you and others were doing at Bletchley. The military crisis gathered together mathematicians and classicists, historians and lawyers, all in a single pool of sharp minds. Is the cross-fertilization crucial to make creative contributions work?*

Hilton: I don't know, because you can also argue the other way: to make progress in an area where there has already been very substantial penetration to a significant depth, you really have to have experts. I think that what I would emphasize in your description of the situation of Bletchley was this sense of common objective and of urgency that was present there and that is so very hard to generate except in a situation like that, in a popular patriotic war. That war was the last war of its kind. Certainly subsequent wars have not been like that. They haven't generated that sort of sense of community and comradeship. At Bletchley there was this strong collaboration between mathematicians and others, because we were, all of us, working on a common objective and also because the methods had not been previously established. To the extent that there was any field you could call cryptography or cryptology prior to the Second World War, it had been the concern of a few linguists. It was perfectly clear that knowledge of the German language was a totally inadequate preparation to decipher German signals enciphered in the very ingenious way they used. The quality that was principally required of us at Bletchley was really the ability to think mathematically. What I fear in our present enthusiasm for applied mathematics is that people are losing sight of the fact that the most important ingredient of applied mathematics is mathematics. There must be something to be applied—you must have studied some piece of mathematics in depth and know what is involved in making progress in it. However strongly motivated you are, as we were—and I think we could not have been more strongly motivated—we could have done nothing had we not had the experience of really getting down to some piece of work, of achieving a deep analysis and executing it. Some of the best mathematicians were the best at it. These were people like Henry Whitehead and Philip Hall, and others. They were absolutely outstanding mathematicians and tremendously good at this sort of work because they could analyze a problem.

So I don't think it matters so much what mathematics you study in order that you should be able to use mathematical thought processes; but you must have really studied mathematics, not just developed a set of skills. I do worry sometimes when people are talking about applied mathematics, that they think that maybe you could replace, say, the study of functions of a complex variable by the study of biology and thereby become a mathematical biologist. But good applied mathematics—was it Rheinboldt who said this first?—is mathematics plus. Far from being an alternative to studying pure mathematics, applied mathematics requires something more. That is one of the lessons I learned from Bletchley.

MP: *Another aspect of the Bletchley work has a modern resonance—chess. Everybody at Bletchley seemed to be a chess player. More recently, the current interest among computer scientists in artificial intelligence has grown in large part out of attempts to analyze chess. Is there something to that? Is chess a good metaphor for that kind of mathematics, or is it just an accident that people who do that kind of work also happen to like chess?*

Hilton: No, I don't think it is an accident that people who do that work like chess. I think that the combinatorial element of chess is very much akin to the combinatorial element present in cryptography and present in many of the mathematical aspects of computer science—automata theory, computational complexity, and so forth. So I think you would expect to find enthusiastic chess players at Bletchley. They were not always the best chess players, but some of them were, of course, extremely good. We had at Bletchley the British chess champion Hugh Alexander, we had Harry Golombek who still writes the chess column for the London Times, and many other outstanding chess players in the British Isles.

We spent a lot of our leisure playing chess. It is a natural leisure activity. It is, perhaps, a little dangerous for mathematicians to get too much involved in chess because it is, after all, an extremely intellectual game. If you take it seriously, it is quite an exhausting game—it can take effort away from your mathematics. Here I am making a value judgment that mathematics is more important than chess. (There is a man at Cambridge, Nunn, who is a professional mathematician and an English Grand Master—in his case you might say that he has not allowed his mathematics to interfere with his chess. He has chosen to make chess his number one activity.)

I believe, in general, that chess is a very seductive game and I do worry about one question brought up by what you say—mainly that you do find this sort of connection between chess and the general interest in the computer. I am a little worried about bright students today getting too much attracted to the various possibilities of the computer, and becoming, as it were, "computer addicts" where they might be outstanding mathematicians with, of course, a strong feel for the positive role that the computer can play. So just as chess can be, I think, a very fine relaxation for the mathematician, so I think the fascination of the computer can be a very good stimulant to the mathematics too. You wouldn't want it to be a diversion, though.

The Beginning of the Computer

MP: *Chess at Bletchley helped nourish the roots of artificial intelligence, an area of applied mathematics that has expanded now to a very broad field. Do you see other elements of applied mathematics whose* *roots go all the way back to the work that you did at Bletchley? Was that the beginning of artificial intelligence?*

Hilton: Yes. It certainly was the beginning of the computer. Turing was the presiding genius at Bletchley Park. He invented and literally constructed for himself a machine, which we called the Colossus, which had as a prototype the Bombe. The Colossus had many features of the computer. In fact, the only thing missing was the long-term memory which we didn't need in Colossus. But otherwise its functions, its binary processing of information, were all there. So the idea that you could use a machine to process and provide information was certainly inherent in our work. Our cryptography was always a combination of the machine and the human being. So the idea of the machine acting as a sort of complement to human intelligence, was certainly there.

I think I would say that we were thinking only of the machines providing information rather than the machines taking decisions. So this notion of artificial intelligence I wouldn't like to say was inherent in what we were doing, but it is true that one of the leading figures in Britian in artificial intelligence, Donald Michie, now a professor of artificial intelligence at Edinburgh University, was one of the team that worked with Turing at Bletchley. So the seeds of this notion of artificial intelligence were certainly there, but I don't think it came out explicitly.

Mathematical Heroes

MP: *Who were your mathematical heroes, not only those whom you knew, but those of an earlier time whose work you admire?*

Hilton: Those whom I could never have known but whom I greatly admire and was much influenced by would be of course the great giants of the past: Newton and Leibniz, of course, Fermat and Lagrange. Coming to more modern times, I would like to feel that two of my strongest influences are Poincaré and Hilbert, Poincaré because of his founding of *analysis situs** and Hilbert in his axiomatic approach to mathematics, particularly in his showing that one can discover through axiomatics, that it isn't just a way of formalizing and codifying mathematics, but actually advancing mathematics. Among those I met slightly and to whom I owe a tremendous debt is Hardy, who exercised such a

Hilton with Kazimierz Kuratowski, Warsaw, 1955.

---

*Poincaré's name for what is currently called topology.

strong influence on British mathematics. Another is Lefschetz, whom I only met on one or two occasions and who had a wonderful understanding of what is important in topology, how it relates to other parts of mathematics. Another person who creates for me great difficulties—I met him once and had a delightful lunch with him—but whom I cannot now fathom is Pontrjagin. I feel that Pontrjagin was one of the great men of the early period of algebraic topology, homotopy theory, and in a sense he really founded cobordism theory in his use of infinitesimal methods. It needed Thom, of course, to turn Pontrjagin's ideas on their head. His book on topological groups exercised a profound influence on me, and I admired his attitude at the 1966 Congress where he spoke of the Congress as the realization of his dream, the real world community of mathematicians. So it is for me a source of tremendous sadness that I now have to identify him as a leading anti-Semite in the Soviet mathematical establishment, along with the late Vinogradov. So I feel a debt to him but a bitterness about him at the same time.

One name I simply have to mention—I knew him and loved him as everyone did: Heinz Hopf. I cannot miss this opportunity to pay tribute to that great, wonderful, lovable man. Henry Whitehead said of Hopf: "For Hopf mathematics was always a question." If you discussed anything in mathematics with Hopf, he would ask marvelous questions. He encouraged young people and was extraordinarily modest, not just in his behavior. He must have had a wonderful, positive influence on so many people he came in contact with. My own closest collaborator, Beno Eckmann, was Hopf's student and learned a great deal from him. Hopf was a marvelous lecturer and he wrote so lucidly. You know the phrase about blinding with science; Hopf did the absolute opposite. Everytime he wrote he gave the impression to the reader, "You could have done this. I'm just setting it out." He was an inspiration. Another thing to remember was that Hopf succeeded Hermann Weyl, and how could anyone succeed Hermann Weyl? And yet everyone came to realize that Hopf was a worthy successor. I think that too is a tribute to Hopf. In Weyl you had a polymath. Hopf, of course was not in any sense, professionally, as broad and yet he came to be recognized as of comparable stature.

MP: *And what do you feel your most important mathematical contributions have been?*

Hilton: Of course, in the context of the people we've just been discussing, my own contributions are very modest. I think I have made more of a contribution as an expositor in algebraic topology than as a researcher, in bringing ideas into good order so that they would be accessible to students of algebraic topology and homological algebra. In my own research I think the best paper I ever wrote is a paper I wrote under the influence of Jean-Pierre Serre on the homotopy groups of the union of spheres. This was, I think, the first time that Lie algebras were used in homotopy theory in an effective way. It was basic and it is a paper very frequently cited. I think then I would have to jump and say that a series of papers I did with Joseph Roitberg and later also with Guido Mislin on questions relating to failures of cancellation in homotopy theory are good papers. There are two sorts of cancellation you can ask about in homotopy theory. You can take the union of two spaces with a single common point—it's like addition—so you're asking about failure of cancellation under addition. We were able to show systematically how to construct examples where one could take two different spaces, and add on the same space to each, so that the two unions had the same homotopy type; and then, what turned out to be a more difficult problem, we showed how to construct examples with the topological product replacing the union. In the course of that work Joe Roitberg and I were able to construct the first new example of a Hopf manifold. And that, I think, began a whole new industry in mathematics. So I was very happy about that paper.

Then I would say the work that the three of us did on localization theory—the new results and the systematization of known results—was a very significant contribution to the whole structure of the subject. I think also that Eckmann and I did significant work in applying categorical notions. Both of us felt these notions were appropriate for looking at concepts and problems in algebraic topology and homological algebra. We were neither of us pure category-theorists and I think that these series of papers we wrote on group-like structures in general categories, and on general homotopy theory and duality, were two very significant contributions to the applications of these categorical notions that suggested ideas and problems. I always feel enormously grateful to Norman Steenrod, that he not only encouraged us, but also when he compiled his list of papers in algebraic topology, he gave Eckmann-Hilton duality a special heading for the papers that had been written in that area. I do think that Eckmann and I did in that way systematize some ideas and we showed how certain ideas are naturally related. Some of these things were intuitively clear to certain people but had not been systematized. By systematizing them, we made them more readily accessible to students, but also we broadened and extended them substantially.

The Decision to Leave England

MP: *Is there anything you would have done differently if you had to go back?*

Hilton: The only thing I can think of is, I would not have accepted the position at Case Western Reserve University. However, I think it's relevant in my case to say that the most difficult decision I had to make was obviously the decision to come to the United States in 1962. I have thought and rethought that decision over and over again, and I conclude that coming was the correct decision, although there have been losses to set against the gains. I feel that coming to the United States has given me opportunities I would not have had and has relieved me of a very heavy burden of administrative duties which would have been my lot for the entire remainder of my career in England. It has given me encouragement to occupy myself, as I have certainly done, with problems of mathematical education at all levels without ceasing to be a mathematician. That I am enormously grateful for, that I have been able to continue actively both in research and teaching at the university level with great encouragement from everybody concerned, and to take a serious interest in problems of education.

Coming to this country, I found one very big difference between Britain and the United States. In Britain, there was always a certain amount of resentment if a university person took an interest in pre-college education. In this country I found my interest very much welcomed. And so I developed that interest further.

MP: *Do you ever feel that perhaps some of your mathematical colleagues feel that this kind of activity is beneath one?*

Hilton: Yes, I think some of my mathematical colleagues are very puzzled. They wonder how it can be interesting because they say that the mathematics that one is thinking about at pre-college level cannot be interesting. That, of course, I completely deny. I think there can be very interesting and open mathematical questions at that level. In fact, Jean Pedersen and I have been doing some work on producing arbitrarily good approximations to regular polygons by folding paper. We can show that you can fold a regular n-gon for any n by a very simple procedure and this work has been a nice combination of geometry and number theory. Our first paper will appear in the May, 1983, issue of *Mathematics Magazine* and we are just drafting the second and are about to submit it. So there has been real mathematical interest here, and that is also very nice. My colleagues also do not wish to concern themselves with educational problems, at any level, preferring to dedicate themselves to research and the training of graduate students. I believe one earns the right to pursue mathematics by one's conscientious concern for good teaching.

MP: *Can you say more about your actual decision to move to the United States?*

Hilton: As I mentioned before, I moved to the United States in 1962. I had been appointed head of department at Birmingham University in 1958, and it looked as if I would be there for the rest of my career. That is to say that I would have approximately another 35 years as head of department. In Britain the universities are really run by the professors. The word professor, of course, has a more specific meaning there than in this country. It is essentially head of department: the professors administer the university.

I was doing a lot of committee work. I was very conscientious and I don't believe unusually proficient, and I realized that it was having a very deleterious effect on my research. I saw no escape, because I certainly wasn't a big enough person to downgrade myself again, say to the rank of senior lecturer, to halve my salary and lose my position. And yet on the other hand I could not see the possibility of my aspiring to one of the very few positions in the country, at that time, where you could have professorial status without departmental and university-wide responsibilities.

At that time English mathematicians were getting invitations from the States. I had a number of them, which I was not at first taking very seriously. But I had visited Cornell and thoroughly enjoyed it. When an offer came from Cornell, my wife Meg and I talked it over seriously. Then we realized that this was really the unique way of escaping from the situation. So we decided to try it experimentally, and it worked very well.

It also worked very well for our sons, Nicholas and Timothy. At that time in England we were suffering under the system of the eleven-plus examination and the subsequent partitioning of students into the grammar school or the secondary modern school. (There was a third possibility of a secondary technical school, but very few went in that direction.) Both of our boys were just about good enough for grammar school. So we envisaged that they would be under constant pressure, and finally they would have great difficulty in getting into the university. We preferred the more relaxed, liberal situation in an American high school.

There were also professional reasons why my wife felt it was a good idea. She is, as I have said, a professional actress, and she might have more opportunities for part-time work in this country than existed in England. So we thought we would try it.

I should mention that I was interviewed about my decision by the BBC, and I had the good fortune to come out with just the phrase for the moment. I was very determined not to appear to criticize Birmingham University, against which I had no complaint. Others had been criticizing English universities and justifying immigration to the States, so I wanted simply to say that I found myself doing this sort of administrative committee work excessively, to the disadvantage of my mathematical work. I hit on the phrase, "I've taken the decision to go now because I really do think it is about time I made way for an older man." This was picked up in the newspapers and many people who don't know anything about topology or homology theory or homotopy theory, know that I am the person who said that.

MP: *You are not only active in research and teaching but also you are involved in a number of writing projects, carry on an extensive lecture schedule, and also serve on many committees and editorial boards and so on. Do you ever feel that you are somewhat over-extended?*

Hilton: Yes, I do. That feeling usually comes to me when I return to my home base and view the pile of correspondence that I have to deal with. But for me it is stimulating and helpful to be able to turn from one thing to another. When I was contemplating an informal invitation from Deane Montgomery to spend some time at the Institute for Advanced Study, I thought, "Yes, it would be nice, but on the other hand I like the idea that if my research is not going well, I can turn to something else that prevents me from being a total drone." So I wouldn't really like a position where I did research only, and on the other hand, I wouldn't enjoy a position where I had so many duties I could not pursue research. I like this multiplicity of tasks. I enjoyed very much recently the fact that, as external examiner for the Honors Program at the National University of Singapore, I apparently became the natural person to inverview candidates for positions at the National University. It's a pleasant experience to interview someone for a position at another university. You learn a lot and you don't feel that you have to make the decisions. Things of that kind, though they are time-consuming, are very rewarding.

I seem somehow to be able to put such things to good use. For example, one of the young men I interviewed was able to tell me some things in the area of general cohomology that I didn't know and that I was happy to learn. So I could learn in a way far pleasanter than by reading articles.

I always tell myself I'm going to do less, and I am doing a bit less, but I enjoy so many of the things that I do. Things I don't enjoy so much are, unfortunately, the things I am paid for.

People think that my interest in mathematical education must be a great intrusion on my time and it is true, I have to be very careful about my time. It's a very serious problem but, fortunately, I'm a fairly well-organized person. Other people, my wife in particular, help me maximize my opportunities to keep everything going at the same time. I believe some people think that I do all this because I realize that I am no longer going to do anything significant in research so I had better have some other activity. Well, there is a grain of truth in that, in the sense that I have said to myself from time to time that when I can no longer do research, I still want to be useful. Therefore, I want to have established some understanding of this territory if I am going eventually to concentrate on it. But I like to think that the time has not yet come for me to give up research. I know that one prominent mathematician with whom I have been in public dispute quite often, has referred to me as 'Hilton, the ex-topologist'. I don't regard myself as such. I hope that I am continuing to do work at approximately the same level as work that I have done in the past. I don't expect the work I do today will be absolutely in the forefront. Continuing, as I have done during my career, to collaborate with young people, I think that I can in some sense marry my technique with their awareness of what are in fact important contemporary problems and I can then achieve something worthwhile. As long as I send my papers to refereed journals, I will trust the judgment of the referees as to whether they are worth publishing.

So it's true that many are surprised and some even take a rather jaundiced view about my interest in mathematical education, but I think that I have on the whole gained very, very much more than I have lost. I have come to enjoy and appreciate the contacts I have made there.

Hilton contemplating the next step of a proof in his office at the Batelle Research Center, Seattle, 1974.

MP: *You receive very frequent invitations to lecture at either mathematical colloquia or at meetings of mathematicians. Do you regard this as an important part of your professional life?*

Hilton: Yes I do. Of course I enjoy it very much—as I said earlier, there is much of the ham in me and I enjoy the sense of being on stage. But such lecturing is something which I think we should take seriously. I sometimes have the feeling that the art of the general, or colloquium, lecture is being lost—too many colloquium speakers, in fact, give seminar talks. If we are justified in describing ourselves as mathematicians this means that we do believe we share in common an interest in a unified discipline, and we should be able therefore to talk to other mathematicians on an aspect of the discipline in an interesting way. I believe also that there is a subtle compromise to be made with respect to the degree to which a colloquium talk should be prepared in advance. Over-prepared, the talk will be artificial, glib and highly misleading; under-prepared, it will be confusing and frustrating.

MP: *You have a reputation for doing a lot of traveling. Isn't this rather exhausting? What are the benefits to you?*

Hilton: I do very much enjoy finding myself in different milieux. The travel itself is tedious and tiring but the rewards are great. When I travel I always try to identify myself with the problems and concerns of my hosts, so that I do not stay on the surface of their lives but enter into them.

It is also pertinent to remark that when one obtains a certain seniority in the academic profession there are very distinct advantages in not being where you are supposed to be. In your office you can be reached by telephone and by mail and you are constantly being asked to carry out important but very time-consuming tasks on behalf of colleagues or students. For example, I send off approximately five hundred letters of recommendation a year. If I am to remain active as a mathematician it is absolutely essential to have some respite. Others, less conscientious than I, achieve their respite by timely neglect of such duties; I find it necessary to place myself from time to time in a situation where it is difficult to be asked.

There is one other point that I am happy to have the opportunity to make. My reputation for traveling is exaggerated! It is true that I travel more than most but it is also true that, during the teaching year, I spend at least 75% of my time at the institution at which I teach, and, of course, I ensure that my courses are covered in my absence. But being where you are and refusing invitations to give guest lectures make neither news nor gossip. Consequently people will talk to each other about my traveling and will ask me where I am going next simply because this is the most conspicuous fact about me. A study of the spread of the story would make an excellent example in the application of differential equations to the spread of rumors.

MP: *I know that you were very much concerned about the Congress in Warsaw and disapprove, with good reason, of the government in Poland. At the same time I know that you have visited South Africa and, I am sure, disapprove of the government there as well. I am curious about the distinction you draw between these two governments.*

Hilton: This has of course been for me a very delicate issue. It is striking that I find myself in this situation this summer, of accepting an invitation to re-visit South Africa and deciding against going to the Warsaw Congress and making very public my reasons in the *Notices of the American Mathematical Society*. And I am losing the opportunity to deliver a one-hour talk at a symposium that is being held alongside the Congress. My reasons for making the distinction are the following: when I go this time, as I went last time, to South Africa, it is perfectly clear and I have made it perfectly plain that I am going there to help black, Colored and Indian students to receive an effective education in mathematics, because I know I will have the opportunity to make contact with people in South Africa with the same objectives as myself. I will be able to say, while I am in South Africa, that those are my objectives, and I feel confident that my opinions will be published as they were last time. There is a strong opposition press that makes courageous, fearless attacks on the government's apartheid policy. In Warsaw, on the other hand, there will be no opportunity to do anything like this. Indeed, if I try to make contact with the people, feeling as I do about the Jaruzelski regime, I would be compromising them very, very seriously. It would, in fact, be quite improper for me to attempt to do so. So this is why I draw the distinction; I feel that I have the opportunity in South Africa to make my small contribution to a change in the positive sense. I feel that everybody who goes to the Congress in Warsaw makes a small contribution, ipso facto, in the negative sense because, by going, they confer a measure of respectability and legitimacy on a regime that is oppressing the Polish people. Now in conversation with a close friend of mine, a topologist, living and working in Poland, I said, "Yes, I will be willing to accept your invitation to visit Poland in 1984, to come individually as a mathematician, to make contact with Polish mathematicians, which to me is a totally different thing from going to an International Congress where one is making contact with other mathematicians but will have no special contacts with the Poles." I can envisage a most horribly embarrassing situation where the visitors who have dollars to spend will be able to go and eat in restaurants and the Poles won't even be able to accompany them, unless the Poles agree to go as their guests, and I know the Poles. I know their pride and I can imagine how they would feel having to be guests in their own country of their own visitors. I think it's a terribly delicate issue. I think what I'm doing is right but nobody can be sure.

With all its dreadful limitations, there is in South Africa a functioning democracy. It's not just the press—there is an opposition party represented in parliament that hammers away against apartheid all the time. Apartheid is an absolutely appalling system but I can go to South Africa and say so. I cannot go to Poland and say that the regime is appalling and get myself heard.

Of course, in certain aspects, the regime in South Africa is worse than the regime in Poland, but the Polish regime is in a transitional stage and therefore subject to some influence; and positive influence there might be achieved by staying away and expressing disapproval. The South African regime is subjected to a lot of influence from within and I believe we can help that influence from within, by going there. I used to support the boycott; I do not support it now. The Soviet regime is a regime where I feel the derivative is zero. The regime is at its minimum and I think there is nothing that we can do to influence them one way or another. Any decision we take about visiting the Soviet Union should not be based on the hope of effecting change. I think you make your gestures and try to do what you can where you feel that the derivative has a sufficiently large absolute value that you can do something.

MP: *Finally, what projects do you have underway and what plans do you have for the near future?*

Hilton: I look forward enormously to the International Conference on Algebraic Topology being held in August [1983], under the auspices of the Canadian Mathematical Society, which will mark my 60th birthday. I also look forward to continuing my research on finitary automorphisms in group theory and homotopy theory which I have initiated with Joseph Roitburg and Manuel Castellet. I plan to continue various writing projects and feel particularly enthusiastic about the second volume of the three-volume project which Jean Pedersen and I are undertaking to refurbish the teaching of precalculus mathematics and of the calculus itself. I also hope to continue my efforts to improve the quality of mathematics education through my writing (I am an author of the basal program, *Real Math*, published by Open Court) and through my teaching. And I confidently hope to continue to enjoy being a mathematician.

JOHN KEMENY

JOHN KEMENY

Interviewed by Lynn A. Steen

John G. Kemeny, co-author of BASIC and co-developer of the Dartmouth Time-Sharing System, returned to full-time teaching in June 1982 after completing eleven years as President of Dartmouth College.

In 1979 he was selected by President Jimmy Carter to chair the Presidential Commission to investigate the Three Mile Island accident. The commission's report, submitted in October 1979, was highly critical of the nuclear power industry and its federal regulators.

A member of the Dartmouth faculty since 1953, John Kemeny was inaugurated thirteenth president on March 1, 1970, at the age of forty-three. He served as chairman of Dartmouth's Department of Mathematics from 1955 to 1967, building it into one nationally recognized for leadership in both undergraduate and graduate instruction. Deeply committed to teaching, John Kemeny continued during his term as president to teach two courses each year.

As President of Dartmouth, John Kemeny moved the all-male institution to coed status; renewed the College's founding commitment to educating significant numbers of American Indians; began the Dartmouth Plan for year-round operation; and initiated a program of continuing education in liberal studies for business and professional people known as the Dartmouth Institute.

As chairman of mathematics at Dartmouth, John Kemeny helped guide Dartmouth to national leadership in educational uses of computing. He also introduced finite mathematics as an important alternative to calculus for students in the social sciences.

A native of Budapest, Hungary, John Kemeny came to the United States in 1940. During World War II, while still in his teens, he interrupted his undergraduate study at Princeton to work on the Manhattan Project in Los Alamos, N. M. Later, as a graduate student at Princeton, he served as research assistant to Albert Einstein. He received both his bachelor's and doctor's degrees from Princeton, both in logic.

MP: *You are now returning to the classroom after eleven years as President of Dartmouth.*

Kemeny: Eleven and a third years, almost to the day.

MP: *Are there any special projects on your agenda? Are you going to write a book, or develop a new programming language?*

Kemeny: I am thinking of two different kinds of things, one having to do with teaching. At Dartmouth, although we do have a number of options, I think we have to work on our introductory mathematics sequences. I want to play a role in that, most particularly to bring computing more into the sequence. Eventually I'd like to do battle also on the kinds of things that the Sloan conference talked about, that science students shouldn't be introduced only to calculus. Those are my teaching priorities.

In research, although I did quite a bit of reading during my year off, I have not yet made up my mind where I will do my research. I probably will go back and do some work in probability theory, which is the last field I worked in before I became president. And I would like to get active in computer science. Although I enjoy talking about it, I have not yet gotten active in it. I talked to my colleague Steve Garland, chairman of the computer science program, who is planning for this coming academic year a seminar on computer science problems. I suspect that in the beginning I will be more a listener than an active participant, but I hope out of that will come some problems that interest me.

MP: *Is computer science at Dartmouth still a program within mathematics, rather than a separate department?*

Kemeny: Yes it is. That is a matter of controversy. The committee has come up with five different models. My guess is that the one that will carry is that in which we stay joined, but that the name of the department will be changed to Mathematics and Computer Science. This is more than just a symbolic change. It may mean certain reorganization. The department has a chairman and a vice-chairman and we may wind up having a chairman and two vice-chairmen, one for mathematics and one for computer science. It may be, for example, that on tenure decisions in mathematics the mathematics members will have more say-so, and that for computer science tenure the computer science members will have more say-so. It may be sort of a federalist system.

MP: *At big universities, of course, they tend to be separate departments and at small colleges they tend to be the same. Dartmouth is right in the middle.*

Kemeny: We are always in the middle. The overwhelming argument still is that if we split up, for the foreseeable future, computer science would be fairly weak. That's because some of their allies who help teach computer science would stay in the math department. I can't imagine, for example, either Don Kreider or myself leaving the mathematics department. An even bigger argument is that the best way to avoid duplication of courses is to stay within one department. Then you don't get into arguments as to which department teaches which courses.

The Origin of BASIC

MP: *Let me go back to twenty years ago when you were working on creating BASIC and time-sharing. With hindsight we can see that this was a really revolutionary development, that has had a dramatic effect on computing. What were your thoughts when you were beginning it? Did you anticipate its effects?*

Kemeny: Let me give you a little bit of history on that. It is very important that Tom Kurtz should be mentioned in this connection, because he is a very modest person and I am not. I seem to have received 90% of the credit, when the effort was strictly 50-50. Actually the initiative was taken by Tom. We only had a small computer and he came to me when I was chairman of the math department and said, "Don't you think the time has come when all liberal arts students should know how to use the computer?" I said, "Sure, Tom, but there is no way on today's computers that we can teach 800 students." Tom said that he was thinking of a different kind of system, and he vaguely outlined what is now called time-sharing.

He won me over fairly fast, and we designed the first time-sharing system together. It was my idea then to say, while we are at it, can't we design a language better than FORTRAN? Remember this was the FORTRAN of 1963 which was a horror compared to the FORTRAN of today. Tom said, "Yes, but what's the use of teaching a language to Dartmouth students that they will never be able to use anywhere else except at Dartmouth?" Tom was normally farsighted, but that was his famous incorrect prediction. He loves to tell about it, since, as you probably know, BASIC is now the most widely used language in computers.

We did do both together. We both were absolutely convinced that the time would come when any intelligent person had to know how to use the computer. That does not mean that we foresaw everything that happened since then. I can best point to the year 1966 when we dedicated the building for the computer. Up to then it was in some horrible basement. On the dedication of the Kiewitt Computation Center, I gave the main speech in which I predicted what computers would be like ten to twenty years later.

If one looks back at that speech, it has some remarkable farsightedness in it, and some major shortcomings. I did predict in that speech that within twenty-five years computers in the home would be as common as television sets were in 1966. And I predicted some of the things computers would be used for in the home, predictions that have turned out to be remarkably accurate. In 1966 everyone who heard me thought that I was just making up things.

Kemeny examining the insides of one of the machines in Dartmouth's first large computer system.

Where I was totally wrong in my predictions was that I did not foresee the coming of microcomputers. All of my predictions were based on terminals being connected to a central system. I also did not foresee graphics—pictures then were terribly primitive. I sort of foresaw the software revolution, but underestimated dramatically the hardware advances.

MP: *Does it bother you at all now that computer science departments are trying very hard to get people away from BASIC and to use structured languages like Pascal?*

Kemeny: Let me tell you what bothers us at Dartmouth. When people think of BASIC, too often they think of BASIC as it was in 1966. BASIC at Dartmouth is a totally structured language. As a matter of fact the International Standards Committee in BASIC is about to report and they are reporting a highly structured version of BASIC. We all agree that structured languages are far superior. We haven't used the nonstructured BASIC in many years at Dartmouth. The problem, however, is that the versions of BASIC that are implemented on microcomputers tend to be the old BASIC. Therefore people tend to think of BASIC as it was ten years ago.

MP: *That [ten-year-old BASIC] is what is being taught in the public schools.*

Kemeny: You are right. But we have all been won over to structured BASIC, and we have had one at Dartmouth for six or seven years now. Look, I completely agree that structured languages are far superior and that is what should be taught. I just wish they wouldn't say that if it's structured, it's not BASIC.

Computing and Active Students

MP: *One of the really special things about computing is how active students are. They are creative, they take leadership roles, they really get involved. Why is computing so special? Why don't they do that in mathematics or in writing or in history?*

Kemeny: In the development of the original time-sharing system, it was not just Tom and myself—we were highly part-time. It was twelve undergraduate students, and believe me they worked incredible hours. We have endless stories which I won't go into; I'm sure you know similar stories of the number of hours students are willing to work at computing.

Equally important, computing attracts not just students who naturally drift towards mathematics. It is true that students inclined to mathematics are often good at computing—although not all of them are. But the converse is not true. There are large numbers of students who would never have come into math. I even know some who hated math but who fell in love with computing. In an article I have just drafted I tell a story of a woman religion major at Dartmouth who almost didn't graduate because she refused to take one more science course. She was a friend of my daughter, who persuaded her to take my introductory programming course almost over her dead body. Then she became a computer scientist—and a very good one. Of course that is an extreme case, of a person who absolutely hated anything mathematical.

While there is some correlation between mathematical talent and computer talent, the correlation is far from perfect. There are major exceptions. There is something about the fascination of the computer. I don't know a good analogue in mathematics to playing games on the computer. It is a very good way of attracting attention.

I think what's special about computers is not just that they give you a great deal of power—so does mathematics—but that you have to learn an awful lot of mathematics before you have any power. You can study math for years and years before you feel, "Gee, I can really do something." After only three months' experience with computers you can do all kinds of terribly useful things with it. That I think is a very big difference.

MP: *That is probably something that can be turned to the advantage of education generally. As teachers in other departments begin using the computer, they can harness students' natural enthusiasm for computing.*

Kemeny: That has happened here. The latest figures I remember—they may be out of date—is that one-quarter of all undergraduate courses use the computer. These are courses in which the faculty require use of the computer. The students may use it in other courses on their own. These figures do not include the use of the computer for word processing, but genuine uses of computing. Remember that as an undergraduate school Dartmouth is like any liberal arts institution—maybe larger than small ones. But the distribution of courses is the same: the largest number is in the humanities, the second largest in the social sciences, and the science division is the smallest. So one-quarter of all courses is a lot of courses.

MP: *One of the social issues that people comment about frequently is that at the early ages in junior high school, when children begin working with computers, boys outnumber girls by about 4 or 5 to 1.*

Kemeny: This does not surprise me. I bet that in most junior high schools you can't use the computer without fiddling with the hardware. I would be no good at that. The typical American boy learns to fiddle with all kinds of gadgets at an early age. I hope it doesn't have something to do with masculinity because I did not have that kind of upbringing. In Hungary you didn't do those things. An American boy knows how to fix a car. I have never learned how to find out what is wrong with my car, partly because most Hungarians didn't have cars. When it comes to fiddling with gadgets, our society is certainly prejudiced towards boys.

Now at Dartmouth, I don't have to know how my terminal works. I do have to know about software, and how to write good programs. But I bet you that at the typical junior school something is not quite right about the equipment, and unless you know how to turn knobs here and there, nothing will work.

For example, in my classroom the connection between the television set and my terminal is very complicated. People haven't worked out a simple interface yet. So I always try to get a volunteer to take a lesson from the one member of the department who knows what gets hooked to what. That volunteer has always been a boy. On the other hand, the assistants that I have had have been an equal mixture of men and women.

I think this has to do with worrying about gadgetry. So it will be terribly important to get equipment in the schools where you don't have to know a thing about the hardware.

A lot of the microcomputers are, I believe, still where the big computers were when I got into it. You have to understand something about the hardware in order to make the thing work. The nice thing on a modern system is that you don't have to think about those things. Open a file and tell it to stick something in it, and you don't have to care where the computer puts it.

The Future of Books

MP: *Let me ask about the future. There is a lot of talk about putting textbooks and even whole libraries on computers. Is it likely, say in the next ten years, that there will be major changes in the way schools deal with books?*

Kemeny: Let me tell you my other major prediction—it actually came earlier than time-sharing. This was a talk that I gave at the MIT centennial called "A Library for 2000 AD" where I predicted that the research and reference portions of a library will all be using computers. I am sure I predicted

Kemeny teaching a class during his Presidential years.

all the wrong technology about how it would happen, but the logical structure was right. You would have computer-organized searches of abstracts. Mathematical journals, for example, would not be published at all. The editor, after having accepted the article, would put it on the computer. If the Library of Congress was in charge, you would submit volume 17 on tape to the Library, and anybody who has access could just call it up.

The technology is just about here now. My prediction was that it would take 10–20 years for the technology to make it possible, and another 10–20 years for people to get around to doing it. Not only is the technology now available, but there is a tremendous economic motivation. I predict that there will be major savings except for the start-up costs. Putting all this equipment together involves enormous conversion costs. But even if it is done only prospectively, the economic incentives will never be greater than they are now. You know what research journals cost—they are totally unaffordable. There is so much junk published that you can't find what you are looking for. I think we are going to be forced to go in that direction.

Of course, I did not foresee technology like this at all. Now it should be possible, if the Dartmouth library is hooked in, for me to have an extension to it on which I can search through the Library of Congress from my office. As a matter of fact we are taking a very small step in that direction. We have a project that Dartmouth is doing for the National Research Libraries to produce one of the first on-line catalogues. They very cleverly sponsored two experiments—we are one of them. Then they can compare which one is better. It is something you cannot afford to do as a single institution. But twenty-five members of the National Research Libraries jointly sponsored these two experiments. We all share in the costs, and if it works we all share the benefits; if it doesn't work, we all save the money of not having each one of us go the wrong way.

We now have a new nuisance in our terminals. When I first turn my terminal on I must type where I want to be connected. That's because there are now two computers here, but it is really preparation for the fact that there are going to be four very soon because there will be a separate one for administrative uses and, more importantly, there will be one for the library.

So instead of typing C for connect, then D1 for the first Dartmouth system, this fall I should be able to type C L to get to the library computer. And then the catalogue will be available with its retrieval system. I haven't tried it, but some of my colleagues have. It is far from perfect, but they say that it is not bad.

What we have done is to start the computer catalogue from some date on. The plan is that everytime someone checks out a book, the process of checking it out will add it to the system. We have a million and a quarter volumes. Going back and cataloging them all is just crazy. But this is a natural self-correcting system—when something gets checked out, it gets added to the catalogue. If a book never gets checked out, then there is not much point in adding it to the system.

This system is very nice, terribly useful, but it means extra expense. The system I am proposing will be very nice, terribly useful, and will save money. By putting the actual contents on line, you don't force every library in the country to have a copy of, say, the *American Mathematical Monthly*.

MP: *Do you think individuals—mathematicians, philosophers, people who now read journals—will really sit down and read things on a screen?*

Kemeny: I certainly would use it for research, for deciding whether I wanted to get an article or not. Certainly the technology is here so that if you want, the same terminal (if not the one in your office, then one in the library) can produce hard copy for you. It is a modern version of demand printing. You can search the system, and if there is one you want to study carefully, then you get a hard copy of it.

This is why I was very careful to say that this is for research and reference. I don't see any advantages to having a novel in there unless you want to do textual studies. But that is a different thing. For reading Shakespeare or a history book, there is really no point in having it on computers. On the other hand, you know the statistics on research journals. Hardly anyone ever reads any particular article. You subscribe because in a given year there are two or three articles that you cannot afford to miss. For this you have to pay whatever horrendous sum the journal costs, not to mention what it costs the society to publish that journal. It just doesn't make sense.

MP: *Some of the costs are in the paper supplies and in the mailing, but other major costs are in the editorial process and these would have to go on.*

Kemeny: I was flooded with letters about infringement of copyright. They wanted to know how this thing would be financed. Frankly, I hadn't thought of it when I gave the talk. But there is a very simple solution. Computers do keep track of things, so you could pay a royalty by use, not to authors (that would be too messy) but to the professional journal itself. Every time the journal is called up, you are charged 10 cents or whatever is appropriate. Certainly there will be user charges for the system—some to have access, some of it for connect time, and a portion for the journals themselves. The journals might get interesting feedback as to whether anyone is reading them.

MP: *Let's turn to your personal background. Your first book was called "A Philosopher Looks at Science," and your first appointment was in philosophy. Are you a philosopher or a mathematician?*

Kemeny: My first full-time faculty appointment at Princeton was in philosophy. That actually was an accident. All my degrees are in mathematics, but philosophy was my hobby, which I continued in graduate school. I audited courses, and I almost got a master's degree in philosophy. I had everything but the general exam. That was the year I became Einstein's assistant, so I didn't have the time to study for it. I had roughly the equivalent of a master's in philosophy, and my Ph. D. thesis was in logic which is often taught in philosophy departments. So I wasn't that far away from a Ph. D. in philosophy. But I never thought of it as anything but a hobby. When I looked for a job in 1951 the job market was about what it is now—good jobs were almost impossible to get. I was looking for jobs in mathematics; it never occurred to me to apply in philosophy. To my total surprise the only good job offer I got was from the Princeton philosophy department. So I moved 100 yards and became a philosopher. I was very happy there, until Dartmouth came along and asked me to join the mathematics department. Even then, I asked whether I could teach some philosophy, and did so for several years.

You are right that my first book was in the philosophy of science. It is essentially the lectures I developed for the philosophy of science courses I taught at Princeton and for a number of years at Dartmouth.

The Hungarian Connection

MP: *What about the Hungarian connection? Why are so many great mathematicians Hungarian?*

Kemeny: It is very, very hard to understand. There are a few fields in which there are an inordinate number of Hungarians—mathematics, theoretical physics, and Hollywood. I forgot who the producer in Hollywood was, with a big sign on his desk saying "Being Hungarian isn't enough; you must also have talent." Really there were an inordinate number of Hungarians that you would never have guessed were Hungarian. Take, for example, the person my wife and I always thought was the epitome of British acting, Leslie Howard. When he was killed tragically in World War II, the story of his life was printed in all the papers. It turns out that he had been born in Hungary but went to England when he was a small child.

I don't quite believe Gail Young's theory that the Hungarian language is so hard that only the brightest children manage to survive. Certainly, for mathematicians and theoretical physicists, the school system in Hungary was very good. No, that's not true. The school system in Budapest was very good. And Budapest had about 10% of the population of Hungary. Secondly, there are so many temptations for Americans to go into all kinds of fields, many of which just did not exist in Hungary. There is no way you could become a great industrialist in Hungary—there were no great industries. I don't know if there were any Hungarian millionaires at all. Medicine was very strong in Hungary, but a lawyer just didn't have the kind of opportunites you have here. And you couldn't get involved in politics. There were just fewer areas, so a larger percentage of the talented people went into fields like mathematics and physics.

But the school system had a lot to do with this. Let me give you an example. I went there through $7\frac{1}{2}$ grades. The Hungarian split is 4 and 8 (4 elementary, 8 gymnasia), not 8 and 4 as in the United States. For the last $3\frac{1}{2}$, from fifth to the middle of eighth grade, I had a mathematics teacher—it happened to be the same one—who would have been well qualified to teach at a good college. He just did an enormous amount to strengthen my interest in mathematics.

I liked mathematics before that. I don't know where I got interested in mathematics—it went way back. But being interested and knowing something is very different. This teacher was better than any teacher I had in high school in the United States—really, significantly better.

There is another interesting story about this teacher. There was a mathematical contest for high school seniors in Hungary that was a very big thing. If you were talented, I'm told, all through your last three years they would drill on practice tests. It was a great honor, not just for yourself, but for the school.

When we left for the United States, my whole class came out to the train to see me off, and my math teacher did too. It was really nice. He said something that has stuck with me all this time. He said that he was terribly happy for me that I was leaving for the United States, because he was worried about the future in Hungary. On the other hand, he said, he had only one regret. He had never had a winner in the math competition.

Look, for God's sake, I was four and a half years away from that exam, and he was already thinking that maybe I could make it to the top in the competition.

MP: *Certainly the tradition of competition has continued in Hungary, and in all of eastern Europe. The International Olympiad started there. Now the West is participating, and doing quite well. But the initiative for having the contests was with the Eastern countries.*

Kemeny: Let me contrast my experiences this way. New York City had a competition when I was in school—it was the Pi Mu Epsilon Contest. We happened to hear about it purely by chance. We heard that at other schools people got help in practicing for this exam. In our high school of 5000 students we could not get one math teacher to help us with it. So two of us went and took the exam in our junior year just to find out what was on it; then we worked the next year drilling each other.

I think I came in third. Considering that I didn't have any coaching, I thought that was pretty good. That sort of thing makes a great deal of difference. But it is incredible that in a school of 5000 students there wasn't a single teacher willing to help, let alone encourage you to take it.

MP: *Certainly the high school exams are better organized now than they were then. In New York City there is a very active league—the Atlantic Regional Mathematics League.*

Kemeny: Contests aren't everything. They are just symptomatic of the status of mathematics teaching. Problem solving is only one type of mathematical talent. I happened to be good at that, but there are very good mathematicians who are not problem solvers. I think that a system that encourages problem solving is in effect showing that mathematics is important. The Hungarian exams go way back, very far back.

MP: *Are there specific things that you think the United States should be doing? We have a real crisis in mathematics education now—few teachers, few that are well prepared, low pay, low morale. You know the litany of the problems.*

Kemeny: It is horrible. I once was Chairman of the MAA committee on teacher training. We totally bombed out on elementary school training, but we thought we made significant impact on high school teacher training. I have been away from that for a long time now. We did provide a strong program for high-school teachers, but I suspect that most of them are no longer teaching high-school mathematics. They can get much better jobs elsewhere.

I did not get really caught up until this spring when I was at a meeting and heard from the chairman of the Northeast Section of the MAA all the horrible statistics. It is now much worse than what I thought was a terrible situation twenty years ago. In between, I think there were temporary improvements with the NSF institutes and other things. But the situation now is probably as bad as it has ever been during my professional life.

What you can do, I haven't got the foggiest idea—except to train more teachers. I think one has to give differential pay to teachers. But most school systems are reluctant to do that. They are much more likely to take a gym teacher who has some free time and train him to teach mathematics.

MP: *Even with differential pay, it would take a long time to develop a large pool of trained teachers.*

Kemeny: Money would help. While industrial jobs are very attractive, as you know there are a great many people really dedicated to teaching. But the salary differential that exists today between what that person can command in industry and what he could get in teaching is so enormous that it is unfair to his family. The gap doesn't have to go to zero. But if it were narrowed somewhat, I think a certain number of people might return to teaching.

And equally important, there would be much more motivation for college students to go into high-school teaching. In a way, if the pay were better, this would be a good time because there are so few jobs available in colleges. This might be a time when the kind of person I had in Hungary—a person well qualified to teach in college—might go into high-school teaching. But not with the kind of salaries available in high school today.

MP: *Does Dartmouth train students for secondary education in mathematics?*

Kemeny: We have always felt that our major task is to give them the mathematical training. But we do have an education department and cooperative arrangements with several local schools where students can do practice teaching. You can't major in education at Dartmouth, so we are not in the elementary training business. But the students can get the minimal amount of education and practice teaching in order to get a teaching certificate.

MP: *Do you know what the numbers are like in recent years in mathematics education?*

Kemeny: We have never trained a very large number of teachers for secondary school. But I have the impression that they have gone up slightly in recent years because of the decline of opportunities in college teaching. But the numbers are not big. There is a certain degree of self-selection in this. The kind of person who gets into Dartmouth, paying the kind of tuition we have (whether through parents or through borrowing), usually wants to get a job that pays somewhat better. We have traditionally trained quite a large number of college teachers, but high school teaching just hasn't had that kind of appeal. But we do have some around.

We have a special Master's program—a Master of Arts in Liberal Studies, which is a terrible name. It is supposed to be the opposite of the Master of Arts in Teaching. The MAT program was for graduates who had the subject matter background, but needed to be certified as teachers. We did the opposite. The MALS program was for teachers who had lots of education courses but who felt weak in their subject matter. They come here four summers and get a Master's degree. Half their courses are in what they teach, and half in other things. In effect we let them take undergraduate courses toward the Master's degree, as long as it is more advanced than what they had had. It is a very successful program —it has gone on for about a decade. At least half of the people in the program have nothing to do with teaching. For people who live in the region and who would like to take courses at Dartmouth, it is a natural thing to do.

MP: *I suppose many of the teachers now are coming back to get computing courses.*

Kemeny: That is part of it. In the two courses I am teaching I have several MALS students. The elementary probability course is a natural. You don't have to be interested in mathematics teaching to be interested in probability.

Logarithms at Age Nine

MP: *You mentioned the influence of your teacher in grades 5–8. Were there others, perhaps teachers, perhaps not, who were very influential in your early years?*

Kemeny: Let me go way back. My father had a one-person import/export business. He had a male secretary/assistant, and his office was in our home. He may have been the earliest influence on me. The business didn't keep him that busy, and this young man seemed to be talented in mathematics. He would chat with me, and I seemed to be very curious about what he was doing. I still have one thing he gave me—a seven place table of logarithms. It is an old, horribly worn copy. That's not important. What matters is that at age nine he taught me how to use logarithms. I thought they were marvelous.

That is the earliest influence I remember. Next was my Hungarian mathematics teacher. My high school I would say was a negative influence. It is the momentum I had from Hungary that carried me through high school. Princeton was a revelation. Whatever advantage I had in Hungary in the gymnasium I lost in George Washington High School in New York City. I really did not know what mathematics was until I came to Princeton.

I started at Princeton during World War II. The mathematics department had just decided that you did not have to take analytic geometry before calculus. So I signed up for calculus, and I had A. W. Tucker teach me first term calculus. I found that Princeton was not as hard as I thought, and I was nervous about what I might be missing from analytic geometry. So I signed up for analytic geometry. During the war all the young faculty were drafted, so we got senior professors. Professor Chevalley, the great topologist, was teaching analytic geometry. I am sure it was the only time in his life he taught it. He was an absolutely terrible teacher at that level. I remember there was one exam, with a 30 point score. I got 28, and the next highest score was 14. Everybody was terrified. Someone asked how many students did he normally flunk in this course. He mentioned some number, say 6. So the student said: if we get six volunteers to flunk, will you give the rest of us all D's? Of course he was a very kindly person in terms of grading. But two-thirds of his problems were terribly hard.

But imagine having gone to high school as I did, having Tucker and Chevalley both in the first semester of college. I entered in the spring, because they had high-school graduation twice a year. Chevalley had another, more important influence on me. I was one of the few who cared for his course, and he soon discovered that I was abysmally ignorant of anything mathematical. Before summer vacation he asked where I lived; we lived on Long Island. He said, "Why don't you come

Kemeny was one of the first people to have a terminal connected to a timesharing system in his home. This photograph shows his daughter Jennifer (then a junior in high school) using the terminal. Jennifer has become a computer scientist.

Mathematical People

once a week to Princeton. I'd like to teach you some mathematics." So that summer I went once a week to Princeton. He lectured to me on sets, cardinality, and point set topology. That summer was the first time I saw what mathematics was all about.

The next term I doubled up on integral calculus and differential equations. And that term something important happened to me. Let me tell you a funny story.

My interest in philosophy was strong at that time. Bertrand Russell was lecturing at Columbia, and I went to hear him. Later I would hear him several times when he lectured at Princeton. But then I went to Columbia to hear him. I sat next to a Columbia student, and we starting talking. I said that I was going to be a math major, and he asked what I was going to specialize in. I said that I didn't know, but since I had been doing a bit of reading on my own, I thought I might study logic. He said, "You lucky dog, having the greatest logician in the United States at Princeton." I said, "Oh really? Who's he?" (What would a freshman know about such things?). He said, "Alonzo Church, of course." I said, "Oh that's a strange coincidence. I am taking integral calculus and differential equations, and he is teaching both of them."

At Princeton you have to write a small junior paper and a senior thesis. I did my junior paper, my senior thesis, and my Ph. D. thesis under Church. So the single mathematician who had the greatest influence on me was Church.

MP: *You have been department chairman, and president. I am sure you have talked to students who have different reactions to teachers who are very good for the brightest students, but who are not so good for others. What's you reaction to that situation from the point of view of an administrator as compared to your reaction as a bright student?*

Kemeny: My reaction is really very simple. The only problem is with faculty who are not very good at teaching anything. I can think of very valuable faculty members at Dartmouth who are just as you described. The trick is to use them only for advanced courses. We also have some splendid faculty who are really superb at teaching the large lectures in introductory courses, but who are not so good in advanced courses. Occasionally you are lucky—Don Kreider is an example. It doesn't matter what Don teaches—he is spectacular at anything he teaches. Therefore, first of all, at a school like Dartmouth, you try not to have any faculty members who are bad at every form of teaching. And you try to get the others in a good mix. It is hard to do that in a small department, but in a large department like mathematics, as long as you have a good mix, you do fine. You try to use people where they do best.

"Einstein Did Need Help in Mathematics"

MP: *Later in your education, as a graduate student, you were Albert Einstein's assistant. But you did not go into physics. What kind of influence did he have on your career?*

Kemeny: Einstein's assistants were always mathematicians, not physicists. Obviously they were mathematicians who knew a certain amount of physics. I had taken all the undergraduate courses in physics, and had a couple of graduate level courses. I really am not a physicist, but it turned out that all of the advanced subjects in physics, the one that most fascinated me was relativity theory. I had done a lot of reading on my own.

People would ask—did you know enough physics to help Einstein? My standard line was: Einstein did not need help in physics. But contrary to popular belief, Einstein did need help in mathematics. By which I do not mean that he wasn't good at mathematics. He was very good at it, but he was not an up-to-date research level mathematician. In fact, some of the things he achieved were miraculous because he had to do original mathematics, and much of it he did the hard way.

His assistants were mathematicians for two reasons. First of all, in just ordinary calculations, anybody makes mistakes. There were many long calculations, deriving one formula from another to solve a differential equation. They go on forever. Any number of times we got the wrong answer. Sometimes one of us got the wrong answer, sometimes the other. The calculations were long enough that if you got the same answer at the end, you were confident. So he needed an assistant for that, and, frankly, I was more up-to-date in mathematics than he was.

The influences he had on me were of two kinds. I tell you, I was a little worried in graduate school because many of the first-rate mathematicians were a little bit peculiar. One gets a bit of a hang-up, that you have to become peculiar in order to be a great mathematician. At least I know I had that hang-up. The same was true of some theoretical physicists. Then I met two people who changed my mind completely, and I met them fairly close to one another. One was von Neumann, certainly the greatest mathematician of that time, and he was not peculiar at all. The other was Einstein. Einstein was the kindest, nicest human being I ever met in my life.

He also once gave me terribly important advice which saved me from going the wrong way. Having worked at Los Alamos, I was terribly worried about nuclear war. I am still worried about nuclear war. But at that time I was working with the World Federalists to try to educate people about the dangers of nuclear war. They had asked me to become executive director. This happened the year I was Einstein's assistant. Einstein absolutely talked me out of it, on these grounds: with that kind of movement, he said, if you are a paid employee nobody will pay any attention to you. If you ever want to make a contribution, get to be a first-rate mathematician or get to be first rate at something. Then people will listen to you on other issues as well. But the worst possible thing you can do is to work as a paid employee for a group like that.

Also when Dartmouth had an opening for a chairman—the math department sort of went out of business here—the dean talked to a lot of people. Al Tucker was deeply involved, but the two that were most influential in the final choice were von Neumann and Einstein. They really changed my life. As it turned out, they changed the history of Dartmouth College as well. The Dean got more than he bargained for.

Mathematical Literacy

MP: *One last topic: mathematical and scientific literacy in the United States. There have been several recent reports on this matter, from the President's Commission, from the National Academy of Sciences. One that I remember is from the report on the Commission on Public Information of the Three Mile Island Investigation. That contained a lengthy, devastating discussion of the problem that reporters have in dealing with scientific information.*

Kemeny: Absolutely. The Commission as a whole spent a remarkably long time talking about that. I don't really enjoy talking about nuclear power, but that particular incident is a very good example of why it is dangerous for average citizens to be as ignorant of science as they are.

I'll never forget our discussion of reporting on radiation. A very high percentage of public statements were unacceptable. For example, they found no acceptable statements in the New York Times—and it is supposed to be a fairly good paper. Someone on the Commission asked: "You mean they got the numbers wrong?" And the answer was no. Strangely enough, they got the numbers right but the units wrong. It turned out that they got the units wrong not only by having them, say, one thousand times as big or one thousand times as small, but they often had the wrong kind of units.

For example they did not understand the difference between total amount and rate of radiation, which was terribly important for one incident when they had indeed detected a 1200 rem radiation. The coverage had two major things wrong. First of all, that amount is what was measured right at the top of a smokestack; it is not what was measured off the island. Second, everyone said that 1200 rem was given off, which is not true at all. Actually 1200 rem *per hour* was given off for less than a minute. And that's a very big difference. Therefore, with those two mistakes the story becomes dangerously wrong.

This happened over and over again. Either some major fact like the location or the units was missing, or they garbled the units completely. It is the kind of mistake people make when they use the term light-year as a measure of time rather than of distance.

MP: *What should educators be doing about this?*

Kemeny: If mathematics teaching is in bad shape, science teaching is really in horrible shape. A survey the New York Times did a couple of years ago contained some depressing statistics. I can't remember them exactly, but roughly they found that a large majority of high-school graduates never had a science course beyond general science. And those who do have something beyond tend to take biology. Almost never is it a physical science. That's madness. I can't believe there is any other western country that educates its citizens in science as badly as we do.

I'm not talking about them becoming experts in physics. But a few elementary things you have to know just to be an intelligent person. I am writing an article for *Daedelus*, the journal of the American Academy of Arts and Sciences, on the case for computer literacy as part of a special issue on scientific literacy. The editor said that others are reporting all the terrible statistics, so I did not have to do that. In this article I lead off with C. P. Snow's two cultures, not because I think his essay is that great, but his basic point is fundamental. It is not just that we are split into humanistic and scientific cultures—this is more applicable to the United States than to the England for whom he wrote it—but the terribly dangerous thing is that most scientists admit that a well-educated person should know literature, or music or whatever, while the humanistic culture is not willing to concede that understanding science is part of being a cultured individual. I think that is where the great danger comes.

MP: *That relates to the "New Liberal Arts" theme that the Sloan Foundation is talking about.*

Kemeny: At a recent Sloan conference on the New Liberal Arts, I argued for a slightly different position. I think the program is basically excellent. But I don't think they have to win over a large number of faculty to actually teach this stuff. It seems to me that students who take most of their courses in the humanities must see science, mathematics and computing. But all we can hope for is that we can get the humanities faculty to where they are not antagonistic to this material. I guess I am too pessimistic to believe that you are going to get any significant number of humanities teachers to change their teaching. But there are many humanities teachers who can come to appreciate it.

When we got our first computer, one of my very dear friends in the English department was denouncing the coming of computers. But he got the strangest kind of punishment for this behavior. Once we got the time-sharing system, we put a terminal in the high school. It turned out that in the first group of students who had access to it, by far the most brilliant programmer turned out to be his son. As a result of this, he is now extremely knowledgeable about computers, and has completely changed his mind.

Of course he is not to the point that he will teach this stuff in his courses. And he shouldn't be. If it is useful, he (and his students) will use it.

MP: *You are relatively unusual among mathematicians for having served as president of a major university, and now coming back into mathematics. I wonder if there is any general advice that you would give to the mathematical community, things that you see differently now as a result of your experience as President of Dartmouth.*

Kemeny: Can I answer a slightly different question? I don't see anything different as far as mathematics goes. However certain trends are developing that I may see more clearly because I was President, trends that weren't there twelve years ago. There is a very strong trend away from mathematics. I knew this was happening, but I was shocked to see Gail Young's report [of the CBMS study] on the nationwide statistics. At Dartmouth the trends are much less noticeable.

I would have to say to mathematicians that if they are not going to learn the important applications of mathematics—not only in physical science but in the growth areas of social science and computer science—if they are not going to learn something about those areas, they are going to lose most of their students. I happen very much to believe in strong mathematics departments, so I would hate to see this happen. I don't believe it is a good thing for social scientists to teach their own mathematics. I don't think it is a good thing if computer scientists decide to go their own way and teach their own mathematics.

By all means, mathematicians should learn all the pure mathematics they want. But also they must learn applications. Get to be an expert in either the social sciences or in computer science. That's the secret of survival for mathematics departments.

MORRIS KLINE

MORRIS KLINE

Interviewed by G. L. Alexanderson

Morris Kline is Professor Emeritus of the Courant Institute of the Mathematical Sciences at New York University and has recently been Visiting Distinguished Professor at Brooklyn College of the City University of New York. He received his doctorate at NYU and began a career of teaching and research in 1930. Between 1936 and 1938 he was at the Institute for Advanced Study at Princeton. Though trained as a topologist he is known as a research mathematician mainly for his work in differential equations and applied mathematics. For many years he directed the Courant Institute's Division of Electromagnetic Research. He has been awarded numerous academic honors—a Guggenheim Fellowship, a Fulbright Visiting Lectureship, various visiting professorships, and a Great Teacher Award at NYU.

He has written many research papers as well as papers on mathematics education. His publications range from books on mathematics (*Electromagnetic Theory and Geometrical Optics*, Interscience, 1965; *Symposium on the Theory of Electromagnetic Waves*, NYU, 1950, Dover, 1965) to books on the role of mathematics in society (*Mathematics in Western Culture*, Oxford, 1953; *Mathematics and the Physical World*, Crowell, 1959), on the history of mathematics (*Mathematical Thought from Ancient to Modern Times*, Oxford, 1972), and on mathematics education (*Why Johnny Can't Add: The Failure of the New Math*, St. Martin's, 1973). In addition he has written several texts: *Mathematics for Liberal Arts*, Addison-Wesley, 1967; *Mathematics: A Cultural Approach*, Addison-Wesley, 1962; and *Calculus, an Intuitive and Physical Approach*, Wiley, 1967; among others. He also edited the popular *Mathematics in the Modern World*, Freeman, 1968, a collection of articles from the *Scientific American**.

Early Opposition to "New Math"

Professor Kline is no stranger to controversy. He was an early and outspoken critic of the "new math" and in the 1950's and 1960's he wrote a series of articles in *The Mathematics Teacher* ("Mathematics Texts and Teachers," vol. 49 (1956), pp. 162–172; "The Ancients vs. the Moderns," vol. 51 (1958), pp. 418–427; "A Proposal for the High School Mathematics Curriculum," vol. 59 (1966), pp. 322–330) and in the *Monthly* ("Logic Versus Pedagogy," vol. 77 (1970), pp. 264–282) in which he attacked what he viewed as excessive rigor and a failure to deal with the applications of mathematics. For example, in his 1966 article in *The Mathematics Teacher* he stated: "Instead of presenting mathematics as rigorously as possible, present it as intuitively as possible. Accept and use without mention any facts that are so obvious that students do not recognize that they are using them. Students will not lose sleep worrying about whether a line divides the plane into two parts. Prove only what the *students* think requires proof. The ability to appreciate rigor is a function of the age of the student and not of the age of mathematics. As Professor Max M. Schiffer of Stanford University has put it, 'In teaching never put logical carts before heuristic horses.'" This same admonition was echoed in his 1970 *Monthly* article, this time aimed more at college teachers.

In his book, *Why the Professor Can't Teach: Mathematics and the Dilemma of University Education* (St. Martin's, 1977), Professor Kline takes on the university establishment and argues forcefully for a renewed concern for quality in undergraduate teaching. This book has generated some excited discussions, much the way his earlier works stirred up debate on the "new math."

---

*Since this interview, his *Mathematics: The Loss of Certainty*, Oxford, 1980, has appeared.

"Back-to-Basics" versus "New Math"

MP: *Professor Kline, having read your many books, books covering a wide range of mathematics, historical, cultural, and pedagogical topics, I hardly know where to begin. One could, I think, plan an interview on any one of them. But the plans for this interview were prompted by the appearance of your book, "Why the Professor Can't Teach," so I do want to get to some of the issues raised there. Let me start, though, with a few questions on the "new math." Now that the "new math" seems to be on the way out, are you encouraged?*

Kline: I am pleased that teachers around the country have recognized that the new mathematics was not an improvement in mathematics education. However, I am unhappy about the fact that a real improvement is not in sight. I am also bothered by the fact that prominent, intelligent mathematicians could ever have believed that that curriculum would be suitable for young people. But perhaps no one should make an *a priori* judgment about education.

Kline upon his graduation from Boys' High School of Brooklyn, New York, in 1923.

MP: *I know that you would like to see more motivation from the physical sciences in the mathematics curriculum. Do you see signs of this being introduced as the schools abandon more and more the emphasis on structure and axiomatics?*

Kline: I certainly would like to see physical problems introduced as motivation and application in many of the mathematics courses. Of course these applications must be carefully chosen to suit the level of the course and they should not require a background of physics. They need not. I believe that professors are not introducing these physical problems because the typical Ph.D. has no background in physical science and fears that he will be unable to answer questions from students in the area. Actually the fear is groundless because so little physics, astronomy, or chemistry is involved. There is much talk these days about applications, but I fear it is just talk.

MP: *What do you think of the current trend to a back-to-basics curriculum? Might this not herald an* *even worse problem in the schools than was caused by the sometimes absurd nit-picking about language and notation?*

Kline: The back-to-basics movement, in part a reaction to the new math, is not the solution to decent mathematics education. It means to me the meaningless drill in techniques that was common twenty and more years ago. That type of education failed, as is evidenced by the attitude of most intelligent, educated people toward mathematics. It will almost surely fail again. It may not be worse than the new math but it is surely not better.

MP: *I thought you made an interesting point about hand calculators in your book. You point out that to be able to do algebra one must understand the skills of arithmetic. One is, after all, only doing with letters what one has presumably already done with numbers. The hand calculator cannot, by itself, provide the necessary experience with arithmetic. Would you care to remark on the ways the hand calculator can be used in the classroom at various levels?*

Kline: I have heard prominent mathematicians say that we do not have to teach arithmetic any more now that hand calculators are available. Hence I pointed out in my book why that assertion was wrong. However, in the teaching of arithmetic the hand calculator can be used as a check the students can make on the operations they have performed with the usual arithmetic processes. Also I believe that laboratory materials are desirable. The calculator is not strictly a laboratory device, but it is a novelty and something students can handle. It is intriguing. At the secondary and higher levels the hand calculator is a minor aid. It can be used in place of the slide rule or logarithms. But this does not mean that logarithms should not be taught, because the log function is important in the calculus. The calculator may also arouse interest in computer science.

History—A Guide to Pedagogy

MP: *You remark at one point in your book that literal symbols for numbers did not come into use till around 1600. And in an earlier remark you point out that a logical foundation for the calculus was developed in the 19th century, roughly 300 years after Newton and Leibniz. Yet we expect young students to take to algebra and beginning calculus students to appreciate the need for ϵ's and δ's. Do you think those who develop courses in mathematics should stick more closely to the sequence of historical development of the subject.*

Kline: I definitely believe that the historical sequence is an excellent guide to pedagogy. The introduction to the calculus should not involve ϵ's and δ's. This rigor belongs in advanced calculus. One need not follow history literally, but if great mathematicians had difficulties with some creations, our students will also.

MP: *Would you advocate a history of mathematics course for every prospective teacher?*

Kline: Every teacher of secondary and college mathematics should know the history of mathematics. There are many reasons, but perhaps the most important is that it is a guide to pedagogy.

MP: *One of the problems I saw in the writing of text materials for the "new math" was a tendency on the part of teachers, who were at that time receiving extensive retraining in summer and academic year institutes, to assume that the material they were being taught (for example, rigorous courses in the foundations of geometry) could be taken back almost unchanged to the high-school classroom. There was a failure to distinguish between the experience of the teacher who already knew geometry, say, and who could appreciate the refinements of a carefully arranged course, and that of the young student encountering the subject for the first time. Are there ways of avoiding this type of problem in future curriculum reform projects?*

Kline: You are certainly correct that professors who could appreciate rigor did not realize that youngsters would not. The movement to teach rigorous mathematics to youngsters was and is a mistake. They are not prepared to appreciate the need for the rigor. I tried to make this point at some length in my article "Logic Versus Pedagogy." [See earlier reference.] The first approach to any subject should be intuitive, even though the teacher knows that from a rigorous standpoint the approach is faulty. For about two thousand years the best mathematicians regarded Euclid's *Elements* as rigorous. Here again history is a guide to pedagogy.

Does Research Affect Teaching?

MP: *Let me move on to more specific issues raised in your book. You use some strong language. On the new math, for example, you say:* "Of course, the new math was a disaster at both the elementary and secondary levels . . ." *There are those who will certainly disagree. And your remarks on research in mathematics are sharp; for example:* "No doubt much worthless research is done in all academic fields. But remoteness and pointlessness are far more prevalent in mathematics." *Has the book prompted a strong response, even a counterattack? I have seen the two articles by Hilton and Hochstadt in "The Mathematical Intelligencer."*

Kline: My attack on research in mathematics was not relevant. Actually I can defend this attack and may someday do so. But what I should have stuck to in *Why The Professor Can't Teach* is that participation in modern, highly specialized research does not improve the ability to teach undergraduate mathematics. The criticisms of my statements on research by Peter Hilton and Harry Hochstadt in the *Mathematical Intelligencer* attributed to me statements I never made. For example, Hochstadt said I wish to abolish research. I never said any such thing. Neither man took the trouble to read the book. They read only the excerpts in the first issue (of the *Intelligencer*). This is hardly fair. But fortunately I have received dozens of letters from teachers who not only agree as to irrelevance of the research for teaching but resent the pressure to do research.

MP: *You have no trouble convincing me that there is some very bad teaching in the universities and that great researchers can be pretty awful in the classroom; but I have the impression from your book that you scarcely admit the exception, the good research mathematician who is also a good teacher. Aren't there some who are good at both?*

Kline: The good researcher who is a good undergraduate or even a graduate teacher is really an exception. One reason is not that more research people could not be good teachers, but that if one is to excel in research, he must keep pace with so much literature, attend conferences, and spend so much time solving any significant problem that he cannot spend time meeting the multiple demands on a teacher. Hence only a person with extraordinary energy can do both.

MP: *One of my problems in reading your book is that I am not sure what a good teacher is. I recall a teacher of mine who was an internationally known research mathematician. He would dwell at great length on ideas that were rather simple, and as soon as we got to something I thought was hard (a series of Tauberian theorems, as I recall) he passed right over the proofs, saying that they were too hard. At the time I thought the course was a disaster and I had notes from the class that were just about worthless. Yet the instructor passed along an enthusiasm for the material that has stuck with me to this day, and I look back on it as one of my best courses. My point is that teachers affect their students' thinking in subtle ways, and superficial student evaluations of an instructor's "performance" don't tell the whole story. Would you care to comment? And how would you define a good teacher?*

Kline: There is no definitive characterization of good teaching. A teacher who stimulates his students to learn is good. Of course he should be doing more but he is doing something vital. A teacher who is boring in class but fully aware of the difficulties students have, presents the material clearly in class, and meets them after class to provide additional help is good. Even the person who is neither stimulating nor especially careful in his presentations but gets to know his students and makes them feel that he is their friend to whom they can come for any kind of help or advice is a good teacher. Every good teacher must know the average student's background and prepare his lectures and choose a text accordingly. Student evaluations must be taken with a grain of salt.

MP: *As I said before, though I think there are exceptions, I would agree that great researchers are often not good teachers, and are, in any case, not sufficiently accessible to undergraduates. But in your book, you are pretty hard on teaching assistants. It's true they have other priorities, but I have observed that undergraduates who are taught by teaching assistants are often the lucky ones. Wouldn't you agree?*

Kline: Just as it is the rare researcher who, for reasons cited earlier, is a good teacher, so, I believe, it is the rare graduate assistant who is a good teacher. That students often learn more from graduate assistants than from professors means to me that bad teaching is better than no teaching.

Kline in May of 1930, the month in which he received his B.Sc. degree from New York University.

MP: *You feel, apparently, that research and teaching for the undergraduate mathematics instructor are almost antithetical. You suggest in your recommendations that the undergraduate and graduate departments be split apart with separate funds and separate faculties. How many schools have done this and how successful has it been? Does it result in better teaching at the undergraduate level?*

Kline: I stick to my recommendation that undergraduate and graduate education be completely independent of each other. The graduate schools feed on the money brought in by undergraduates. Instead of using that money to support first-class teachers, the universities use graduate assistants and large lecture classes to teach undergraduates. The large universities have been doing this for thirty or forty years. I know from personal experience at New York University (I was chairman of undergraduate mathematics for eleven years) that when the undergraduate college was independent the teaching was far better. It was done in small classes by mature, full-time, generally tenured faculty. But New York University has gone the way of Harvard, Princeton, Chicago, Illinois, Michigan, Berkeley, and, to my knowledge, all the major universities. In a recent "Op-Ed" article in the *New York Times*, the president of a major university deplored the impending decline in undergraduate enrollment because less money would be available to support the research.

MP: *At one point you refer to the fact that federal intervention in the form of laws and the withholding of grant money has been effective in exerting pressure on universities to end discrimination and that similar federal pressure could force universities to change in order to improve teaching. Are you advocating such intervention and what form should the pressure take?*

Kline: I would not like to see much federal intervention in education. But now state university systems are the biggest institutions. The money to run these is provided by the states (sometimes cities). The legislators should make sure that ample funds are provided for sound undergraduate education and see to it that the funds are so used. This is the first priority. Money for graduate education and research should be provided, but only after the undergraduate needs are provided for. The National Science Foundation and hundreds of private foundations help graduate students and research projects. I am not oblivious of the value of research, but it should not be supported by cheating the undergraduates.

MP: *On page 235 you say:* "*In the educational field the universities' insistence on research as the qualification for appointment and tenure of professors . . . , large lecture classes, the use of teaching assistants on a wide scale, and inadequate textbooks are all highly detrimental to the progress of mathematics and to the effectiveness of education.*" *There are certainly those who feel strongly that research as a criterion for appointment and tenure is essential for the progress of mathematics. Are you referring here only to appointments to the undergraduate faculty? Researchers are surely necessary for the graduate faculty, unless one drops the Ph.D. program for the D.A. (Doctor of Arts).*

Kline: My statement on page 235 of my book to the effect that research should not be a qualification for appointment and tenure was intended to apply primarily to undergraduate teachers. However, even graduate students, especially in their first year or two, need good teaching. Also many graduate schools provide service courses to, say, engineers who want to learn more mathematics but are not Ph.D. candidates. In the next paragraph of the book I was clearer that we must recognize scholarship as well as research. Of course research professors are needed to advance mathematics and to train prospective research people.

A Doctor of Arts Degree for Teachers?

MP: *You devote some pages to a discussion of the D.A. degree. I certainly agree that it could be and probably is in many cases a more appropriate degree for undergraduate teaching than the Ph.D. Do you think it will ever really catch on and be regarded as something more than a second-rate degree compared to a Ph.D.? And what do you think of the D.A. as an appropriate terminal degree for the two-year college teacher?*

Kline: The D.A. degree is slowly gaining acceptance. I believe that the clamor for good teaching will help. Though for a while this degree will be regarded as second-rate, I believe it will acquire status. It certainly is the proper degree for the two-year college teacher. Now about fifty percent of the undergraduates are in two-year colleges. These students must have good teachers, and I expect that the young American Mathematical Association of Two-Year Colleges will be exerting pressure for training teachers.

MP: *I would like to move now to some discussion of what mathematics should be taught. On page 206 [*"Why the Professor Can't Teach"*], you object to some topics often included in courses for liberal arts students or for prospective elementary school teachers:* "*the logical development of the real number system; set theory; transfinite numbers; Boolean algebra; truth tables; abstract mathematical structures such as groups, rings, and fields; finite geometries; and a heavy emphasis on axiomatics and proof.*" *Now there are certainly attractive topics in all of these fields (well, almost all) but, for the student not drawn to mathematics naturally, they make a pretty dreary list of topics, overall. I know that you would suggest topics more closely tied to science. Aren't there topics, though, from pure mathematics, for example, from number theory or from geometry, that would be closer to the student's experience and to his interests than the topics above?*

Kline: As for the material taught to liberal arts students and prospective elementary school teachers, the topics currently taught are, in the main, worthless for them. However, I do not exclude some purely mathematical topics. My *Mathematics for Liberal Arts* gives one possible version for liberal arts students. I would not advise the book for prospective elementary-school teachers. They need a far better understanding of arithmetic, the elements of algebra and geometry, and of applications at this level. These students can study a number of topics, such as bases of our number system, which would enhance their knowledge and make them better teachers.

How to Motivate: Pure versus Applied Mathematics

MP: *I get the feeling throughout that you are a bit hard on pure mathematics. Surely topics from number theory, combinatorics, and geometry can be very appealing without being tied to practical problems and can be used to motivate students to acquire good mathematical skills. How else does one explain the popularity of Martin Gardner's books and columns? You do admit in your book that games are effective in the early grades. Cannot the upper grades benefit from mathematics that is appealing and fun?*

Kline: I am hard on pure mathematics as the basic diet in courses. A bit of it can be attractive, intriguing, and even fun. But the mathematics that is most important and, in my experience, most attractive to students is the kind that applies to their world (at the respective ages). The popularity of Martin Gardner's columns and books is undoubtedly due in the main to the puzzle aspect. I would use at any level puzzles and games that are truly instructive. But many are insignificant. Let us keep in mind the popularity of crossword puzzles. To what extent should these be the substance of English courses?

Kline in 1936 after receiving his Ph.D. from New York University.

MP: *I have a problem with the suggestion about more examples from the physical sciences or astronomy. The fault may well be mine, but I have trouble conveying to students my excitement about such problems. All too often they seem to remind my students of unpleasant experiences in physics classes. Then, too, they often involve extensive computations and end up looking too much like real work. And astronomical calculations, though they may be of interest to some students caught up by the space program, are, it seems to me, just about as applicable to everyday life as the Königsberg bridge problem. Would you care to comment?*

Kline: Physics is no better taught than mathematics. And it is true that some students are as much repelled by physics problems as by mathematics proper. If mathematics teaching is improved, the physical problems can be presented attractively. Moreover, by selecting problems that do relate to the student's world we can arouse interest and even overcome antagonism to physics. For a high-school student, physics applied to athletics (there are books on this) might be one suitable application, though not perhaps the most significant. Extensive computations should be and can be avoided by rounding off numbers. We must experiment to find what applications do appeal at the respective age levels.

MP: *Have you seen some recent compendia of applied problems (for example, the MAA–NCTM publication, "A Sourcebook of Applications of School Mathematics") that have been put together by various groups? I have read through some of them looking for problems for my classes. For the most part, they look dull to me and I'm afraid to try them on my students. They seem to me to be scarcely more interesting than the artificial problems they would replace. There is always the problem of finding really good materials from which to draw examples. Most authors, as you point out, do not have adequate background to write well on applications. Even people with the background and stature of Pólya have trouble if one is to accept the review of "Mathematical Methods in Science" by Robert Karplus in the Monthly." Are there any short-term solutions to this problem?*

Kline: I have not yet examined the MAA–NCTM publication, *A Sourcebook of Applications of School Mathematics*, but I have seen others. Usually they are disappointingly dull. The trouble with such projects is that the members write what they already know. And what they know is limited. I believe that the mass and variety of applications are so great that we can find good ones. I myself have notes on hundreds of applications. But good ones must be unearthed through hard digging in various places. And I stress that they must be tried. Pólya is a splendid mathematician and really interested in pedagogy. But his ability to write for high-school students is questionable. Like other European-trained mathematicians—some are my colleagues at NYU—Pólya is really not at home with lower-level United States education.

Expository Writing for Teachers

MP: *As a co-author of several texts, I have to admit I squirmed a bit when I read your chapter on textbooks and their authors. Let me quote you. "The writing in mathematics texts is not only laconic to a fault; it is cold, monotonous, dry, dull, and even ungrammatical." I'll admit to some of these charges but not willingly to the last! Your comments apply, of course, not only to many texts but even more to journal articles. Opaque mathematical writing seems to have become a virtue, in the eyes of many, instead of a vice. Is there any reasonable hope of changing the style of mathematical writing so that inclusion of motivation and historical background will not be considered bad form?*

Kline: My condemnation of texts as of research people as teachers is general. It was not meant to exclude any exceptions. All mathematicians, but especially D.A.'s, should be trained in expository writing. At present, mathematicians receive no such training. Writing is a skill and an art and is not automatically acquired. If we train and find places for good teachers, they will seek out the well-written texts and so force others to improve if they wish to gain adoptions. As for research articles, I see no hope of improvement. The American Mathematical Society could do a lot but won't.

MP: *I recall visiting a small liberal arts college some years ago and looking at the textbooks in the bookstore. I know what kind of students they had. Where I expected to find Granville for calculus I found Apostol. Now surely Apostol's book is a fine text—for some students. I know you have something to say about such practices. Would you care to comment?*

Kline: Apostol's *Calculus* is far more an *advanced* calculus text. It should not be used for an introduction to the calculus. Many professors adopt books of interest to themselves but totally unsuitable for the students.

MP: *You make a strong case for recognition of scholarship, rather than research ability. I assume you mean by scholarship, ability to synthesize, evaluate critically, and put work in a historical framework. Surely ours is one of the few fields, if not the only field, where such scholarship is eschewed and only the creation of something new counts. Recognition of such worthwhile activity is what is required if the D.A. degree we discussed earlier is to succeed and have any impact on the profession. I only hope that as scholars we would not be measured by your scholarship. I found "Mathematical Thought from Ancient to Modern Times" an amazing tour de force.*

Do you include historical remarks in your classroom lectures?

Kline: My *Mathematical Thought* is not a text. It is for occasional reading and general background for professionals. But history can play a great role in pedagogy. For example, if students taking introductory calculus were told that even Newton and Leibniz, despite notable predecessors, did not have a good grip on the concepts and that about two hundred years of effort were required before mathematicians began to pin down the concepts, the students would not feel baffled if they don't grasp the concepts properly at the outset. Instead they will gain courage to continue. There are many other pedagogical values of history.

Scholarship should also include good expository articles and critical articles.

MP: *Do you feel there should be more separate courses in the history of mathematics offered in mathematics departments?*

Kline: Every mathematics department should offer a course in the history of mathematics, on both the undergraduate and the graduate levels. I could write a long article on the values to be gained from such a course.

Life at New York University

MP: *You have been close to some important recent history yourself. You were at N.Y.U. for a number of years, during the Courant years of which we have read so much recently in the Reid book.* Earlier you were at the Institute for Advanced Study—in the late 30's, wasn't it?—when Einstein was there, along with some other rather good people. Could you give us a glimpse of what life was like at NYU and at the Institute during those exciting times?*

Kline: It would require a book of the size of Constance Reid's to describe life at NYU during Courant's administration from 1934 to 1958. All I can say briefly is that Courant was the wisest and most able administrator I have ever met and that he built an insignificant department into one of the greatest. Working for him gave me insights I could never have gotten elsewhere. The two years, 1936–1938, that I spent at the Institute for Advanced Study were also very valuable but only for the acquisition of mathematical knowledge. Einstein, von Neumann, Weyl, Morse, Veblen, and Alexander were the mathematics professors. I was a research assistant to Alexander. I state reluctantly that the limitations of these men—not in creativity or knowledge—were also very apparent.

MP: *Did you always have a strong interest in applied mathematics? Who influenced you in this, any particular teacher or any figures in the history of mathematics or science? Are there are mathematicians in history for whom you have an especially high regard?*

Kline: As a high school undergraduate, and graduate student of mathematics (before Courant came to NYU in 1934) I hadn't the least idea of what mathematics was all about. I could do the required work and make good grades and so I preferred it, for example, to English. I believe I was a victim of the poor knowledge and poor teaching that was prevalent during the second and third decades of this century. My doctoral degree was in topology, and this is one reason I got the appointment of research assistant to the topologist James W. Alexander. But when I returned to NYU to work for Courant, he convinced me that the greatest contribution mathematicians had made and should continue to make was to help man understand the world about him. And so I turned to applied mathematics.

Though Courant was certainly the ablest administrator I have gotten to know, and an unbeatable judge of men and ideas, the broadest and wisest man in *mathematics proper* was, in my opinion, Hermann Weyl. Even his articles in *American Mathematical Monthly* are worth reading and rereading. If we go back in time, then my hero is Leonhard Euler.

---

*Constance Reid, *Courant in Göttingen and New York: The Story of an Improbable Mathematician*, Springer-Verlag, New York, 1976.

MP: *As an applied mathematician you must be following the current controversy concerning catastrophe theory. Would you care to comment?*

Kline: I do not know enough about Thom's work in catastrophe theory to evaluate it. I suspect it has substance, but when a word such as catastrophe is attached to it, it gets more attention than it may warrant.

MP: *I first met you in 1958 at Stanford when you were out one summer to teach in a special program for high-school teachers. You were, of course, at that time writing on teaching and have written a number of texts and other works concerned with teaching. When did you develop this strong interest in teaching? Did you always have it?*

Kline: When I started to teach as an instructor in 1930 at New York University, teaching was still regarded there as the most important activity of the faculty. Though for various reasons I did turn to research in applied mathematics, worked for the U.S. Army on applications during World War II, and then founded the Division of Electromagnetic Research at the Courant Institute, I still believed that teaching was at least as important, and I continued to pursue teaching interests. We met at Stanford in 1958. I was there again to teach in 1961 and 1966. When research began to take precedence over teaching in this country, roughly about 1945, I became incensed about the shoddy treatment of undergraduates; and even though I was heavily involved in research I resolved to spend some time in

Morris Kline, champion of applications and an early opponent of the "New Math."

efforts to argue for good teaching. I hope to continue these efforts as long as I am able to. Fortunately Courant was sympathetic and in fact appreciated teaching, and so I did not have to face personal hardships at NYU.

Advice to Teachers

MP: *Of course, you have a reputation as a great teacher. Are there any hints you can pass along to the rest of us? Can people be taught how to teach or are some born with the ability, others not?*

Kline: I believe that almost anyone can learn to be a good teacher. The exceptions are people who, probably because of influences in their early years, develop withdrawn or introverted personalities. Perhaps the person with a lively temperament is born with that quality. But as I noted earlier, this feature is not essential. There are various kinds of good teaching. However, there must be a will to teach and a will to learn how to teach. It is not something one does well merely by "knowing his stuff." Somehow one must learn that motivation is vital, that some students think rapidly and others slowly but that both can learn and even create equally well, that the background and interest a student brings to class must be taken into account and that these backgrounds and interests differ considerably, that an attitude of friendliness and even camaraderie on the part of the teacher is essential, that humor has a place in teaching, and that one must constantly search for the best materials and presentations. There are many qualities that good teaching demands, but they can be learned. However, as long as research is the measure of the man and the qualification for personal progress, it is a natural reaction of human beings that the will to teach will be suppressed.

MP: *I believe it's Pólya who tells the story of Hilbert, who, when asked what question he would want answered if he could come back, like Barbarossa, after five hundred years, said, "Has anyone proved the Riemann hypothesis?" If you could come back after five hundred years, what would you ask?*

Kline: If I could come back after five hundred years and find that the Riemann hypothesis or Fermat's last "theorem" was proved, I would be disappointed, because I would be pretty sure, in view of the history of attempts to prove these conjectures, that an enormous amount of time had been spent on proving theorems that are unimportant to the life of man. I would hope that medicine would have made progress over the five hundred years at the rate that mathematical physics made in three hundred years. When medicine has discovered how to cure or prevent cancer, heart troubles, birth defects, mental disorders, and other diseases, I would be so overjoyed that I would be more tolerant of even that large part of mathematical research that is useless. It is not easy to define human happiness and, as the poet Archilochus put it, "Each man must have his heart cheered in his own way." But good health is the first prerequisite. Perhaps research in medicine had to wait for some of the progress made in mathematics, physics, and chemistry. But I believe that a far greater amount of research in medicine could have been undertaken two hundred years ago. It is deplorable that Harvey's discovery that the blood circulates through the body, Descartes' experiments in biology, and John Bernoulli's and his son Daniel's proposal to apply fluid dynamics to the flow of blood in arteries and veins were not followed up until very recent times, and biological research generally received little attention. If their work and that of others had been followed up, I might have my wish fulfilled today.

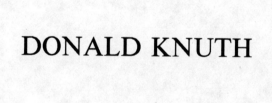

DONALD KNUTH

Interviewed by Donald J. Albers & Lynn A. Steen

As a high school senior in Milwaukee, Donald E. Knuth had doubts about his ability to graduate from college. Four years later he received his B.S. in mathematics, *summa cum laude*, from the Case Institute of Technology in 1960. His work had been so distinguished that by a special (unprecedented) vote of the faculty he was simultaneously awarded an M.S. degree. In 1963 he received his Ph.D. in mathematics from the California Institute of Technology. Over the years he has received many prestigious awards. In 1979, at age 41, he was awarded the National Medal of Science by President Carter.

By any measure, Don Knuth is a remarkable man. He is generally regarded as the preeminent scholar of computer science in the world. He also is an accomplished organist, composer, and novelist.

He is a prolific writer on a host of topics, and he has contributed to an unusually large number of publications. His first publication was for *MAD Magazine*. Since then he has written for *Datamation*, the *Journal of Recreational Mathematics*, the *American Mathematical Monthly*, and dozens of other mathematics and computer science journals such as *Acta Arithmetica* and *Acta Informatica*. He is best known for his monumental series of books, *The Art of Computer Programming*, which has been translated into several languages, ranging from Chinese to Russian. Three of a projected seven volumes in the series are now completed, with part of the fourth in preprint form. His progress on the series has been slowed by a four-year-long diversion on computer-assisted typesetting.

"I Got Headaches from Drawing Those Graphs"

MP: *You are a computer scientist, and yet you started out in mathematics. When did your mathematical interests first emerge?*

Knuth: In my freshman year of high school I got very interested in mathematics. In fact, I think I ruined my eyes drawing hundreds of graphs on orange graph paper with dim lighting. I started to get headaches from drawing those graphs, but I was fascinated with them. The typical graph would be some function like $y = \sqrt{ax + b} - \sqrt{cx + d}$, where I would fix b, c, and d and vary a, in order to see what would happen to the shape of the graph. I had hundreds and hundreds of such graphs where I wanted to see the behavior of functions.

MP: *Did you have an outstanding teacher along the way?*

Knuth: The mathematics teacher that I remember most and who inspired me the most was in college. My high school senior teacher also introduced me to things like binary numbers and encouraged me to do recreational things. During my senior year in high school, I entered the Westinghouse Science Talent Search. I made two entries. One was sort of physics oriented, and the other was a number system based on π. I had thought about imaginary number bases and irrational number bases, and I played around with the kind of logarithm tables that would result from such bases. I didn't win the prize, but I do remember having a lot of fun thinking about number systems. I also played around a lot with absolute value functions. When I learned about the absolute value function, I started making another set of graphs. I worked out a system so that if somebody gave me a pattern of connected straight lines, I would be able to write down a function whose graph gave that pattern. I was absolutely fascinated with graphs in mathematics. My physics teacher was my favorite teacher in high school. I was torn between physics and music, but I enrolled in college as a physics major. I had done a lot of piano playing and some orchestrations, so I didn't know whether I should major in music or physics. My choice of majors was due essentially to the different scholarships I got. The college I chose was better in physics than in music. If I had gone to Valparaiso instead of Case, I would have majored in music.

MP: *Is there any other music or mathematics talent in your family?*

Knuth: My dad was a church organist and now I am.

MP: *Do you think there is much to the suggested connection between musical ability and mathematical ability?*

Knuth: There is definitely something to it. Go, for example, to the Mathematics Institute at Oberwolfach in Germany. Every week mathematicians come there for conferences, and music is the main recreation. They have a tremendous music library and many people will come with their instruments. Chamber music fills the halls almost every night as the mathematicians get together. In our department now we also are surrounded by chamber music. I just was talking to the administrative assistant of our department about this. She had previously worked in the law school, and she said in the law school one professor out of 20 might go to a concert once in awhile. They weren't that much interested in music. Here in the Computer Science Department she felt that more than half the people were musicians themselves.

MP: *Do you have a theory as to what you perceive as a real connection between music and mathematics?*

Knuth: No, I really don't understand why. I guess Euler liked music, but I can't say how many great mathematicians of bygone days were really good musicians. I haven't studied that. There is definitely a correlation, certainly at Stanford.

I also found this in my friend, Professor Dahl from Norway, who carries piano duet music all of the time with him in his briefcase. No matter where he goes, he finds someone to play with him. He was the one who introduced me to the beauty of four-hands piano music, and now I have quite a collection of it built up over the last ten years.

MP: *I wonder if we could get back to your early schooling experiences from a different angle, namely, writing. When did your writing interests first emerge?*

Knuth: Our grade school was very good in English grammar. I remember one of the most interesting things for me in the 7th and 8th grades was to diagram sentences. A bunch of us would get together after class to try diagramming. We could diagram all of the sentences in the English book, but we couldn't diagram many of the other sentences we saw around us, especially the ones we saw in the hymnal. We couldn't figure out what was going on. They just didn't fit any of the rules we learned. We worked hard on this, and it was a big thing for us at the time. In high school, I found out that everyone from our grade school was whizzing through the English classes because of what we had learned; so it wasn't just me.

MP: *I was going to ask you if this was a public school.*

Knuth: It was a Lutheran school. My dad was a Lutheran school teacher, and I attended a Lutheran high school in Milwaukee. My grade school education in writing was really good. I'm trying to do that for my kids now. I have them write an essay every week. If they want permission to watch television the next week, they have to turn in their essay the previous week. When I got to college, I found out that writing was almost 50% of what I had to do well, and the other half was mathematics.

The other strange thing I remember about grade school occurred when I was in the 8th grade. There was a contest run by the manufacturers of Ziegler's Giant Bar in Milwaukee. The contest consisted of trying to find how many words could be made out of the letters in "Ziegler's Giant Bar." This contest appealed to me very much, and I told my parents I had a stomach ache so that I didn't have to go to school for two weeks. I spent all those two weeks with an unabridged dictionary finding all the words I could get from the specified letters. I wound up with about 4,500 words, and the judges had only 2,500 on their master list. Afterwards I realized that I could have made even more words if I had used the apostrophe! My dad and mom helped type up the answers I had written out when they saw how interested I was in this project. The prize was a television set for the school. So our school got a TV set in the classroom, and we got to watch during class that year (1952) one of the first live transmissions from San Francisco across the country. We also got a Ziegler's Giant Bar for everyone in the class.

Professor Guenther was my freshman calculus teacher at Case, and he first exposed me to higher

Knuth is an accomplished organist and composer. "I want to write some music for organ with computer help. If I live long enough, I would like to write a rather long work that would be based on the book of Revelation. The musical themes would correspond to the symbolism in the book of Revelation."

mathematics. Paul Guenther died about five years ago, but he was a great teacher for me mostly because he was so hard to impress. Every time I made a suggestion, I was put down, but he would grudgingly appreciate it when I finally came up with a good one. I don't know why I got so excited about him. In addition to his unimpressibility, he had a good sense of humor, and he seemed to know as much physics as my physics teachers and as much chemistry as my chemistry teacher. That impressed me: it seemed that mathematics was a little better somehow.

Also, I was scared stiff that I wasn't going to make it in mathematics. My advisors in high school told me that I had done well so far, but they didn't think I could carry it on in college. They said college was really tough, and the Dean had told us that one out of three would fail in the first year. In high school, I did have the all-time record for grades. We graded not on A, B, C, D, but on percentages in every course. And my overall percentage in classes was better than 97.5%.

"I Always Had an Inferiority Complex"

MP: *Didn't that instill in you a great deal of confidence?*

Knuth: I always had an inferiority complex—that's why I worked so hard. I was an over-achiever probably.

At Case, I spent hours and hours studying the mathematics book we used—*Calculus and Analytic Geometry* by Thomas—and I worked every supplementary problem in the book. We were assigned

Donald Knuth

only the even-numbered problems, but I did every single one together with the extras in the back of the book because I felt so scared. I thought I should do all of them. I found at first that it was very slow going, and I worked late at night to do it. I think the only reason I did this was because I was worried about passing. But then I found out that after a few months I could do all of the problems in the same amount of time that it took the other kids to do just the odd-numbered ones. I had learned enough about problem solving by that time that I could gain speed, so it turned out to be very lucky that I crashed into it real hard at the beginning.

I started as a physics major, but my turn towards mathematics came in my sophomore year. I took a course in abstract mathematics from Professor Green, who is still teaching at Case. He had written his own textbook for the class, where he would give axioms for Boolean algebra, logic, etc. All of a sudden I realized that it was something I liked very much. And he gave a special problem without telling us whether it was possible or not. He said that if anyone could solve the problem, they would get an automatic "A" in the course. So, of course, none of us tried it. It was obviously hopeless, for he had quite a reputation. As far as we were concerned, there was no way to do it.

I was in the marching band that fall. But I missed the bus and had nothing to do, so I decided to kill time by working on that impossible problem. By a stroke of luck I was able to solve it, so I handed a solution in on Monday. He said, "Okay, Knuth, you get an 'A' in the course." I cut class the rest of the quarter, and he lived up to his bargain. I felt guilty about cutting classes afterwards, so I became the grader for the course the next year. Instead of doing the homework, I was grading it.

In physics I was having a terrible time in the welding lab. I never was very good in laboratories, either in chemistry or in physics. My experiments just wouldn't work. I would drop things on the floor, and would always be the last one to finish. Once I had to report an experimental error of 140% in chemistry lab. I objected to their formula for experimental error, since I thought nobody could be more than 100% wrong, but they wouldn't listen to me.

In welding it was even worse. I was too tall for the welding tables, and my eyes weren't good enough; I couldn't wear my glasses underneath the goggles. Everything would go wrong, and I was terrified by what seemed like hundreds of thousands of volts of electricity! I just wanted nothing to do with it, yet physics majors were required to do this lab work.

On the other hand, I had Professor Green's course in abstract mathematics, which seemed very appealing to me. Just for fun I had made up sort of random axioms for what turned out to be a ternary logic, something that looked a little like Boolean algebra. The idea was to see if those axioms would lead to any theorems. I was working hard, trying to get something to follow from those axioms, so hard that I found my other grades were going down, so I had to stop working on it. I had set up—it's probably pretty trivial now, I suppose—some operation that would be analogous to a truth table, and I finally proved a theorem that went something like this: "*The absitive of the posilute of two cosmoframmics is equal to the posilute of their absitives.*" I just made up words for certain abstract concepts, and it appealed to me that I could prove a theorem that was analogous to De Morgan's Law.

All the way through my student work I had been joyfully stuck in Chapter One of my math books, thinking about the definitions of things and trying to make little modifications, seeing what could be discovered and working from there. I really enjoyed that, but the physics labs were killing me: The combination resulted in my switching to a math major at the end of my sophomore year.

"I Discovered Computers in My Freshman Year—Before Girls"

MP: *Somewhere along the line you began to work with computers. Was there some point where it became clear to you that you would work with them for a long time?*

Knuth: Between my freshman and sophomore year, I had a summer job drawing graphs for statisticians at Case. In the room next to where I worked was a computer.

MP: *So the graphing continued?*

Knuth: Yes, I could draw graphs.

MP: *Was this compulsive?*

Knuth: Well, yes; to some extent, METAFONT (the system for computer-assisted typography that I recently developed) probably reflects my love of graphing.

The Statistics Department at Case was located right next to a new computer, a wonderful machine with flashing lights. Early that summer, someone explained to me how it worked. Pretty soon I was hooked. I spent a lot of nights, all night long, at the console of the computer. Nobody else was there. I discovered girls in my sophomore year. This was before that; I had computers first.

I still have my first computer program. It factored numbers into primes. You would dial a ten-digit number into the console, and it would punch the factors on cards. The program initially was about 70 instructions long, and as I recall, by the time I finished it, I had removed more than 100 errors out of 70 lines. In other words, I made a lot of mistakes, but I always felt I learned from those mistakes. The program wouldn't work, and I kept on fixing it, and finally it worked. My second program was to do base conversions.

My third program was to play tic-tac-toe, and it also would learn *how* to play tic-tac-toe. I worked hard on this one. I developed a learning strategy where every position in a game the computer won would be rated as a little bit better; but if it lost the game, every position would be marked as bad. This was a memory that would adapt itself. Each position in the game had a number from zero to nine representing how good it was thought to be; the neutral value was four, so if it was a drawn game, the position ratings would tend to go towards four. I wrote another program that would play tic-tac-toe perfectly, and then I had these two programs playing each other. After 90 games, the learning program learned how to draw against the good one. In another experiment, after 300 games, two learning programs starting with blank memories learned to draw against each other. They played a very conservative game, not very exciting, but it was interesting to see "the blind leading the blind."

That was my first month of learning to program. Those things were really fun. The next thing I did was somewhat different: I wrote an assembly program for the machine. I started to read the code of other people, and I got especially interested in programming because most of what I read wasn't very well written. I could look at programs and say here I am, only a college freshman, and I can do better than these professionals. I didn't know that lots of people could do better than those professionals. The standards of software at that time were pretty bad. I began to think that I had a special talent for it. Maybe I did, but I don't think now that it was as special as I had once thought. Then I read the code for Stan Poley's assembly program, which I thought was truly beautiful, a masterpiece of elegant programming, so I was inspired to carry his ideas one step further.

MP: *Feelings of inferiority were certainly not present in that work.*

Knuth: Well, part of me was anxious to prove to the other part that I was OK! So, I was always motivated by seeing publications by someone who was apparently an expert, where I thought I could do a little better. Now I always tell my students when I make mistakes in class that I'm just trying to motivate them.

"Students Aren't Learning How to Write"

MP: *The kinds of experiences with computers you had as a college sophomore, with minor modifications, are going on in junior high school now. What effect do you see that having on the future?*

Knuth: The students aren't learning how to write. That's serious. They don't know how to spell "mnemonic." They're specializing too early. There's a danger that people aren't seeing the other side of the coin, which is writing. That's what I worry about. As long as people keep open to a lot of aspects of life, then it's good, but if they become involved too much with one subculture, then it's going to limit them later.

MP: *What is it about computers that makes people so compulsive, either pro or con? A lot of people just can't stand programming. There are others who just get consumed by it.*

Knuth: It's partly a strange way of thinking. There are so many different modes of thinking, not really understood yet by psychologists. Teachers of computer science regularly find that two percent of the people who enroll in their courses are natural-born computer scientists who really resonate with computer programming. There seems to be a correlation between that and mathematical logic. I said I enjoyed the abstract algebra course where I first really studied axioms and Boolean algebra. If you look in math departments, the faculty who have traditionally been the closest to computer science have been the people in logic and combinatorics. Conversely, the mathematicians who are best at geometrical visualization tend not to enjoy a discrete universe like the computer world.

Another difference is between finite and infinite mathematics. I used to say to Peter Crawley at Cal Tech that he and I intersect at countable infinity, because I never think of anything more than a countable infinity, and he never thinks of anything less than a countable infinity. Higher infinities involve a kind of reasoning and intuition that doesn't apply very much to computers at all.

There are different flavors of mathematics, based on what kinds of peculiar minds we have; we aren't going to change that. People find out what things are best for them. And as for their mentalities, physicists are different from mathematicians, as are lawyers from doctors. Each of these fields seems to have predominant modes of thinking, which people somehow recognize as best for them. Computer science, I am convinced, exists today in universities because it corresponds to a mode of thinking, a peculiar mind-set that is the computer scientist's way of looking at knowledge. One out of fifty people, say, has this peculiarity.

Historically, such people were scattered in many walks of life; we had no home to call our own. When computer science started out it was mostly treated as a tool for the existing disciplines and not of interest in its own right. But being a useful tool is not enough in itself to account for the fact that computer science is now thriving in thousands of places. For example, an electron microscope is a marvelous tool, but "electron microscope science" has not taken the world by storm; something other than the usefulness of computers must account for the rapid spread of computer science. What actually happened was that the people who got interested in computers started to realize that their peculiar way of thinking was shared by others, so they began to congregate in places where they could have people like themselves to work with. This is how computer science came to exist. Now we can look back in old writings and see that certain people were really computer scientists at heart. It was latent back in Babylonian times, and throughout history.

Computer Science vs. Mathematics

MP: *Are computer scientists really different from mathematicians, then?*

Knuth: I think you can recognize a difference.

MP: *But how would I recognize a computer scientist if he walks in the door?*

Knuth: By the thought process. I've been trying to answer exactly that question. In order to get a handle on it, I tried to study several works on mathematics to discover the typical paradigms of good mathematicians, using a random sampling technique. What I did was to take nine books that would represent mathematics, and I looked at page 100 of each book. I analyzed that page very carefully, until I understood what kinds of things were there on that page. It was interesting to see what aspects of mathematical thought processes were involved. I asked myself: "If I had to write a computer program to discover the mathematics on page 100, what capabilities would I have to put in that program?"

I found that one of the most striking things distinguishing mathematicians from computer scientists was their strong geometric reasoning and reasoning about infinity. The things that were common to both computer science and mathematics were primarily things like the use of abstractions and the manipulation of formulas. The main thing that was prominent in computer science that wasn't in mathematics was an emphasis on the state of a process as it changes, where it changes in time in a discrete way. In computer science when you say n is replaced by $n + 1$, the old value disappears and the new value takes over. We know how to think about an algorithm that is half-way executed; it has a state consisting of the current values of all the variables, and the state also specifies what rule to apply next. In order to formulate this for most mathematicians, it requires putting subscripts on everything. Traditional mathematics doesn't have this notion of a process in highly developed form, but it is vital in computer science.

The other striking difference was that computer scientists are willing to deal with diverse case analyses. The more pure the mathematician, the more he or she instinctively likes to have a clean formula that covers everything in all cases. But computer scientists are able to reason comfortably about things that have different cases, where we do step one, step two, then step three. A mathematician likes to have one step that you can apply over and over again.

MP: *What you are saying reminds me of an argument by Edsger Dijkstra, that computer scientists have now learned enough about the process of mathematics—in terms of how formulas are manipulated and how things change—that it should feed back into the process of teaching mathematics. Does the computer scientist actually now know enough to develop a science of mathematics that would be useful in teaching?*

Knuth: Since people have different modes of thinking, I doubt if any one way of teaching will be simultaneously the best way to reach different types of students; and I also doubt if many people can design educational plans that work for students having a different mind-set from the educational planners. So I can't be confident that a method best for me would be best for the world. But certainly

President Carter presenting the National Medal of Science to Knuth.

Dijkstra's proposal would be the best way to teach mathematics to a natural-born computer scientist. From my own perspective, I feel that I have really learned some subject of mathematics at the point when I understand how it works in an algorithmic formulation.

For example, consider Volume 2 of my books. I think every theorem of elementary number theory is in there somewhere, but it's in the context of an algorithm that somebody needed because of a computational problem that had to be solved. I believe that the original discovery of these ideas was because of the need for such algorithms, so I presented it that way. It's a different aesthetic from mathematics, you see, from what is mathematically "clean" to what is not elegant in the same way. It's a different way of thinking, and I can't argue that one is better than the other.

Donald Knuth

Dijkstra and I are natural-born computer scientists. We found that out after we got older. Such ways of organizing knowledge are not going to be for everybody, but for our subset of the population, an algorithmic approach works best.

The knowledge that we have a computer scientist mentality is also a challenge because we have to do our best for the other 49 out of 50 people who don't think as computer scientists; computers are affecting everybody's lives. We have to find a way to make it comfortable for other people to use computers, even though we don't really understand the way they think any more than they understand the way we think. You need people who are halfway between the different modes of thinking to bridge these gaps.

Of course, there is no definite boundary that you cross in going from computer scientist to mathematician to physicist, etc. There tends to be, in a multidimensional space of different kinds of abilities, a focus around the place that is most representative of computer scientists, and another one that is more typical of mathematicians. Maybe musicians are also close. Who knows? But it is a continuous thing.

Computer science departments thrive because there are a lot of people near our focal point in "thought space." In past ages, some of the people we now would call computer scientists were called mathematicians, physicists, chemists, doctors, businessmen. Now they have a home. That's what is holding the field together. But to develop our field well, if we want other people to make use of computer science, we have to realize that we aren't able to do it all ourselves. We need others who can understand the other modes of thinking.

Maybe Dijkstra would argue that our mode of thinking is more powerful somehow, that it includes the other ones. I'm not quite so bold yet to do that, but the reason I raise this possibility is because I once met someone who had been a computer science major in graduate school and then went on to become one of the main advisors to President Lopez in Mexico. He told me that his training in computer science was of great value to him in working with all the people he had to deal with. Even though he wasn't programming computers any more, he felt that his approach to knowledge enabled him to understand lots of different people who were talking to him but who couldn't understand each other very well. The computer science view seemed to be more powerful in its models of reality. This might be true because computer scientists are accustomed to models that can handle a variety of cases. The real world breaks down into cases, and mathematical models are better or worse depending on the uniformity of what they're modeling. The computer scientist dealing with less uniform models is perhaps able to cope with more general things. On the other hand, when it gets to something that is truly uniform, a computer scientist will not be able to go as deeply as a mathematician. Uniformity might really be the most striking difference between the fields.

MP: *A few have suggested that mathematics may be a part of computer science. Others say that computer science is a part of mathematics. What do you think?*

Knuth: I really think that they are two things, although they are related. There is a lot of overlap, but also I think I can tell when I'm in my mathematics mode, wearing my mathematician's cap, and that I can almost feel the changes when I go into a computer-science mode. I can't exactly say why; maybe I'll never know. But I can definitely feel when I'm behaving as a computer scientist.

I remember once giving a lecture about a number-theoretic problem that seemed to straddle the two fields. I started with the mathematical definitions and took things as far as I could using traditional mathematical tools. Then I said, "Now let's tackle this problem as a computer scientist would." And I carried on for fifteen minutes with an algorithmic viewpoint, after which I said, "This is as far as the computer scientist is going to get. Now let's be mathematicians again for a while." And I could feel strongly that this was true, that I was sometimes doing what a mathematician would definitely do, while at other times I knew I was doing something that a mathematician wouldn't do.

Neither the mathematician nor the computer scientist is bound by a study of nature. With a pencil and paper we can control exactly what we are working on. A theorem is true or it's false. The fact that we deal with man-made things is common both to mathematics and computer science, but the nature of the thought process is sufficiently different that there probably is a justification for considering them to be two different views of the world—two different ways of organizing knowledge abstractly that have some points in common, but in a way they have their own domains.

MP: *In addition to these intellectual contrasts between mathematics and computer science, there is getting to be a lot of social and educational concern about their interaction. The growth of the computer field, for example, is drawing so many people into bachelor's level computing that there aren't very many* *people going on into advanced work in mathematics or computer science or into high school mathematics teaching. Computer science and mathematics appeal to the same limited group of people, and at the moment, the momentum seems to be running pretty well in favor of computer science.*

Knuth: But that's an economic consideration. Right now bachelors in computer science are getting better salaries than anybody else. This is our problem because people are going into it who aren't natural-born computer scientists; they're just going into it for the money, and not for the love of it. This trend can actually be to some advantage to mathematicians because they've now got motivated students in their classes instead of people who are just there for some external reason.

I don't agree with your statement that "computer science and mathematics appeal to the same limited group of people," but our disagreement is probably due to the fact that I have been thinking mostly of extreme cases, the differences between the best computer scientists and the best mathematicians. According to the law of large numbers, there will of course be very few people who rate a 10 on a scale of 1 to 10, under almost any ranking criteria, and there aren't many 9's either. So let's consider somebody who is an 8 at mathematics and a 6, say, at computer science. That person is somewhat likely to go into computer science instead of mathematics, nowadays, because the salary is better; and perhaps society will be better off since there is a pressing need for computer scientists. I agree that economic pressure is causing mathematics to lose a lot of its 7's and 8's (if you'll excuse my callous use of numbers in place of human beings); but I hope that the differences between computer science and mathematics will be well enough understood that you don't lose the 9's and 10's. Their mathematical abilities are vital to people in all other fields, including computer science.

MP: *One of the things that we keep hearing today is that good instruction in mathematics is reversing dramatically and perhaps in other subjects as well. Many say the reversal is due to the influence of computers and calculators.*

Knuth: I have been quite disappointed in the mathematics textbook my son has as a sophomore in high school. But his teacher is very good and compensates for it. The worst excess of this book is the pedantry. The second-worst is that its three authors seem to have written three different kinds of chapters without much awareness of what the other authors had done. But the book has some good points, too, like its emphasis on graphs. My suggestion for teaching mathematics at the young levels is to draw graphs! I was glad to see there was much more use of graphs here than in other books I had seen.

I read an article a week ago where someone said he had started to employ older textbooks, and the older the textbook the better the students would do on exams. He went back to about 1900, and he said that was the Golden Era.

Clearly the Hungarian educational system has been the most successful for pure mathematics; it's a model that ought to be studied very carefully because it works. It produces so many good mathematicians *per capita.*

As I see my children learning mathematics now, I find that the books have too much emphasis on flashy things and on memorizing formulas, rather than on what an idea is good for as a general tool, or on how to reconstruct a formula from a few basic principles instead of from memory.

After reading these books, you don't remember what the degree of a polynomial is, or how to answer various precise questions. My son had learned algebra, but it was not clear to him how to add fractions; that had been glossed over and needed in only one homework assignment. He would remember for a few weeks what the distributive law was, but that would come and go. The high-level wording for things wasn't being put to any use. So it disappeared from his mind. In my own case, I got along fine without knowing the name of the distributive law until my sophomore year in college; meanwhile, I had drawn lots of graphs.

MP: *Do you think students should be introduced to computing earlier in the formal curriculum?*

Knuth: I would like to see an approach by means of algorithms, but I really hesitate to say much about it for fear somebody will believe me when I really haven't thought it through. The only experience I had personally with doing something on an algorithmic basis was at a higher level. At Cal Tech I taught an introductory abstract algebra course. We were studying matrices and there was a question on canonical forms of matrices. The book was very obscure on that point. We wanted to know when one matrix was similar to another. How can you tell? Well, you have a canonical form that says that if two matrices are similar, they have the same canonical form. Books rarely even say that. They just say here is the name of the form, and here is the definition of it.

So we said then, how can we determine if two matrices are similar? There is a little operation you can do on a matrix, something like this: "Subtract a multiple of the third row from the second row, and then add the same multiple of the second column to the third column." That operation preserves similarity: although it is very simple, it can be used to get another matrix similar to it. So let's take a look at what it does. We start with a matrix and try to zero out most of the first row and the first column, and we see that we can almost always do this by simple transformations that preserve similarity. But every once in awhile we get stuck.

The point is that we are solving a problem. We are trying to do it step by step with an algorithm; pretty soon we find this canonical form. Of course, we might fail because some special case might happen. Well, in that case, we won't be able to make so many elements zero. This, I believe, is the way these canonical forms were first found. But later on there was a tendency for people who had learned about them in a concrete way to present them in a high-level, abstract fashion. For example, they would look at the way linear operators behave on subspaces. This is elegant, but pedagogically unsatisfactory because it conceals the method of discovery.

I think you need both views, especially when you are learning a subject at first. The algorithmic point of view tends to be at a more intuitive level. It helps very much in the early stages of teaching. But as I say, I haven't been teaching these courses myself for a long time, and my own intuition might not correlate well with that of the majority of students. Careful experiments should be tried, since an algorithmic approach might well be very successful.

MP: *It was my impression that you were perhaps very excited by your tic-tac-toe learning program. Now that's using the computer for playing games. What do you think of games as an introduction to computing?*

Knuth: When children are young, they get to a stage where they like to make up rules of games. They enjoy arguing with each other. They argue incessantly about the rules because they like to make up rules to games. I think this might be related to making up computer programs. There is probably in children a tendency that can be well exploited to encapsulate these kinds of rules.

I'm way out of my expertise, of course, when thinking about elementary education. But I would like to see an approach where the students learning some concept not only learn how to follow the rules but also how to explain the rules. Suppose you are teaching somebody to add. We carry this out by giving a bunch of examples and tell children that you go from right to left and you carry, and so on. If there were only a simple enough computer language for these second and third graders, it might be good to teach addition by having them write little programs in this language. The teacher would say: "Here's the way to add numbers and we're going to teach it to this goofy machine." I think it's definitely worth a try. Students learn quickly to make up computer programs that will draw pictures. It might be that similar skills would work on things like arithmetic.

I like to see the rules and their exceptions. Other people are very comfortable without the rules; maybe people who are good with language just absorb dictionaries very quickly and don't even make rules out of things. To them, everything is an exception. So I suppose an algorithmic approach will be of no help to them; we need a variety of teaching methods.

MP: *I have been impressed with a toy called Big Track that uses a language like TURTLE, which comes out of Papert's laboratory. The rules are very explicit and very tight.*

Knuth: That's very good for teaching algorithms. I think arithmetic is another thing that people should still keep learning even though we have calculators. It's nice to be able to add and to multiply numbers by hand, in a pinch. Not only nice: it introduces important patterns of thought.

MP: *Are you very enthusiastic about calculators?*

Knuth: Well, no. Computer scientists were the last to get enthusiastic about calculators. One of my colleagues remarked the other day that this is probably the only department at the university where nobody has a pocket calculator at the faculty meeting. This may be because we have all these super-computers here in our offices. I can do my symbolic calculations at M.I.T., too, three thousand miles away, using a computer network.

Basketball and Computers

MP: *Can we turn to some other games for a minute? One of your former teachers at Case tells a good story about you. I don't know whether it is true or not, but it is a good story. He says you were absolutely instrumental in the success of the Case basketball team in 1960. Is that true?*

Knuth: Well, it would be nice to say that. We did win the league title that year, and I'll be glad to take all of the credit for it. But here's what really happened: As the manager of the team, I worked out a formula, a rather complicated jumble of symbols that I really don't believe in any more. It was a system that would rate each player with a magic number. The magic number would tell how much each player really contributed to the game, not just the points he scored. If he missed a shot, there was a certain chance that our team would not get the rebound, and the other team would get the ball. Thus, he would lose something for each shot missed. Conversely, when he stole the ball, that was very good because our team got possession.

In fact, it is interesting to watch a basketball game and imagine that possession of the ball counts one point. Such an assumption isn't true in the last seconds, but during a good part of the game it isn't out of the question to say that if your team has possession of the ball, it's worth a point. In other words, you look at the scoreboard, and if it's 90 to 85, you add one to whichever team has the ball then. It tends to give you a better feeling of the current score. If you look at basketball this way, then when someone makes a field goal, the score hasn't really changed because his team has gained two points but lost possession of the ball, while the other team has gained possession. The two points sort of cancel each other out. If the other team just goes back and makes another field goal, we're back to where we started. But if somebody steals the ball, he is really making the previous field goal count.

According to my magic formula, possession of the ball turned out in most games to be worth about .6 of a point, so you got some credit for field goals too. As I said, I don't really believe in that formula

Student Knuth and Coach Helm putting player ratings into an IBM 650 computer. In 1960 Don Knuth was a student manager of the basketball team at Case Institute of Technology. Knuth developed a formula for rating each player. Helm used Knuth's formula, and Case's Rough Riders went on to win the league championship.

Donald Knuth

any more, but it did include all the statistics like steals, fumbles, etc., and you could plug into it and get a number. The coach liked these numbers. He said it did correlate with what he felt the players had contributed; and the players on the team, instead of competing for the most points, tried to get a good score by this number. The coach thought that was a good thing. It was written up in *Newsweek Magazine*, and it was on Walter Cronkite's Sunday News. They sent a cameraman out to take a picture of me taking the statistics and feeding them into the computer.

MP: *Did you receive offers to become a consultant for professional teams?*

Knuth: There was some talk about the Cleveland Browns wanting to use computers already in those days, but I never was involved with that.

Origin of "The Art of Computer Programming"

MP: *What was the inspiration to launch your series ("The Art of Computer Programming") in the first place?*

Knuth: *The Art of Computer Programming* developed when I was a second-year grad student at Cal Tech. I had been working as a private consultant writing compilers for different machines. In those days a software firm would ask for hundreds of thousands of dollars to write a compiler. But I didn't know that, and I had written one for $5,000. I guess the word got around that I knew how to write compilers. A little later, Richard Varga, who was an advisor to Addison-Wesley, suggested that they ask me to write a book about compilers. They came to me in January of '62 and said: "How would you like to write a book on writing compilers?" It occurred to me that I really did like to write. It sounded good; I decided that it would indeed be nice to write such a text.

I had gotten married in the summer of '61, and I wonder now how I broke the news to my wife that I was suddenly planning to write a book. Surely neither of us knew how much this was going to change our lives. Anyway that day I went home and sketched out twelve chapter titles that I thought would be nice, and then a little while later I signed a contract for a book about compilers. I got a chance to teach a course at Cal Tech during the fall of 1962, while still a graduate student, with the idea that the class notes would develop into the first three chapters.

Right after getting my Ph.D. in '63, I started to work hard preparing the chapter on sorting. I knew hardly anything about sorting, but I thought it would be nice to read up on the subject and to toss a chapter about sorting into a book about compilers, especially because the LARC Scientific Compiler had just come out and it was reputedly based on the idea of sorting the data in unusual ways. I found that sorting was really interesting, and pretty soon I found myself digging into lots of technical articles

The main thing that struck me was that the literature was so spotty. Computer science was a very new field, without an identity of its own, and standards of publication were not terribly high, especially when quantitative aspects of algorithm performance were concerned. A lot of the published articles were just wrong, so you had three possibilities: the wrong answer by the wrong method, the right answer by the wrong method, and the right answer by the right method. You had about a one-third chance on any of these possibilities. The literature on computing was already large but very unreliable, so it was clear even in 1962–63 that it would be nice to have a summary of the right parts of the literature. Publications were so bad, in fact, that people didn't even bother to read them, and the good ideas were being rediscovered because people found it easier to do this than to sort them out from the bad ones. Thus, one of my big motivations was to clean up the story that had been presented badly in the literature.

I guess I have an instinct for trying to organize things. At that time, everybody I knew who could write such a book summarizing what was known about computer programming had discovered quite a lot of the ideas themselves. It seemed to me that they would slant it to their own perspective, which would present only one side of the coin, their own particular part. By contrast, I really hadn't discovered anything new by myself at that point. I was just a good writer. That was very prominent in my mind: for example, I was the only computer scientist I knew who hadn't discovered how to compile arithmetic expressions by the precedence method. Ten people independently discovered it, yet the problem had baffled me; I hadn't seen my way through it. So I felt not only that the story needed to be presented, but that I could present it from a less biased viewpoint than these other people who seemed to have done their work more in isolation. I had this half-conceited and half-unconceited view that I could explain it more satisfactorily than the others because of my lack of bias. I didn't have any axes to grind but my own. (Then, of course, as I started to write things I naturally discovered one or two new things as I went, and now I am just as biased as anybody.)

But you asked about my original motivation. My original motivation was to write a text about how to write compilers, so I began drafting chapters. I was seriously planning to finish the book before my son was born. (He's a sophomore in high school now, so I'm currently trying to finish before he starts writing his own books!) I recently found copies of letters I wrote in 1965 saying, "I wish I could come to visit your university this summer, but I can't because I just have to finish this book-writing project." It was going to be just one book.

As I said, however, I started to get interested in sorting, after I had been writing for a while. There were so many interesting things on sorting. So the book was growing rapidly.

I eventually wrote a letter to Addison-Wesley saying: "Do you mind if this book is a little long? I would like to give a fairly complete presentation of the material." They said: "Don't worry, go right ahead, write whatever you want." So I kept gathering more and more stuff. In June, 1965, I had finally finished the first draft of the twelve chapters. It amounted to 3,000 hand-written pages of manuscript.

To me this was something of a longish book, yet only one volume. I thought I knew about books, and the printed letters in books seemed to be a lot smaller than the ones I write by hand. So I figured that about five pages of my handwriting would be about one page of book. Then I went ahead and did chapter one, typing it from the handwritten manuscript and sent it off to Addison-Wesley, just to see if it was okay. This was October, 1965. They hadn't heard from me for quite a while, and were wondering what was going on. I felt good, for at least I had finished chapter one. Incidentally, that chapter one was pretty much the same as the chapter that was eventually published.

Immediately I got back a letter from Addison-Wesley, from a person very high up in the company. It turned out he was one of the people who had originally talked to me in '62, but meanwhile he had been promoted three times. From the length of chapter one, the book was now estimated to be almost 2,000 printed pages, and they said: "Don, you said you might be writing a longish book, but do you

At the age of three, Don Knuth was already attracted to keyboards.

realize that one and one-half pages of typing is one page of book?" I thought to myself that it can't be so. "I've read a lot of books; these guys don't know what they're talking about." So I took one of their books, Thomas's *Calculus*, and I sat down at my typewriter and typed up a page of it. Lo and behold, they were absolutely right! Then I knew why it had taken me so long to get that first chapter finished —three and a half years to write the first chapter is not too good.

It gradually dawned on me how large a project this was going to be. If I had realized that at the beginning, I wouldn't have been foolish enough to start; I wouldn't have dared to tackle such a thing. But by 1965, of course, I was hooked because I still felt the need for these books, and I still felt the project ought to be done. I still believed that I was fairly unbiased and could try to be a spokesman for the people who were making the discoveries. I had collected so much material that I felt it was my duty to continue with the project even though it would take a lot longer than I had originally expected. I had done all of the background work and it would have been very hard to transfer it to anybody else.

Addison-Wesley took another look at the 12 chapters. At first it appeared that the material would fill up two volumes, then three volumes. Publishers have horror stories about ponderous tomes like that. So my guess is that they showed these things to some consultants, and the consultants recommended that seven of the 12 chapters would sell pretty well as individual volumes. So they suggested combining the remaining five chapters with the good seven, hoping to sell seven books this way. I think this was probably the motivation for the present plan. At any rate I saw that it was possible to reorganize the chapters so that they would fit together reasonably well in seven separate volumes.

Volume Two, for example, was the combination of my original idea for Chapter Two on random numbers and Chapter Six, which was originally called "Miscellaneous Utility Routines." Suddenly I noticed that all but one of these miscellaneous utility routines was really about arithmetic. So I decided to call that chapter "Arithmetic." That little change in title suggested one or two sections that I ought to add; and as I added them, I came to a marvelous realization that there was this book out there waiting to be written, bringing together what is known about arithmetic from a computer scientist's point of view. As a result, that chapter almost wrote itself, and led me to fascinating things in a variety of journals that had just never been put together in book form. Having the title "Arithmetic," and having it bound in the second volume, is what turned out to add a lot of unity to the subject, a unity that probably hadn't been realized before. Most of the articles I found in the literature were written by people who were not aware of many of the other journals, nor of the relation between their ideas and others.

I got so excited about writing Volume Two that I started working day and night on it. As a result, I got a serious attack of ulcers, and had to change my whole life style. By the middle of Volume Two I kept thinking I was going to finish it soon, and I had something of a breakdown of my health in the summer of 1967. About the middle of Euclid's algorithm is where I broke down. That happened on what is now page 333 of Volume Two out of 688 pages. So I still had a lot to go at the time. I knew it, but I wouldn't admit it to myself.

I always underestimate time, too; otherwise, I never would have started writing these books. If only I had a better estimator of time! Because of my writing, I have now resolved not to give any lectures away from Stanford until 1990. All of book writing comes out of spare time. When I go on a trip to give a lecture, one day wipes out at least five days of book writing because when I come back I still have to do everything else that I was supposed to have done when I was gone. So the spare time disappears.

MP: *Your field, computer science, has been growing at great speed. Is it still possible to capture it? Is it growing faster than you can write?*

Knuth: Yes, perhaps. I'm thinking of the novel *Tristram Shandy*, a fictional autobiography whose supposed author goes through Volume One of his memoirs covering just the first year of his life. But I still believe it will be possible for me to finish. Volume Four, Combinatorial Algorithms, has exploded the most. Volume Four is the one that I'll return to immediately after I finish my typographical research. I think it is going to become Volumes 4A and 4B.

MP: *So it really is exploding?*

Knuth: In fact, the chapter called "Combinatorial Algorithms" that I planned on that first day in 1962 was thrown in almost as an afterthought. There were very few combinatorial algorithms at the time, but I liked that sort of programming so I thought it would be nifty to have a chapter about it. Almost none of the present material in Volume Four was known then. So when people talk about "combinatorial explosion" the words have a significant double meaning for me. At one point, three years ago, I think seventy percent of the journal articles being published in computer science were about combinatorial algorithms. Volume Four will surely be the hardest, because of the explosion.

MP: *Is the organization you worked out twenty years ago still pretty well holding its shape?*

Knuth: There are new topics, but I won't live long enough to include those. For example, I never promised to write about operating systems. Therefore, I'm very happy whenever I get a journal in the mail, if most of its articles are about operating systems.

The Discipline of Writing

MP: *How do you combine the discipline of writing with spontaneous creativity?*

Knuth: When I'm writing a book, I surround myself with that subject and nothing else. I read exhaustively on one area, and then after I finish that section, it goes out of my head, and I bring in another. That's what computer science calls "batch processing," as opposed to continual "swapping in and out" or "thrashing."

I don't read the literature as it happens. I only read the titles and abstracts to know where I can put articles on the agenda of things to read later. I do the same thing with quotes. I file the quote. Here is a wonderful one from the Beatles' songs: "There's nobody in my tree." That's just perfect for branch-and-bound methods. I hope to live long enough to finish just because I have so many quotes on file that are great.

When I'm working on a topic, I may have to read sixty papers on one subject. The first two I'll read slowly, but with the next fifty-eight I know what to do already. When I read the first two, I use the strategy of trying to figure out the problems before I look at the answers. Then I am ready for the vocabulary and ideas that are going to be occurring in the other papers.

I keep a little notebook, too; every day I write a summary of what I've worked on that day. It helps me to schedule myself a little bit and it helps me to realize how hard things are so that I can plan ahead. If too many days go by where I said I was just too tired and went to bed, or if it reads, "Today I goofed off," then it helps me to make a little more rational schedule.

For the seven-volume book project, I have to cross a threshold every day to get started when I am writing. I have to get psyched up for it. It is a long, on-going process; and I know that even after the end of the day I won't have finished. Every morning I wake up and say: "Another day, and the book isn't finished." I still feel a strong need for the book in the world, and that it is filling a necessary role, but all these logical arguments aren't going to make me get started. On the other hand, once I've started, then I'm excited about it, and it's hard for me to stop again. I have to force myself to stop and not just stay up all night. So I always read a variety of things—detective stories, or more serious works of fiction, or history, sort of rotating between them—at bedtime.

MP: *Do you do most of your writing at home?*

Knuth: Yes, always; it is not part of my Stanford job. It's all spare time.

MP: *Do you do it in long hand?*

Knuth: I can't compose at a typewriter. I can't even compose a letter to my relatives on a typewriter, even though I'm a good typist. I went to secretarial school during the summers of my high-school years, and I learned to type eighty words a minute. I learned machine shorthand, and I learned Gregg shorthand. But I can't compose in any of those modes.

MP: *Why would a prospective physics major take those courses in high school?*

Knuth: I had summer jobs doing secretarial work, and I thought it would help me in college taking notes. But all I learned were the abbreviations for "Dear Sir" and "Yours Very Truly", and that didn't help very much in my chemistry class. I kept making up new abbreviations, sitting in the back of the class with my stenograph machine; afterwards I couldn't figure out what I had put down, so I gave it up.

The Roots of METAFONT

"Mathematics books and journals do not look as beautiful as they used to." With those words, Don Knuth introduced his 1979 article, "Mathematical Typography," to readers of the *Bulletin of the American Mathematical Society*. His statement clearly reveals an aesthetic concern about the physical appearance of mathematics.

This concern has resulted in his inventing TEX and META-FONT. TEX is described by C. Gordon Bell, Vice President of Engineering, Digital Equipment Corporation as follows:

"Don Knuth's Tau Epsilon Chi (TEX) is potentially the most significant invention in typesetting in this century. It introduces a standard language for computer typography and in terms of importance could rank near the introduction of the Gutenberg Press."

METAFONT is a system that makes use of classical mathematics to design alphabets. Knuth's aesthetic concern is clear when he says: "Of course it is necessary that the mathematically-defined letters be beautiful according to traditional notions of aesthetics. Given a sequence of points in the plane, what is the most pleasing curve that connects them? This question leads to interesting mathematics, and one solution based on a novel family of spline curves has produced excellent fonts of type in the author's preliminary experiments. We may conclude that a mathematical approach to the design of alphabets does not eliminate the artists who have been doing the job for so many years; on the contrary, it gives them an exciting new medium to work with."

Four years after he had started his research on METAFONT, Don's mother sent to him the alphabet book that he had enjoyed as a boy of two or three. A page from that book is reproduced on the facing page. Note the *x*'s that Don placed by each serif in the K and L. The 7 inside the K is a count of the serifs of that letter. Clearly, his aesthetic and mathematical interests in alphabets go back to his early childhood.

K

kitten

L

lighthouse

"Surreal Numbers"

MP: *How did you come to write "Surreal Numbers"*?*

Knuth: I wrote it in one week, while on sabbatical in Oslo, Norway. It hasn't been a best seller, but it's been steady and translated into lots of languages. I'm glad for that. Writing *Surreal Numbers* was probably a once-in-a-lifetime experience for me. I got inspired to do it, and I guess there was a muse sitting behind me telling me what to write. The book just fell together, and I don't think I could do it again. That week was one of the most exciting periods of my life.

It was December, 1972, and I was in the midst of writing *The Art of Computer Programming*, when suddenly I got the idea for *Surreal Numbers* in the middle of the night. I woke my wife and I said, "Jill, you know how this series of seven volumes—the books I started on after we had been married for only six months—is affecting our lives? Well, it turns out that there's another book I would like to write too. But I don't think it will take me very long to finish this new one." I said that I thought I could write this other one in about a week, if I just worked on it and nothing else. To my great pleasure, she was also delighted by the idea. She said, "This is the best time in your life for you to do such a project."

We planned it so that after the new year I would get a hotel room in downtown Oslo near where Ibsen wrote his plays, so that I might be able to pick up some of the nuances of his art. Then I could work on this book and also she would come to meet me twice. (We always wanted to have an affair in a hotel room.)

During the three weeks or so before I started *Surreal Numbers*, as I would be walking along or skiing, I would be going through the first page or two of the book in my mind. But I didn't go any further than that, at the time, because I wanted the rest of the book to be fresh as it was being written. I wanted it to be a faithful recording of mathematical discoveries, so I didn't want to do any of the mathematics in advance. I only vaguely remembered what John Conway had told me at lunch a year before.

I got to the hotel and started to work. Fortunately, I didn't go through the scene you frequently see in the movies where the guy types the title of the book, stares at the page for awhile, and tears it up. I didn't have to go through that. I could write out the first page and most of the second, since I had that memorized.

Every day of that week had pretty much the same scenario. I would start out in the morning with a very leisurely breakfast. Students from Saint Olaf College happened to be staying at this same hotel, and I eavesdropped on their conversations to see what phrases they were using. Then I would go to my room and work for about three or four hours. Then I would get to something I wouldn't know how to handle, and I wouldn't have any idea what to do next. So I would go for a walk around Oslo for about two hours. Maybe I'd go to the library, but usually I'd just walk around watching people. Then the solution to the problem would present itself. I would go back to the hotel, and after two more hours of work, I would get over the hump, and I would magically be able to move a little further. Then I would have a nice relaxing dinner, watch Norwegian television for about an hour, go back to the room, write some more, and put out the light.

The reason I think I had this muse was that the book would seem to write itself and fall into place. Things seemed to work out too nicely. I am, of course, very biased. But after I turned out the light, the next page would flash into my head, and I would have to get up quickly and write it down. The thoughts would come so fast that I would only have time to write the first letter of every word. Then I

---

*In 1974, Addison-Wesley published a novelette by Don Knuth. Its title is *Surreal Numbers: How Two Ex-Students Turned on to Pure Mathematics and Found Total Happiness*. It contains a development of a remarkable new way to construct numbers. The new construction had been found by John Horton Conway of Cambridge University. One day over lunch in 1972, Conway briefly explained his system to Knuth. Knuth was so taken by this revolutionary approach that he was motivated to write a book about it. Martin Gardner says: "I believe it is the only time a major mathematical discovery has been published first in a work of fiction."

In a postscript to *Surreal Numbers*, Knuth explains his purpose in writing the book: " . . . my primary aim is not really to teach Conway's theory; it is to teach how one might go about developing such a theory. Therefore, as the two characters in this book gradually explore and build up Conway's number system, I have recorded their false starts and frustrations as well as their good ideas. I wanted to give a reasonably faithful portrayal of the important principles, techniques, joys, passions, and philosophy of mathematics, so I wrote the story as I was actually doing the research myself."

could turn out the light and sleep like a log. The next morning I would have to figure out what the first sentences were from the first letters of all the words. Every day the same pattern repeated.

The day I finished was the happiest day of my life. Oslo was so beautiful; there was a hoarfrost on all of the trees, more than an inch thick. I walked around in the gardens of the king's palace after having been to a movie with my wife. The frost-encrusted trees by the palace were magnificent. The midnight sky was a perfect, deep blue. I spent an hour gaping upwards in the park, marvelling at the patterns of trees against the sky, and then went back to the hotel. That was one of the greatest times I can remember. I knew that I was just one or two pages away from the end of the book. Then I finished the final chapter, except for a few unimportant mathematical details that I knew I could work out, and I relaxed into sleep.

The writing of *Surreal Numbers* had taken six days; so on the seventh day I rested. In fact, I still had the seventh day to tidy up the last page, which I did. Then I wrote "The End."

I couldn't write a word after that. I tried to compose a letter to Phyllis, my secretary at Stanford, telling her how I wanted this book to be typed. I would get into the middle of a sentence, and I could not figure out what verb to use. Suddenly, I couldn't even put simple things onto the page. I had just gone through a week where everything was sort of flowing out, and all of a sudden it was gone! That's why I love this book. It was a part of me that had to be expressed. I wish everyone could have a chance like this—some inspiration that could touch them.

**Knuth working at the computer terminal in his study.
He refers to his study as his "book factory."**

MP: *At the conclusion of "Surreal Numbers" there is a "dear teacher" letter, in which you suggest that students using the book for a course should do a project and write it up. In your last paragraph, you say that two of the major problems in teaching mathematics are a lack of experience in writing and also a lack of experience with creative thinking. Do you feel the same way today, several years later?*

Donald Knuth

Knuth: Oh, absolutely. I try to do that with our graduate students in computer science. We try to minimize the competition. We encourage working together on problem-solving, and discussing problems with each other. Creativity seems to be encouraged until about the fifth grade, at least in the education of my children. There were a lot of creative things that they once were asked to do; but after fifth grade, schools seemed to say: "We haven't got time for that anymore, no time for you to discover anything for yourself. From now on you will have to absorb all this information the world has been developing."

That's wrong. We should teach people things with an emphasis on how they were discovered. I've always told my students to try to do the same when reading technical materials; that is, don't turn the page until you have thought a while about what's probably going to be on the next page, because then you will be able to read faster when you do turn the page. Before you see how to solve the problem, think about it. How would you solve it? Why do you want to solve the problem? All these questions should be asked before you read the solution. Nine times out of ten you won't solve it yourself, but you'll be ready to better appreciate the solutions and you'll learn a lot more about the process of developing mathematics as you go.

I think that's why I got stuck in Chapter One all of the time when reading textbooks in college. I always liked the idea of "why?" Why is it that way? How did anybody ever think of that from the very beginning? Everyone should continue asking these questions. It enhances your ability to absorb. You can reconstruct so much of mathematics from a small part when you know how the parts are put together. We teach students to derive things in geometry, but a lot of times the exercises test if they know the theorem, not the proof. To do well in mathematics, you should learn methods and not results. And you should learn how the methods were invented.

Artificial Intelligence

MP: *Occasionally today we have been getting close to the burgeoning area called artificial intelligence. Has your attention been attracted to that area at all?*

Knuth: I enjoy reading about it. Many of the algorithms in Volume Four are used in artificial intelligence to solve interesting problems. They turn out to be in one-to-one correspondence with things that electrical engineers use for other purposes, and again I enjoy bringing together two literatures that are talking about the same thing. The shortest path problem, for example, is something that arises in many different disguises. In artificial intelligence, we find algorithms for theorem proving and problem solving, expressed in a different language than other people will be using for the same kind of problems that they are encountering in electronics for wire routing, or something like that.

I'm not a specialist in artificial intelligence, but I think the most interesting thing about it is something that I can paraphrase from a book by Pamela McCorduck (*Machines Who Think*). She points out that the question used to be "Can computers think?" By now, however, everything that has been associated with thinking has been done by computers; and the only human accomplishment that computers *can't* do well are things that people do *without* thinking! This is so true. The things we do without thinking are the things that computers have never done or hardly done, like walking. To control a robot to walk like an ant walks, or to program a computer so that it will recognize a face when someone has grown a beard, is extremely difficult. Children can talk languages; computers can't even translate languages very well. All the things we do subconsciously are the things that artificial intelligence hasn't been able to do. That's the most striking thing about the subject now. The big mystery is what goes on when we are not thinking. How do ants do such complicated things with no leader telling them what to do? How do their small brains come to the decision of how they are going to communicate with each other and solve problems? That's way beyond what we know now. I am fascinated by that, but I never promised to write a book on it.

MP: *You're not terribly optimistic then?*

Knuth: I believe the study of artificial intelligence is really important in that we learn much more trying to find out how these things are done than we actually do by having a computer system doing them. Trying to automate something is a great scientific achievement. After you automate something, the important thing is what you learned in the process, not really that the computer can now do a complex job.

While you're trying to explain something to a computer, you have to understand it so well that it's even better than the understanding you get by teaching it to somebody else. The old saying is: "You don't learn something until you have taught it to someone else." Today's saying is: "You don't really know something until you have taught it to a computer." That's the secret of learning. The computer is really a good test of understanding. It doesn't allow you to wave your hands and say: "Now you use some common sense." You have got to understand it clearly—there's no room for wishy-washiness. That's why I believe computer science impinges on education. If students can teach something to a computer, then you know they have got it in their heads.

MP: *How long have you been working on T$_E$X and METAFONT?*

Knuth: Spring of '77 is when I started, so it will be four or five years by the time I'm done. I thought it would be a one-year project.

MP: *Do you think it was time well spent?*

Knuth: Yes, I think the things I've learned are really exciting, and they are causing a lot of good waves in the printing world. I think it happens fairly often that a person from one field, say a mathematician, will stumble into another field. He'll have a different background than the people in the other field, and so he can contribute new insights. Sometimes such people will change fields and change their life's work. Sometimes they just make the contribution. For example, I have heard that Larry Shepp's daughter developed cancer of the brain. He got interested in that problem, and worked on a technique for locating cancers that's actually turned out to be an important breakthrough. You'll see this happen a lot of times for one reason or another: a mathematician will wander into some other area, and he'll see from what he knows that it is possible to apply ideas to that area right away, and that will help those people a lot.

In my own case, some of the things I learned about typesetting seem to be important enough that I've really gotten excited about them, and that is why it is taking me four or five years instead of one. If the ideas hadn't worked very well, I would have just kept them to myself and not bothered telling anybody anything about them; I would just have used them for my own purposes, and that would have been enough. But several ideas from mathematics and computer science now seem important to typography, so I want to do my best to refine them and explain them well. At the same time I am anxious to return to Volume Four before it becomes too big to write.

BENOIT MANDELBROT

BENOIT MANDELBROT

Interviewed by Anthony Barcellos

Benoit Mandelbrot is the inventor of fractals, and presently holds the position of IBM Fellow at IBM's Thomas J. Watson Research Center. The span of his interests is best illustrated by listing some of the positions he has held: Visiting Professor of Economics, Applied Mathematics, and Mathematics at Harvard, of Engineering at Yale, of Physiology at the Albert Einstein College of Medicine, and of Mathematics at the University of Paris–Sud. Fractal geometry has been the unifying theme in all of his work.

Fractal geometry is one of the major developments of twentieth century mathematics. Its characterization of fractional-dimensional objects permits a systematized and unified approach to irregularities in nature. Clouds, coastlines, and fluid turbulence are examples of natural phenomena which are best discussed in the language of fractal geometry.

MP: *Were any people or events particularly influential in your choice of mathematics as a career, and the highly individualistic manner in which you have pursued it?*

Mandelbrot: The most influential person was an uncle. His being a prominent professional mathematician affected me in contradictory ways. The most influential events were the disasters of this century, insofar as they repeatedly affected my schooling. It was chaotic much of the time.

In 1929, when I was five, my uncle Szolem Mandelbrojt became professor at the University of Clermont–Ferrand, and I was thirteen when he moved up to the top, as the successor of Hadamard and the colleague of Lebesgue at the Collège de France in Paris. Therefore, I always shared in my parents' (surprised) awareness that some people lived by and for creating new mathematics. Hadamard, Lebesgue, Montel, and Denjoy were like not-so-distant uncles, and I learned to spell the name of Gauss as a child, by correcting by hand a misprint in a booklet my uncle had written. At twenty, I did extremely well in mathematics in the big French exams, despite an almost complete lack of formal preparation, and my uncle took it for granted that his gifted nephew would follow right in his steps.

However, we had entirely different tastes in mathematics. He was an analyst in a very classical style (he had learned French by studying Poincaré's and Hadamard's works, and he had come to Paris because it was the cradle of classical analysis), and I call myself a geometer. For him, geometry was essentially dead except in mathematics for children, and one had to outgrow it to make a genuine scholarly contribution. It seems I did not like the idea of growing up in this fashion. Therefore, my uncle's plans for me backfired. While he never ceased to wonder what "had gone wrong" and took no interest in my work, we remained friends. But he had a largely negative influence on my work, therefore on my life.

At this point, the influence of my father became dominant. He was very proud of having already helped raise my uncle, who was his youngest (sixteen years younger) brother. My father was a very scholarly person, and the descendant of long lines of scholars. In fact, it often seemed that everyone in the family was—or was expected to become—a scholar of some sort, at least part-time. Unfortunately, many were starving scholars, and my father—being a practical man—saw virtues in a good steady job. His own work was to manufacture and sell clothing, which he did not enjoy, yet he strongly believed in the notion that a scholar's independence and happiness had better hinge on a steady income from a very different source, preferably one that would not be overly sensitive to the world's catastrophes. Thus, in the wake of 1914–1918, he had hoped that his gifted brother would go into the desirable field of chemical engineering (John von Neumann's father had also wanted him to go into chemical engineering). Again, in the wake of 1939–1945, my father feared my uncle's success was a fluke, and preferred to see me make a living as an engineer. Because of my distress at the reported "death of geometry," and because of my distaste for the obvious alternatives in science, I accepted my father's argument, and in particular let myself drift farther and farther away from mathematics.

Eventually, I came back. In fact (most unexpectedly), I made good use of the classical analysis I had read under my uncle's prodding. However, I was never imprinted with the normal way of being a mathematician, a calling whose rules exist independently of what any individual can do, and which provides peers or successful living role models to whom to conform. Those who accept such a calling perceive the normal unpredictability of life as unwelcome perturbations whose effects must be compensated for; it takes more than even a war or other similar catastrophe to change their way of operating. To the contrary, as I allowed myself to drift, I soon came to view the normal unpredictability of life as contributing layers or strata of experience that are valuable, demand no apology, and add up to a unique combination. Hence perhaps the impression that I encounter more than the customary amount of randomness! Looking back I must agree that it is hard to see how I managed to survive professionally and accumulate the proper school ties, without ever settling down in an established career. I made several attempts to settle down, but then accepted the inevitable: none of the existing careers fitted my growing cocktail of interests.

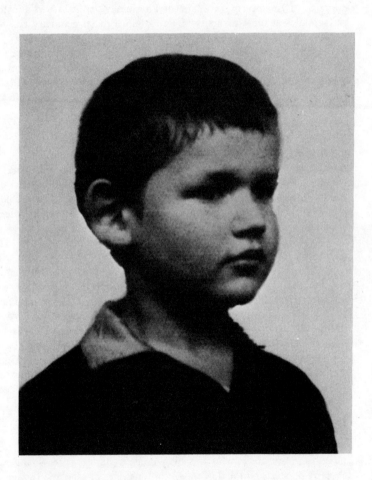

Benoit Mandelbrot, age five, 1929.

A Fractal Orbit

Of course, the reason why you sought me out for this interview is that I brought my interests together eventually, in a way that is attracting attention: I conceived, developed and applied in many areas a new geometry of nature, which finds order in chaotic shapes and processes. It grew without a name until 1975, when I coined a new word to denote it, *fractal* geometry, from the Latin word for irregular and broken up, *fractus*. Today you might say that, until fractal geometry became organized, my life had followed a fractal orbit.

Ultimately, the surprise is not that my manner of practicing mathematics should seem individualistic, but that I should be generally recognized as a mathematician. But I am a physicist also, and an economist, and an artist of sorts, and . . .

MP: *What was the actual course of your studies?*

Mandelbrot: Without ever trying, I did very well at avoiding being imprinted by schools. It all began way back, by my *not* attending grades 1 and 2. My mother was a doctor and afraid of epidemics, so she was trying to keep me out of school. Warsaw, where I was born and lived, had been hard hit by the depression, and my uncle Loterman, who was unemployed, offered to be my tutor. He never forced me to learn the whole alphabet, nor the whole multiplication table, but I mastered chess and maps, and learned to read very fast.

We moved to Paris in 1936, and by 1937, when I entered the Lycée, I was 13 instead of 11 years old. The Lycée was the secondary school that prepared for the Universities. Then World War II came, and we went to live in central France, at Tulle near Clermont–Ferrand. To an older boy from the big city, the Lycée de Tulle was ridiculously easy, but several marvelous teachers from famous schools were

Mandelbrot, rear center, age thirteen, with his school class and teacher.

also stranded there, and they gave me hard work to do. In effect, they tutored me, mostly in French and history. By the end of high school, I had caught up with my age group, which moved on to rather intensive mathematics with a first-rate teacher.

Then, poverty and the wish to keep away from big cities to maximize the chances of survival made me skip most of what you might call college, so I am essentially self-taught in many ways. For a while, I was moving around with a younger brother, toting around a few obsolete books and learning things my way, guessing a number of things myself, doing nothing in any rational or even half-reasonable fashion, and acquiring a great deal of independence and self-confidence.

Benoit Mandelbrot

In French education, attendance was not that important, but exams were vital. So when Paris was liberated in 1944, I took the entrance exams of the leading science schools: Ecole Normale Supérieure (Rue d'Ulm) and Ecole Polytechnique. Normale, which was exclusive and tiny (a class of thirty, half of them in the sciences), prepared university and high-school professors. Polytechnique had classes of about 250 (one out of ten applicants), and led to the top technical positions in the Civil Service and to other extremely diverse careers.

The two sets of written and oral exams take a solid month—a test of physical stamina. I passed both very handily; in fact, I was ranked very near the top in each case. Everybody else had spent two or more years in special preparatory classes, a kind of cramming college, but I had only a few months of that drill, so my passing was considered very unusual.

I did not do well because of my skills at algebra and complicated integrals—in fact, these skills demand training and I had had little formal training—but because of a peculiar inborn gift that revealed itself in my mid-teens, quite suddenly. Faced with some complicated integral, I instantly related it to a familiar shape; usually it was exactly the shape that had motivated this integral. I knew an army of shapes I'd encountered once in some book or in some problem, and remembered forever, with their properties and their peculiarities. More generally, I could find instantly geometrical counterparts for almost any analytic problem. Having made a drawing, I nearly always felt a kind of esthetic lack to it. For example, it would become nicer if one were to add its symmetric part with respect to some circle or some line, or if one were to perform some projection. After a few transformations of this sort, the shape became more beautiful, more harmonious in a certain sense; the old Greeks would have called it more symmetric. At this point, it turned out usually that the teachers were asking me to solve problems that had already been solved by just making the shape more harmonious. Classmates and teachers who watched me play my tricks told me later that it was a strange performance.

You might say this was a way of cheating at the exams, but without breaking any written rule. Everybody else took an exam in algebra and complicated integrals, and I managed to take an exam in translation into geometry and in thinking in terms of geometric shapes. Besides, it did matter to my overall ranking that I was skilled at drawing and that I could write good French, so it did not matter that the answers in physics and chemistry could not be guessed. That's how I got away with my "legal cheating."

At this point, my uncle had returned from the USA, where he had spent the war years, and family and friends held agonizing discussions about which career I should choose, and which school I should go to. We were surrounded by the ruins and the hunger of 1945, which figured significantly in my decisions. I started Ecole Normale (ranked first among those who entered), but with the intention of avoiding my uncle's kind of mathematics.

Unfortunately, the only alternative was to follow "Nicolas Bourbaki." In the nineteen-twenties, my uncle had been among the bright young iconoclasts who founded Bourbaki as a pleasantly jocular club; they planned to write together a good textbook of analysis (to replace the aging treatises of Picard and Goursat). But he did not rejoin them in 1945, when they started a dead earnest drive to impose a new style on mathematics, and to recreate it in a more autonomous ("purer") and more formal ("austere") form than the world had ever known. Thanks to my uncle, I knew they were a militant bunch, with strong biases against geometry and against every science, and ready to scorn and even to humiliate those who did not follow their lead. Bourbaki was one of several conflicting social movements that flourished after the War, when the yearning for absolute values was especially strong and widespread.*

Anyhow, having no taste for Bourbaki, I gave up on Normale after a few days, and went over to Polytechnique. My father was relieved. Since there were to be no electives, I was receiving the gift of time: it seemed that the need to make a firm choice was postponed until graduation, but in fact I was to be never forced to choose.

MP: *Why was that?*

Mandelbrot: Initially, because of a legal mix-up. Polytechnique offered well-defined numbers of several favored positions to its graduates, and the students chose a life career on the basis of their weighted grade point average over two years. Everything was graded, or so it seemed. The competition

---

*See Jean A. Dieudonné, "The Work of Nicholas Bourbaki," *American Mathematical Monthly*, 77(1977), 135–45, and Paul Halmos, "Nicholas Bourbaki," *Scientific American* **196** (1957), 81–89.

for the top slots was ferocious and left no time off. (If France wants to dominate world chess, the easiest approach may be to teach chess at Polytechnique!) I would have competed for the top slot, but the school's legal people thought—wrongly, as it later turned out—that there was a Catch 22 that disqualified me from competing. (To explain it would require a lecture on law and history.) Since my rank did not matter, I allowed it to erode slowly by not studying enough for a few dreadful exams. Instead, I felt a free man. A friend initiated me to classical music—which, he said, I absorbed like a big dry sponge. And I did lots of interesting reading of every kind. Many course notes had lots of appendices that no one else could afford to study.

The Influence of Lévy

Today, Polytechnique is without a permanent staff, and it borrows professors from outside, mainly from various universities. The same has long been true of Normale. But in 1945, Polytechnique had its own professors chosen by its own committees. The pattern was that some were moonlighting scholars from the University, for example, Gaston Julia, some were alumni who had never done any research, and some were research people picked outside of academia's mainstream, for example Louis Leprince-Ringuet and Paul Lévy. Julia was a brilliant teacher, but well past his research prime. I spoke to him once; no one could have predicted that thirty years later I was to help revive his theory of iteration of functions so that it might reach full glory. The professor I was most aware of was the Professor of Mathematical Analysis, Paul Lévy. He was lucky that Polytechnique had hired him for life, when he was a promising scholar in the early 1920's, because his way of doing mathematics and his choice of topics had been becoming less and less popular in the mainstream. He had an extremely personal style, even in his basic analysis course. The course notes given to the students were at the same time rather leisurely and surprisingly brief, and they were the despair of many of my classmates because many facts seemed to be declared "obvious." But I found his course to be profound and in a way very easy; perhaps I was the only one who liked it.

Paul Lévy was nearing sixty. Suddenly, he was becoming famous; it was being "discovered" that he was a very great man in probability theory and that this new field was a branch of mathematics. Its good fortune was to be built on the great Norbert Wiener's work on Brownian motion, and to rest mostly on the shoulders of two very different persons: Lévy and a man in the middle of the mainstream, Andrei Kolmogorov.

Having learned the basic mathematical analysis from Lévy, I was used to his style, and could read his research papers much more easily than almost anybody else. One had sometimes to guess what he meant. Many major difficulties were not tackled at all but swept under the rug more or less elegantly. Many respected Ph.D. dissertations or articles consist in the proper statement and proof of a single "obvious" fact from Paul Lévy. Several years later, a would-be faculty advisor recommended a Ph.D. of this sort, but I never tried. Eventually, as fractal geometry came close to being implemented, I found myself fully involved in observing further "obvious" facts about diverse shapes and configurations drawn by chance.

MP: *Were you Lévy's student?*

Mandelbrot: No. Several people later claimed they had been his students, but Lévy specifically disclaimed having had any students. Besides, it took years before I came to be called a probabilist. Polytechnique requires two years of study, ending roughly at the level of a strong Master's degree in the U.S. During the last term at Polytechnique, I kept looking for ways to use my mathematical gifts and growing knowledge in dealing with real concrete problems in nature. My hopes were thoroughly romantic: to be the first to find order where everyone else had only seen chaos. Someone who heard me say so commented that my dream was to be Johannes Kepler, but that the Kepler days were over. Luckily, someone at Polytechnique felt ashamed about the Catch 22 that I have mentioned, and helped me obtain French and American scholarships for studying in the United States. Also, a professor suggested study under Theodore von Kármán, who was finding order in the chaos of transsonic flight.

MP: *Is that when you went to Caltech?*

Mandelbrot: Yes, for two years. But I found that Kármán had left Caltech, and that the students of transsonic flight had split into a group of engineers building big rockets, and a group of mathematicians doing mathematics. Caltech was home to many people I admire greatly, and many of my best friends are people I met there. But there was nobody at Caltech that I particularly wanted to emulate at that time.

So I went back to France, first into the waiting open arms of the Air Force, which kept me for a year, and then to search for a suitable thesis topic. A book review found in my uncle's wastebasket started me on a task which was extravagant in every way: to explain "Zipf's law," which is a surprising regularity in word statistics. To many people, this topic looked almost kooky, but I saw a golden opportunity of becoming the Kepler of mathematical linguistics. My explanation of Zipf's law received much praise, but a few years later I abandoned this line of work, and later watched it go into a deadend, while mathematical linguistics developed in a different direction.

My doctoral thesis at the University of Paris was defended in December, 1952. It had been written without anyone's assistance—and was poorly written. The mathematical linguistics in the first half was formally a very exotic form of the statistical thermodynamics in the second half (thermodynamics in a space of trees). Some people described the combination as being half about a subject that didn't yet exist (they were right), and the other half about a subject that was no longer part of active physics (they were ill-informed). As the thesis had to be pigeon-holed, we decided to say it was in mathematics, but it was obviously very far from the reigning pure mathematics. For reasons linked to my official advisor's personal ambitions, the committee chairman was Louis de Broglie.

The title was *Games of Communication*, largely because for several years before and after my Ph.D. I was very much influenced by the examples of John von Neumann and Norbert Wiener. Indeed, Wiener's book *Cybernetics* and von Neumann and Morgenstern's book *Theory of Games and Economic Behavior* had come out, and they were very precisely what I wished to emulate one day. Each seemed to be a bold attempt to put together and develop a mathematical approach to a set of very old and very concrete problems that overlapped several disciplines.

Unfortunately, cybernetics never really took off, and game theory became yet another very special topic. Colossal claims had been made when there was little to support them, and they had avoided being shrugged off only because of the authors' renown based on earlier and very different work. It soon became good manners in academia to laugh when someone mentioned "interdisciplinary research." To my bitter disappointment, I had to agree that there was good reason for laughing. I wondered whether things would have been better if von Neumann and Wiener had had the desire and the ability to take an active interest in their progeny.

John von Neumann

I was the last man whom von Neumann sponsored at the Institute for Advanced Study in Princeton, in 1953–54. It was a marvelous year, and again I made many life-long friends. I became aware of the computer—but years were to pass before I became fully involved with it. Unfortunately, von Neumann was not there very often. He was becoming more concerned with defense than with science. But it seemed that he was living proof that one could do science without really belonging to a "guild." In fact, he was under extreme pressure at Princeton. From there, he went to Washington and was not planning to return.

Luckily, von Neumann had realized that by having failed to claim admission to any guild, I was leading a very dangerous life, and a foundation executive told me much later that he had asked him specifically to watch after me and to help in case of trouble. I hope it was a true report.

The papers I wrote during these years were praised individually, and many anticipated later developments in diverse areas. They had a recognizable common style, but they failed to add up, even in my own eyes. Every so often, I was seized by the sudden urge to drop a field right in the middle of writing a paper, and to grab a new research interest in a field about which I knew nothing. I followed my instincts, but could not account for them until much, much later. Anyhow, to work in many fields was not harder than to work in "only" two. I had returned to France, then married Aliette in 1955, and moved to Geneva to attempt a collaboration with Jean Piaget, when the French universities suddenly started expanding very fast and looking for applied mathematicians. They gave me this label, and a professorship at Lille; in addition, Lévy sought me out to help at Polytechnique. But Lévy was about to retire, and Bourbaki was to take over. It seemed I was about to be crushed between them and other big academic blocs that could only think of their own narrow interests.

MP: *Is that when you left academia for IBM?*

Mandelbrot: Yes. I was there as a faculty visitor in the summer of 1958, and decided to take the gamble of staying a bit longer. My wife was not enthusiastic at all, but she agreed to this gamble. I wrote to the Ministry of Education in Paris to request a leave of absence for one year. They failed to respond to my letter, and later told me that my request had arrived a few days before my tenure was to be granted in writing, so they simply dropped me off their lists. They said they would take me back if I wished, nevertheless my "leave" had become open-ended.

It took my wife and me a long time to accept the fact that I had been lucky, and that as an environment to pursue my devouring yet ill-defined ambition, IBM Research was far better than any university department in either country. Much of the practical consulting I did was informal, and some was very exciting and had far-reaching consequences, for example, my work with Jay Berger on transmission errors in telephone links between computers.

In academia, on the other hand, Bourbaki was only the extreme form of a generalized phenomenon. Each of the old departments was working hard at that time to "purify"—that is, to narrow down—its scope, and each of the newly established departments was working hard at finding criteria to define yet another narrow combination of skills which could be rewarded in grown-ups, trained in the young, and endowed with a slice of the job market. The emergence of "pure applied mathematics" and "pure mathematical statistics" brought special discomfort to my life, at times. Finally, each field acted as if it were destined to live forever. For all these reasons, the notion of an academic activity that voluntarily reduces to one man's fancies had become inconceivable.

Everybody's ideal seemed to be sports. Competition is important in life, hence in science, but why should science embrace being dominated by situations such as that in track, where the mile race and the 1500 meters race (only 7.3% shorter!) are often won by different champions? To make things worse, the decathlon survives as an Olympic discipline, but the scientific decathlon that I seemed to practice was not acceptable in academia. The granting agencies were divided in the same fashion. An energetic young fellow could always find support. But mavericks develop gradually and slowly, and maverick enterprises are better off if they develop slowly and solidly. Unfortunately, there was no room in academia for a gradually developing maverick enterprise.

For many years, every group I knew viewed me as a stranger, who (for reasons unknown) was wandering in and out. Luckily, the striking (and often shocking) news I was bringing could not pass unnoticed, and I was acquiring a kind of fame. I became very popular in many diverse departments as a visiting professor, but no major university wanted a permanent professor with such unpredictable interests.

The IBM Fellowship

On the other hand, Ralph Gomory took the gamble of sheltering my one-man project when he joined IBM shortly after me, as the manager of a small group, and again when I returned after two years at Harvard and he became departmental director. Eventually, he assigned a programmer to my project. Then—when Gomory had become Director of Research and my 1975 French book was nearing completion—I was made an "IBM Fellow" and given a small staff. A few dozen IBM'ers are designated as IBM Fellows "in recognition of outstanding records of distinguished and sustained technical achievement in fields of science, engineering, programming and systems. They are given freedom to choose and carry out their work in areas related to their specialization in order to promote creative achievements." Thus, it was stated officially that my work had become widely respected, and that I could proceed in my very own way.

MP: *You said that your whole life has followed a "fractal" orbit until fractal geometry became organized and you mention some of the turning points in the epilogue of your 1982 book. Could you tell us these milestones?*

Mandelbrot: My wild gamble started paying off during 1961–1962. By then, there was no question in my mind that I had identified a new phenomenon present in many aspects of nature, but all the examples were peripheral in their fields, and the phenomenon itself eluded definition. To denote it, the usual term now is the Greek "chaos," but I was using the weaker-sounding Latin term "erratic behavior" at the time. The better word "chaos" came later from others, but I was the first to focus on the underlying notion, and to specialize in studying the erratic-chaotic. Many years were to go by before I formulated fractal geometry, and became able to say that I had long been concerned with the fractal aspects of nature, with seeking them out and with building theories around them.

But let us go back to the year 1961. Starting in that year, I established that the new phenomenon was central to economics. Next, I established that it was central to vital parts of physical science, and moreover that it involved the concrete interpretation of the great counterexamples of analysis. And finally, I found that it had a very important visual aspect. I was back to geometry after years of analytic wilderness! A later turning point came when I returned to questions of interest to those in the mainstream of mathematics.

Economics is very far from what I had planned to tackle as a scholar. However, after I had become bored with the Zipf's law of linguistics, I proceeded to read Zipf's works and became acquainted with Pareto's law of income distribution. I believe my work contributed to the understanding of this topic, but this interest was also giving signs of becoming exhausted when I visited Hendrik Houthakker at Harvard, and saw on his blackboard a diagram that I had already encountered in the study of incomes. On the grounds that such geometric similarity was bound to be the visible symptom of an underlying similarity of structure, I inquired about the problem that had led my colleague to the diagram in question, and was told that it referred to the variation of stock market and commodity prices. I became fascinated with this topic because it involved marvelous examples of unquestionably important quantities whose variation is very erratic, very irregular . . . chaotic. I soon came to distinguish two syndromes in price variation, sudden jumps and non-periodic "cycles," which I later denoted by the expressions *Noah and Joseph Effects*.

Price variation was becoming a source of worry to a few economists, because it was resisting being squeezed into the accepted econometric mold, which had simply been copied from the physics of a gas in equilibrium. On the other hand, I pioneered a radically different alternative approach, based upon self-similarity. This is a widely familiar notion today, largely due to the physicists' work on critical point phenomena, but that work came much later. I showed that the stochastic process obtained via self-similarity generates sample functions that are very rich in configurations, and can account for a great part of observed price variation. During this period, I was doing things analytically, with very few diagrams. The most influential of these diagrams has already been mentioned.

Mandelbrot, age thirty, 1954.

The next major series of related developments concerned noise, then turbulence and galaxy clusters. My investigation of the so-called "excess noises" started with a very practical problem at IBM, but continued long after the problem was settled. Again, my solution was grounded on postulated self-similarity; it necessarily involved random forms of the Cantor set, and its description demanded Hausdorff–Besicovitch dimension. A friend, Henry McKean, Jr., had written his thesis on the Hausdorff–Besicovitch dimension of certain random sets, when we both lived in Princeton. Otherwise I might not have encountered this notion. It was very rarely used at that time, but I discovered that it had an essential application, first to noise, then to turbulence, next to galaxy clusters, and so on to fractals in general.

Incidentally, I met Edward Lorenz in 1963. His work on chaotic behavior in deterministic systems had just been published, but had yet not drawn much attention. Erik Mollo-Christensen predicted that it and my early papers would turn out to concern two faces of the same reality. This hunch is in the process of being confirmed.

In the mid-sixties, however, my message was not getting through well enough to satisfy my ambition. Each context in turn elicited a complaint I had often heard in economics. "Granted that such and such statistical expression is known to converge in all the other fields of science, how can it be that my field (my interlocutor would complain) is alone cursed by the necessity of facing divergent statistical expressions?" "When all the other fields of science can be tackled by proven mathematical methods from familiar textbooks, why should my field necessitate newfangled techniques, for which the only references are dusty tomes written in French, or even in Polish, or incomprehensible modern monographs?

One had to agree that these situations were paradoxical, but I thought this could have a very different origin. I kept observing that, in many applications, these familiar and unquestioned statistical and mathematical techniques had been oversold: in fact, they had failed to come to grips with the truly important problems, as they were perceived by the brilliant but nonmathematical practitioners whom I trusted. This failure could be accounted for, if the "ills" I had already diagnosed in a few fields were in fact of wide occurrence.

How Long Is the Coast of Britain?

In any event, noise, turbulence, and clustering are complicated phenomena, and the repeated experience of unvarying resistance to my increasingly unified approach, and in particular to my use of Hausdorff dimension as a concrete notion, made me wish and search for a simpler illustration. I stumbled upon coastlines, and proposed that the irregularity of a coastline be measured by its Hausdorff dimension. A few (astonishingy few!) scholars had noted in passing that the notion of a coastline's length is meaningless, but no one had done anything about it. Perhaps the finding embarrassed them (for example, Lewis F. Richardson's thoughts on this topic were found after his death among irrelevant unpublished drafts).

MP: *Could you elaborate?*

Mandelbrot: The question I raised in 1967 is, "how long is the coast of Britain," and the correct answer is "it all depends." It depends on the size of the instrument used to measure length. For example, look at this picture (Figure 1). It does not represent a true coastline but a fractal fake, made (years later) as a "model" sharing the significant properties of a coastline. It is clear that, as measurement becomes increasingly refined, the measured length will increase. Thus, all the coastlines are of infinite length in a certain sense. But of course some are more infinite than others. To measure their degree of infinity, I thought of Hausdorff dimension, and indeed I was able to show that the notion of dimension of a coastline is meaningful, and that its value can be measured quite accurately. This is how I came back to my true love, geometry, and went on practicing it in a very strong and intense fashion.

MP: *Do you have any opinion of the way geometry is handled in schools and in research?*

Mandelbrot: Oh, very much so. The old French fashion required years and years of high-school geometry. I found that totally intuitive and childishly easy, which of course is why I considered going into mathematics as a career. So I am distressed by how little geometry there is in American high schools.

For quite a while, geometry was in effect banned from university curricula. Now the availability of computer graphics and, to some extent—I think even to a large extent—my work have made many

Figure 1: An artificial fractal coastline, from *The Fractal Geometry of Nature* by Benoit B. Mandelbrot.

people realize that geometric intuition can enter into seemingly very abstract domains, like the theory of Kleinian groups or the theory of iterations of mappings in the complex domain. In both fields I discovered a number of facts using geometric intuition and computers.

We have entered a period of intense change in the mood of mathematics; increasingly many research mathematicians use computer graphics to enhance their geometric intuition, others cease to hide to outsiders (or even to themselves) that they had been practicing geometry. This return of geometry in the frontiers of mathematics and of physics should have an effect on the teaching of geometry in colleges, high schools, and even in elementary schools, because so much geometry which was before quite impractical now can be easily done with the help of computers.

MP: *It's rather curious that high-school geometry in America was advanced, for the most part, as being good logical training for the mind, emphasizing the deductive process rather than anything having to do with images or visual intuition.*

Mandelbrot: That's also the way it was described in the old French curriculum, but for me geometry was something entirely different. Perhaps I should be careful to use always the leaden expression "geometric intuition," but it would be better to take "geometry" back from those who really do not care about it.

Furthermore, school geometry did instill discipline, but I doubt its value as logical training. To a student, the reduction to the axioms is largely a matter of satisfying the teacher. We were told that certain arguments were "okay" and that others had to be transformed to be made "okay." I never had the feeling that this "okay" was intrinsic, much less that it was the last word in logic, but achieving it gave me good grades, while I was engaged in the truly important task of training my gift to be able to think directly in terms of shapes.

The Rebirth of Geometry

In any event, the geometry of yesterday has become dull and dry, but a combination of Greek and fractal geometry would be alive, attractive, and very useful. The young are very much dominated by the eye, through media like TV and computer games, and a combined geometry would be a way of getting to their minds before one tries to expose them to dry logical constructs.

Establishing such a curriculum will be a very complicated thing, and I don't have precise constructive thoughts, but the main value of the Greek geometry I learned in France was that the problems were hard, but not abstract. When friends ask me whether their fourteen-year-old wonder kids should first take calculus or number theory, I put in a word for old-fashioned books on geometry written about 1900.

Anyhow, I think it is a fact that some people think best in formulas, and other people think best in shapes. A hundred years ago, this was almost a platitude among mathematicians, but people who think in formulas now run the show in every branch of science, and for a while they could not tolerate even one person who proclaims he thinks in shapes.

It may have become true that people who think best in shapes tend to go into the arts, and that people who go into science or mathematics are those who think in formulas. On these grounds, one might argue that I was misplaced by going into science, but I do not think so. Anyhow, I was lucky to be able—eventually—to devise a private way of combining mathematics, science, philosophy, and the arts.

Let me try out a simile. Imagine that a hundred thirty years ago a pandemic virus had wiped out man's ability to sing, but left him with the ability to write music. Great opera scores remained widely available; they became an object of cult and intense study, and great new scores were written by Verdi (Wagner, if you prefer). Generations later, one person, then a few persons, were born, who found they were immune to the virus, and eventually everyone was reminded that if Verdi can indeed be read, he can also be sung. *Everyone* rejoiced! Well, one of the privileges of fractal geometry has been to make the classical hard mathematical analysis of Verdi's time sing out at long last.

MP: *Now that fractal geometry is, as you have written, "taking ominous steps toward becoming organized," at what level of the school curriculum would you consider its introduction appropriate?*

Mandelbrot: Fractals should be introduced first in the presentation of the derivative. This notion first bothered me when I started calculus, and many people should understand calculus far better if they know at the very outset that a continuous function need *not* have a derivative. Until recently, the only counterexamples were artificial and contrived. One could not bother young minds with Weierstrass functions. But I have shown that nondifferentiable functions are essential in very fundamental parts of natural science, and it has become easy to tell a student that a coastline has no tangent and that the components of the motion of a particle along a coastline have no derivative. It is desirable to introduce such notions very early. They will serve as antidote against the ridiculous idea we have already discussed, that the study of geometry is primarily a form of logical training and not a way of learning to reason on shapes.

Secondly, fractals should be taught early in fields like physics and in geophysics, where they are important. For a first step, the best is to try out several short, specialized courses entirely devoted to fractals; this is being done in several places. Later, one will see how fractals should be added to the basic courses on mathematical methods. To my pleasure and surprise, undergraduates accept this stuff very well. Another place where fractals are becoming important is in teaching computer graphics. It seems that every computer graphics demonstration includes fractals.

As to the place of fractals in the training of mathematicians, it raises two distinct questions. Falconer's forthcoming *Geometry of Fractal Sets* is very welcome, but it is a conventional mathematical monograph: its topics are suggested by needs created by my work, but the style is dry, as usual. It will not be the last mathematical monograph on these and related topics, but one cannot make any prediction about the teaching of advanced mathematics, because there are few advanced students at present, and the instruction they receive depends on fashion and the instructor's taste.

A much greater change would occur if the training of mathematicians were to go back to leaving some room for geometric intuition. Fractals may be taught around the theory of Kleinian groups or the theory of iterations of rational functions. These theories became unmanageable quite a long time ago, at least in part because of the lack of intuitive aspects to them. But with the advent of fractals, large parts of these theories become quite easy and widely attractive. They show very graphically how one can or should concentrate on thinking on shapes as live "wholes," and learn to modify an algorithm to affect the shape it generates.

MP: *Geometric shapes are being introduced through computer graphics in quite a few elementary schools through such things as Seymour Papert's LOGO and turtle graphics. What do you think of such things?*

Mandelbrot: I have only a very superficial familiarity with them. It seems that they too overemphasize the algorithmic aspects. For me a circle is not primarily something which is traced by a turtle running a certain course, but above all it is a circle! However, according to friends, it is easy to draw fractals using turtle geometry. This is very welcome and to the advantage of the field.

Computer Graphics

MP: *What role does computer graphics play in your research?*

Mandelbrot: It plays two major roles: to help my work develop, and to help it become accepted. To take an example, I had long known of a very simple algorithm which generates distributions of points in space, and I proposed this algorithm to model the distribution of galaxies, and of clusters and superclusters of galaxies. The first tests, both numerical and visual, were amazingly satisfying. But a long look at these distributions revealed discrepancies that had been much less obvious to the other students of the field, who used ordinary analytic methods. To try to improve the fit, I did not scan any repertory of alternative models, but my repertory of alternative shapes.

The virtue of computer graphics is that it makes it easy to compare a model's imitations to the natural shapes from many viewpoints simultaneously, including some viewpoints that had not yet been formalized or recognized. I encountered the first example when I was interested in the long-term persistence that is observed—since the Bible!—in the discharges of rivers like the Nile. A statistical theory was of course available to represent this phenomenon, but no one had actually thought of looking at the sample functions which this theory generates. It turned out to be surprisingly easy to convince every hydrologist that these samples did not look at all like the records of river levels, whereas the corresponding curves drawn according to my alternative recipe could not be told apart from the real ones, even by the experts, unless they knew the particular river intimately. Thus, computer graphics allowed the elimination of certain theories simply on the basis of the obvious unreasonableness of the shapes they generate. The same trick worked even better with coastlines and mountains. Graphics techniques gradually became better and better, and we could afford to do some fancy stuff.

One must say "we," because I do not program computers myself, but have found ways of working very interactively with several outstanding people: students and assistants, but also colleagues like Richard F. Voss. As a matter of fact I developed a skill for helping "debug" programs that I cannot read, by analyzing the wrong pictures these programs produce.

MP: *Did computer graphics play a role in your work on turbulence?*

Mandelbrot: No, but geometry has been essential. The big question in the study of turbulence is: how does turbulence arise from the differential equations for the flow of fluid? The literature on partial differential equations is tremendous, but it never got even close to tackling the questions raised by turbulence. In 1963, the situation was that Kolmogorov (and other Russian scholars) had written down formulas concerning the intermittency of turbulence. With hindsight, one may say that they were only one step removed from introducing Hausdorff dimension, then fractals. But they failed to take these steps, in fact they resisted them for a while after I took them. At the very same time, and quite independently, I had tackled intermittency in the context of noise, and had developed a very geometric mental picture of it. Upon seeing the facts about turbulence, at a seminar Robert Stewart gave at Harvard in the fall of 1963, I found it obvious that my methods could be translated wholesale in these new terms. Then I tried to explain the validity of the translation. This led me to conjecture that turbulence represented a singularity in the flow of fluids, and that this singularity is concentrated on a fractal. To form my intuition, I had looked at paintings and photographs of turbulence, had looked at records of velocity, and even listened to them (after transposition into the audible range), and used the summaries of measurements (such as power spectra) as further evidence.

This was an entirely new approach to the problem: for many years it was viewed as exotic and even bizarre, and it took ten years to be published other than in abstracts or via allusions. But my hunches on the Hausdorff-dimensional properties of the singularities are in the process of being confirmed in several different ways. It was quite beyond my skills to prove them, and to guess what had to be proven was beyond the geometric imagination of those who provide the demonstrations. Very great minds had tried to tackle turbulence by analytical techniques; they did not succeed, while it seems I succeeded by looking at turbulence via the shapes that it generates.

MP: *Kleinian groups and iterates of rational functions were reputed to be highly technical mathematical topics. When and why did you become involved?*

Mandelbrot: In 1976, after I had read Hadamard's superb obituary of Poincaré (which everyone will soon be able to read—and should read—in an American Mathematical Society book on Poincaré). This obituary made it apparent that my work should be extended beyond the linearly invariant fractals, to which I had restricted myself up to that point. Indeed, the limit sets of Kleinian groups and of groups based upon inversions are fractals also; the latter could be called self-inverse. This forthcoming extension of self-similar fractals was mentioned in a last-minute addition to the 1977 *Fractals*, and then I set out to work, namely to play on the computer in order to acquire a "hands-on" intuition. The payoff comes very quickly, in the form of an explicit construction algorithm for the self-inverse limit sets. It took me longer to ascertain that, to my surprise, I had solved a problem that had stood for one hundred years.

A short step then brought me to some old work of my former teacher Gaston Julia, and of Pierre Fatou. My uncle had once lent me the original reprints of Julia's paper that everyone called "the celebrated Prize Essay" and of Fatou's equally long work on the iterates of rational functions. The "Julia set" of a rational mapping of the complex plane is the repeller set of this mapping, that is, the attractor set of its multivalued inverse. I started playing with fairly complicated mappings, and was amazed to discover that sets that Julia and Fatou had characterized negatively, as being pathologically complicated, were in fact of extraordinary beauty. As they emerge on the computer screen, they seem totally strange for a moment, but one soon comes to feel one had always known them. (An example is shown in Figure 2).

MP: *What about the "Mandelbrot set." Is it also a fractal?*

Figure 2: A Julia set and its interior for the map $z \to z^2 + i\mu$. **Variant of an illustration in** *The Fractal Geometry of Nature.* **Copyright by Benoît B. Mandelbrot.**

Mandelbrot: Yes, it is. A few months of mindless fun with complicated mappings had prepared me for a detailed study of iteration. The best was to start with the simplest mapping, which is the second order polynomial. There is only one significant parameter, and to each parameter value corresponds a Julia set. I drew the set of parameter values such that the corresponding Julia set is not a "Fatou dust," but a connected "dragon." To paraphrase the famous words that J. C. F. Sturm is said to have used when he could find no way of avoiding mentioning the Sturm (-Liouville) equations, this is the set of which I have the honor of bearing the name. My study showed that this set is an astonishing combination of utter simplicity and mind-boggling complication. At first sight, it is a "molecule" made of bonded "atoms," one shaped like a cardioid and the other nearly circular. But a closer look discloses an infinity of smaller molecules shaped like the big one, and linked by what I proposed to call a "devil's polymer." Don't let me go on raving about this set's beauty.

Figure 3: The Mandelbrot fractal set of the map $z \rightarrow z^2 + i\mu$. **From** *The Fractal Geometry of Nature*. **Copyright by Benoit B. Mandelbrot.**

Old Fractals and New Names

MP: *Such things as the Cantor set and nowhere differentiable functions have long been important examples in analysis and topology. Is your term "fractal" well accepted by now?*

Mandelbrot: Yes, but with exceptions. Some mathematicians speak of my work as "generalized Cantor sets" or "concrete applications of the counterexamples of analysis"—these terms underrate the drastic novelty of my endeavor, and really imply that there is no unity to it.

MP: *They still regard them as pathological?*

Mandelbrot: That's one part of it, but mathematicians' lack of perspective can be breath-taking. Some would go so far as to call the whole of physics "a concrete application of harmonic analysis and differential equations!"

One must also be aware that mathematicians have a strange traditional way with words. Indeed, take the theory of "strange" attractors (which are a kind of fractal). To my great surprise, this use of "strange" does not bother anybody. Mathematicians like to take a familiar term, to turn it around, and to use it with a very different meaning. For example, the vocabulary of mathematics is full of terms like "ring," "field," "complex," or "imaginary." The words "distribution," "irregular," and "singular" are used with hundreds of different meanings. Mathematicians rarely coin new words, and the new words they coin are hardly ever graceful.

This may also be why the word "fractal" is already being used in meanings different from mine, and which my books call confusing. For example, a topologist like James Cannon needs a clearly different notion, but the word "fractal" exists, so he uses it. I fear that when this word is accepted it will be as ambiguous as "irregular," but not much can be done about it. One must just let time take its course.

MP: *Despite the resistance to fractals that still exists in certain quarters, they've clearly been accepted and applied in many different fields. You yourself have been involved in most of these branches from the outset. Have there, nevertheless, been recent developments which have surprised even you?*

Mandelbrot: I would say—with regret!—that there has been no major surprise since my work on Kleinian groups and iteration in 1978–79. What is surprising is the fact that fractals have attracted few crackpots, the quality of the applications, the rapidity of their development after the inevitable initial resistance is broken, and the total absence of outright failures. Additional phenomena that were clearly worth looking into but could have been complicated and messy are tackled and prove to be comparatively civilized. Fractal geometry works, which must be a reason why it is becoming so popular. But no new application seems to come out of the blue.

A second negative surprise is that the additional techniques and concepts which are needed keep coming in the same sequence in each study. For a long time, the role of fractal dimension had to be emphasized because it was the principal concept, and I was glossing over difficulties that require additional parameters. Then I introduced a second parameter, "lacunarity." It's very surprising that in many fields the demand for lacunarity materializes shortly after dimension has been fully understood. In this sense things have proceeded alike in different fields. As a matter of fact, scientists' reactions to fractals, both positive and negative, seem to be very much the same in all fields.

MP: *In addition to providing useful tools in various fields, haven't fractals become a field in their own right?*

Mandelbrot: My work was inspired by a strong belief that the division of science has been extremely harmful. On the other hand, new fields keep being created for serious reasons for survival. We must wait and see.

The Antigeometry of Bourbaki

MP: *You mentioned that you weren't attracted by Bourbaki's antigeometric approach toward mathematics. Did you find that the Bourbaki influence posed a significant obstacle to acceptance of your fractal approach?*

Mandelbrot: They mattered greatly in my life in 1945, when I left Normale, and again in 1958, when I left France, but very little since then. They did not prevent me from doing my thing, and for many years my audiences were sheltered from their influence or did not know they existed. After the study of turbulence had inspired me to an isolated bit of "pure" harmonic analysis, there was the right non-Bourbaki mathematician to complete and continue that work.

Furthermore, by fluke, the timing of my books was perfect. They came out when the feeling was beginning to spread that the Bourbaki *Foundations* treatise, like a Romantic prince's dream castle, was never to be completed, and that their old books would never fulfill their proclaimed goal of becoming the universal standard of mathematics. The Constitution phrased to insure that the group would remain eternally a cohesive young rebel was—of course—not working. In a way, the whole enterprise had become boring. The pendulum was therefore beginning to swing from this extreme back to an uneasy but more reasonable balance in mathematics, and my manner was becoming less threatening. If I had formulated fractal geometry much earlier, Bourbaki could have been a major obstacle. But today they do little beyond running a seminar in Paris. As a matter of fact, I may have benefited from the backlash against their old arrogance.

Besides, one of the current leaders of Bourbaki, Adrien Douady, has spent the last several years developing ideas on iteration that I had pioneered; to welcome him has been a treat. Finally, one of the founders of Bourbaki, Jean Dieudonné, has published various demonstrably wrong statements about the meaning of mathematics, which were of great help in making some of my main points. For example, he wrote that a Peano curve is so counterintuitive that only logic can comprehend it and no intuition can be used to understand its properties. That was demonstrably wrong. Today the Peano curve is viewed as completely intuitive, because my work made it so. And I have the feeling that Dieudonné is not hostile but amused.

The Fractal Manifesto

MP: *Why did it take you so long to get what you call your "fractal manifesto" in print?*

Mandelbrot: Today, when the status of fractal geometry is compared with that of other maverick enterprises, like cybernetics and game theory, one is tempted to think that mine has benefited from having developed slowly—granted that I am alive and well to watch its coming of age. Of course, I had intended otherwise, but science is organized into tight branches, and the only assured way to leave a mark on a branch is to visit it in person, so to speak. This demands adaptability on the part of the visitor, and takes enormous amounts of his time.

Linguistics allowed me to mention thermodynamics, but only because linguistics was not yet organized as a modern profession. When working in economics, I was similarly dying to be allowed to make it known in my research papers that my methods were part of a general philosophy, of a certain approach to irregularity and chaos, and that they also mattered in physics. Invariably, the referees asked me to take these statements out, and ultimately I decided to comply. Later on, I went on to study turbulence (which had to resemble the stock market, because the weather and the stock market are equally unpredictable), and again I wanted my papers to appear in the most prestigious specialized journals. Again, the editors forced me to cut out what they scorned as "dubious philosophy," and to give more formulas and more details on the manipulations. In each case, I was pretending to be a technician in the field, which was never completely successful because I always kept a strong "foreign accent," but was necessary and sufficient to get my papers accepted by the good journals.

These papers were excruciatingly hard to write, nevertheless in many cases my foreign accent gave them the reputation of being difficult to read. Also, I did not really learn to write English until tutored by my IBM office neighbor Bradford Dunham in 1968. Anyhow, interpreters invariably followed with their own renderings of my ideas. (One of them called himself a "master at repackaging," and all did very well in the process.)

In the meantime, turned-down, would-be prefaces to my papers were piling up. Moreover, several papers failed to gain acceptance by referees, and the hassle made me accumulate drafts that did not seem worth finishing. Many come in handy these days: when the scientific public becomes interested in a new topic concerning fractals, I often have an old draft that can be revived into a paper. But it is too bad I failed to publish the "tentative fractal manifesto" read at the 1964 Congress of Logic and the Philosophy of Science in Jerusalem. It would have appeared in an unsuitable place, but it should have been in the record. (P.S. I was lately asked to publish it in *Interdisciplinary Science Reviews*.)

Friends started telling me that I could not continue in this fashion: contrary to the cliché, I would perish if I *continued* to publish as I had done. Moreover, my work on galaxies was not to become acceptable until it was known, and would not become known until it was acceptable. Ten years ago, I was fortunate to be on sabbatical in Paris and my uncle had retired, so it had become proper to ask me to give a major talk at the Collège de France. I saw a golden opportunity to present a general manifesto, and to explain how my different interests fit together. Preparing this talk revealed that my work was already more complete and more homogeneous than I had myself known it to be! My lecture of January 1973 was described by a friend as the most autobiographical scientific talk he had ever heard. It was received with much praise and no hostility at all, which made me realize that my

Facing page: A fractal representation of Benoit Mandelbrot

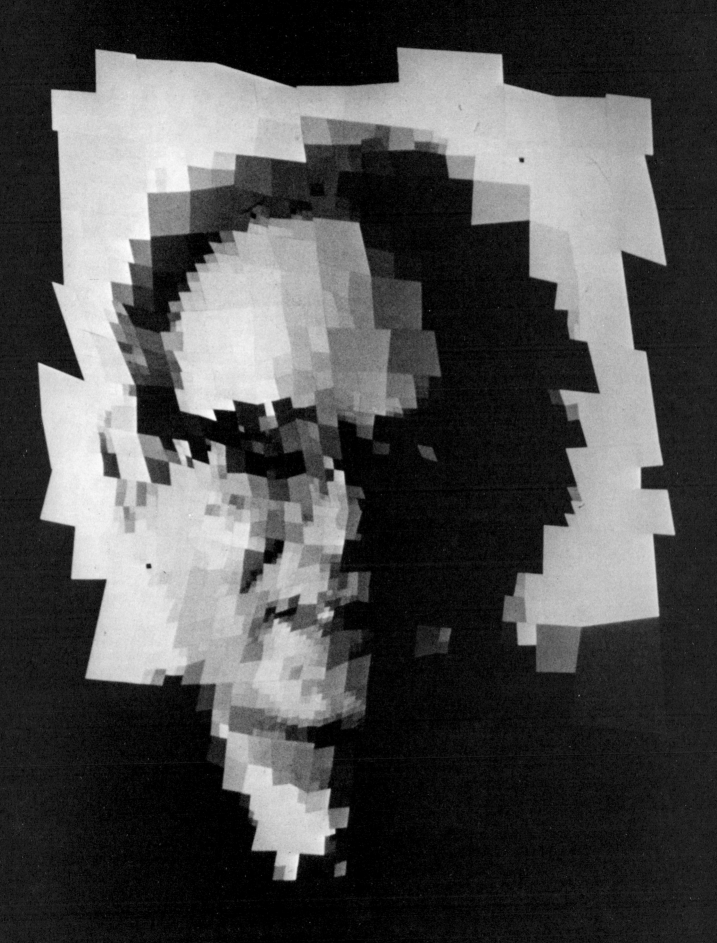

years in the wilderness were about to end. To denote my unified approach, I soon coined the term *fractal*, and the expanded text of this lecture in Paris became my French book, published in 1975 and soon to be reissued in a slightly refreshed second edition.

To summarize, until 1973–1975 my "political" situation as an outsider in all the fields in which I was working was not strong enough to allow me to assert my philosophy and my interdisciplinary approach. Circumstances made me play games which I didn't believe in. The French book marked the change from this piecemeal approach to the present unified approach. Soon afterwards, fractal geometry became organized. My way of life changed profoundly. You may say I have become the slave of my creation.

New Fields to Conquer

MP: *It seems to me that fractal techniques have been embraced fairly readily in the natural and physical sciences like fluid mechanics, astronomy, physics, and geomorphology, but what about other places where you have pointed out applications, like economics or linguistics? Are those techniques being used today by practitioners in those fields?*

Mandelbrot: In linguistics, fractals will not revive. My early work was important to me but was peripheral to the field. However, the mathematical procedures I devised for this purpose continue to survive in other guises.

But in economics, the revival and blooming of fractals can only be a matter of time, because the main two questions which I tackled and discussed in outline are truly unavoidable and cry out for more work. Poincaré wrote that some problems are man-made and other problems pose themselves. Well, the role of discontinuity in economics and the degree of reality of business cycles are problems of the latter variety that will not disappear until they are answered. When a fractal theory really starts moving by itself I tend to become technically underequipped to continue to participate, and it becomes wise to move on. But in economics it is clear that I did not stay long enough. If I come back and show it can be done, this application ought to start moving again.

MP: *You are often referred to as "the father of fractals," and you have been called "tirelessly imaginative." You have, however, always taken great pains in your work to give complete citations to all earlier research which was connected in some way to fractals. One could perhaps obtain from your style of reference the impression that you are cataloguing old results rather than creating new ones. What is your reaction to those who think that you merely pour old wines into new bottles?*

Mandelbrot: This impression is totally without merit, but I understand how it can come to be held by some *very* casual readers. I also understand why it is held by some mathematical extremists who refuse to acknowledge that to build new physics upon existing mathematics is a very creative occupation and *not* merely an exercise in relabeling. There is a price to pay for being called a mathematician, hence for being judged in part by mathematicians.

Allow me a homey comparison. Nearly every theory in my books can be regarded as an evolved model of a "machine" that I conceived. When designing and building the first model of each of these machines, I acted like any other tinkerer, very happy to be able to use many, many existing "spare parts," hardly any of which was to serve the purpose for which it had been listed in the catalogs. Furthermore, many of my contraptions had precursors, in the sense that the legendary Icarus was the precursor of the airplane!

Given the fierce competition that prevails among scientists, the custom under such circumstances is to be content with a casual footnote acknowledging that "an idea—or a tool—somewhat analogous to the author's had also been used for a different purpose in Refs . . . " But my upbringing and my years in the wilderness—when I had no roots in the present, only unconfirmed roots in a distant past—led me to make an arrogant choice. I decided to buck the custom, and to give full catalog references. Of course, these may fail to indicate the ultimate sources, yet I think that giving them helps establish that science is—after all—more than a fast-buck business. Furthermore, I do not mind being a scavenger, and I seek original references for parts I had—quite literally—picked from shelves of remaindered books and from other trash bins of science.

I am pleased to report that the new models of my machines mainly use specially designed parts, and that, among the many reviews my books have won, only one or two brief ones are by casual readers who misinterpreted my gratitude for suppliers of reusable parts.

MP: *It seems that you have not done much work on topics that are not related to fractals.*

Mandelbrot: This is largely correct. The only major exception was work in the statistical foundations of thermodynamics, which I should one day take up again and make better known.

Clearly, this unity of purpose could not have been planned in 1945, or in 1952, or in 1962. Also, everyone has fallow years every so often, and I may have been tempted on occasion—just to keep going—to follow the lead of some other drummer. But I was never tempted; I wonder why.

You must know the line by the Greek poet Archilochus that "The fox knows many things, but the hedgehog knows one big thing." The actual meaning of this line being lost, it is quoted in many contexts. Thus, before a recent lecture, the chairman introduced me as being the hedgehog *par excellence*. I found this very touching and very appropriate.

One should also note that, within this unified thrust, my work has bucked the custom, and has become increasingly *more* "technical" in many ways. My old works seem somehow "lighter" than the more recent ones.

MP: *Do you have any favorites among the fractals, examples that you particularly like?*

Mandelbrot: The unavoidable example is that of coastlines (Figure 1), and my line that "Clouds are not spheres, mountains are not cones, coastlines are not circles and bark is not smooth, nor does lightning travel in a straight line" has gained the supreme accolade of becoming an instant cliché.

The fractal structure of the blood vessels is also a fact that people first find quite astonishing and then very natural. And many people have quoted my assertion that "Lebesgue–Osgood fractal monsters are the very substance of our flesh."

The self-squared dragon curves bring in best the fundamental and amazing discovery that extreme complexity can result from very simple formulas.

But I don't have real favorites, because I tinkered with each of my machines very hard and for a very long time. It's like with one's children. One may be proudest of those who bring the widest renown to the family, but each of my intellectual children has brought equal renown to the fractals family in its part of the scientific world. Anyhow, one can love different children for different reasons, but one cannot really have asbolute favorites.

REFERENCES

Barcellos, Anthony, "The Fractal Geometry of Mandelbrot," *College Mathematics Journal*, March 1984.

Falconer, Kenneth F., *The Geometry of Fractal Sets*, Cambridge University Press, to appear.

Gardner, Martin, "In Which 'Monster' Curves Force Redefinition of the Word 'Curve'," Mathematical Games, *Scientific American*, New York, December, 1976.

Mandelbrot, Benoit B., *Les objets fractals: forme, hasard et dimension*, Flammarion, Paris, 1975.

Mandelbrot, Benoit B., *Fractals: Form, Chance, and Dimension*, W. H. Freeman and Company, San Francisco, 1977.

Mandelbrot, Benoit B., *The Fractal Geometry of Nature*, W. H. Freeman and Company, San Francisco, 1982.

Mandelbrot, Benoit B., "Self-Inverse Fractals Osculated by Sigma-Discs and the Limit Sets of Inversion Groups," *The Mathematical Intelligencer*, 5 (2) 9–17 (1983).

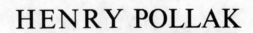

HENRY POLLAK

HENRY POLLAK

Interviewed by
Donald J. Albers and Michael J. Thibodeaux

Henry O. Pollak is Director of the Mathematics and Statistics Research Center of Bell Laboratories. He is responsible for research on the mathematics of physics and networks, communication theory, discrete systems, statistics and data analysis. As Director, he administers a remarkable group of highly skilled mathematicians who not only create beautiful mathematics but who are remarkably adept at drawing inspiration from and linking mathematics to real-world applications.

Pollak is a strong advocate of incorporating applications into the mathematics curriculum at all levels of education. For more than a quarter century, he has worked very hard to improve mathematics instruction. Despite his heavy administrative responsibilities at Bell Laboratories, he has found time to serve professional mathematics associations in this country and overseas. Most recently, he was Chairman of the Fourth International Congress on Mathematical Education and President of the Mathematical Association of America.

Dr. Pollak was closely associated with the School Mathematics Study Group (SMSG), a major mathematics-education reform program that was launched in the late 50's. In the interview that follows, he reflects on both the successes and failures of SMSG, which was one of the first "new math" programs in this country.

After earning his Ph.D. degree from Harvard University in 1951, Dr. Pollak went directly to Bell Laboratories. Since then he has authored over thirty-five technical papers on analysis, function theory, probability theory, and mathematics education. He holds a patent for his work on Interconnected Loop Digital Transmission Systems.

He has been honored for his technical work and his work in mathematics education. He has been awarded several honorary Doctor of Science degrees and was elected a Fellow of the American Association for the Advancement of Science in 1971.

Since the time this interview was held, the breakup of the Bell System has become a reality, and Bell Laboratories is being split along with it. Pollak has left Bell Laboratories to become Assistant Vice President for Mathematical Communications and Computer Science Research for Bell Communications Research, Inc. Ron Graham has become Director of the Mathematics and Statistics Research Center at Bell Laboratories.

The Roots of Pollak

MP: *How and when did you choose mathematics?*

Pollak: I began in it. I got interested in mathematics apparently at a very early age. I say apparently because my memory is very peculiar—I have absolutely no memory at all of myself before the age of five or six, not even isolated scenes, and very little before the age of ten. People tell me that's unusual, but that's a fact. In my attic, there are a lot of old papers, and among them is a composition I wrote in German, in the fourth grade. This composition says, "My father wants me to become a lawyer, but I want to become a mathematician." I was nine years old at the time. I am told that in my parents' diaries there is evidence of my having pattern-searching instincts before going to school. I am told that I would do things like look at a newspaper before being able to read numbers and say, "Hey, the numbering in this must be wrong. It doesn't follow the right pattern." I don't know about that, but I certainly know that by the vicinity of fourth grade, I knew that mathematics was what I liked. I don't think I had much of an idea of what it was, but whatever they were teaching in school that was called mathematics was nice stuff.

MP: *Was mathematics what you started out with as a major?*

Pollak: Yes. What makes you think differently? I never had any doubts in my undergraduate years that I was going to go into mathematics. It is true that I had an offer to do graduate work in Germanic Literature near the end of my undergraduate career, but I didn't do that. It is also true that I enjoy classics very much. I took Latin all through high school and tried to take Greek, but the public school felt that there were too few customers for that, and they couldn't afford it. In my freshman year of college, I took first-year Greek and fifth-year Latin; and in my sophomore year, I took second-year Greek and sixth-year Latin. In my junior year, when I signed up for Latin, Greek, and Sanskrit, somebody in the Mathematics Department at Yale called me in and said: "Pollak, you have got to quit *circumequining* (which is from the Latin and means *horsing around*). You are going to take one of those three, and you are going to take complex variables." So, I took the third year of Greek, and I didn't get my seventh year of Latin which I regretted because that's when you get to the stage where you can read it at 20 to 25 pages an hour, which I never did get to, and I never have learned any Sanskrit.

MP: *Other mathematicians have exhibited strong interests in both languages and mathematics. Do you think there is a connection—a talent for languages and a talent for mathematics?*

Pollak: I don't know how serious a connection there is; I always have felt a whimsical connection in that I have maintained that Latin is an excellent mathematics course because if anybody can learn to live with and obey the consequences of that ridiculous axiom system, they can do any mathematics you could imagine.

Mathematical Heroes

MP: *This question is a little on the schmaltzy side perhaps, but do you have any mathematical heroes, dead or alive, who have influenced you in one way or another?*

Pollak: Schmaltz of course, means goose grease; you aren't supposed to put that on a baseball when you throw it, and you have thrown a curve ball with some extra schmaltz on it. As I think about people, I believe that probably the effects on me that I am conscious of are much stronger in the education process than in the research process. I certainly remember a number of teachers and professors and particular points from them that came through, either directly or subliminally, that made a lot of difference to me.

I had a junior high-school teacher of algebra in Stamford, Connecticut, by the name of Mrs. Shahan, who had a particular pedagogic technique that was very good for me. She had developed a very thorough ability of lip reading. When she gave work for the class to do, she had the problem of keeping the fastest students amused; and how does one keep them active without spoiling the lessons for the slower ones? The way she solved that problem is that when you thought you had the answer you could mouth it. Nobody else was good enough at lipreading to be able to read it in the classroom, but she would nod her head and say, "Yes," and you could go on to something else. She also kept me amused by teaching me the old-fashioned method of extracting n^{th} roots. We did have the old method based on the binomial theorem for taking square roots, although it was never explained why it worked. She taught me the method of extracting the cube roots and fourth roots, and this was good because that certainly allowed me to understand what the square root rule was. I enjoyed that. It was very good for me to get that additional work and built up in me a feeling that I have always had: Enrichment is better than acceleration if you have a choice between the two, although that's not something that everybody will agree with, or that I might agree with myself on another occasion.

Mr. Farrell taught me high-school geometry, and I learned something else from him. He used a particular device to keep me quiet: He had a habit of asking students to go up to the board to do a particular geometric proof. I didn't realize it right away, but I did afterwards. With other students he would give them problems, let them sit in their seats for a moment, and then have them go up to the board to work. With me, he would tell me to come up to the board and then tell me what problem he wanted me to do. That was a technique that I appreciated—afterwards.

My second-year algebra teacher's technique was to provide you with a choice. You could work with the regular class and cover the year's material at the pace which she set; or if that was too slow for you, you could study a chapter on your own. For every chapter, she had three 5″ × 7″ cards. She would give you one of those, and that was the test on that particular chapter. Then if you did that successfully, you could go on to the next one. You could either stay with the class, or run at your own pace. Incidentally, success meant that you had to get 100% on the test for that particular chapter. If you didn't get 100% on the first card, you could try the second one. I finished the year's work in a quarter by that particular procedure, and then they had to find some other things for me to do. They gave me some advanced algebra and various other things. That was an experience that I remember.

Cross-Country Mathematics

At the college level, I certainly remember Ed Begle. What I remember, particularly, and I don't know how to put it in the very positive sense in which I felt it, is that he came to class sometimes not completely prepared. You had a very interesting time watching him struggle, inventing the proofs and trying to think about the right way to do it. I learned a lot more mathematics that way than I might have if it had been a perfectly polished lecture; and I think already at that time I developed my feeling that I like cross-country mathematics. Mathematics, as we teach it, is too often like walking on a path that is carefully laid out through the woods; it never comes up against any cliffs or any thickets; it is all nice and easy; there are beautiful views, and at five o' clock every day, you come to a well-stocked hut where you stay overnight and start out on the path the next day. I like cross-country mathematics from time to time. You take a rough compass heading, and you get lost. I learned the excitement of that because Ed, intentionally or otherwise, occasionally did that.

Henry Pollak as a Harvard graduate student, 1950. Pictured left to right are Alex Blair, Pollak, Bob Osserman, and the late Bill Turanski.

Another experience that I remember, as a graduate student at Harvard, was Georges deRham teaching a course on distributions on manifolds. In the first lecture, he defined continuity of one of these currents as they were called then. He went on for 20 minutes or so and proved various things about them and then came up with something that was pretty absurd—some property that just didn't make much sense. Of course, we realized at that point that it was the wrong definition of continuity.

The counter-example gave you the idea of how you wanted to change it, and the other definition is what we worked with ever after in the course. Now that kind of open-ended teaching was undoubtedly done to me many times, but that's the first time I noticed it, and I say that's a good way to teach. You get something across that way, and you understand something that way, but I certainly wouldn't have understood it as well if he had just started out with the right definition of continuity the first time around.

Pollak as Teacher

MP: *Your interest in mathematics education is remarkable among applied mathematicians. What prompted it and what keeps it going?*

Pollak: It is basically like hit-and-run driving. In the real world hit-and-run driving is illegal, but I can come out of industry and pontificate about teaching, and never have to face up to anybody about it. I never have to listen to my dean in the morning about what horrors I said about teaching, and I just go away again and have some more fun.

MP: *Have you ever thought seriously about being in a university as a teacher of mathematics and mathematics education?*

Pollak: The last time I thought seriously about teaching was when I taught, which was in my last year of graduate work at Harvard in the academic year 1950–1951. That experience convinced me that there was no way I wanted to be a teacher of mathematics.

MP: *What convinced you not to become a teacher?*

Pollak: As a graduate student, I had a class of 25 students, and there were usually 25 ridiculous, arcane reasons why they didn't understand what I was saying at the moment. If I had had several hours apiece with those students, one at a time, I could have dug back into their backgrounds and probably have found out what was misunderstood, what was missing, and what would then get them over a particular difficulty. It was also quite clear that the university educational system was not at all interested in my spending three to five hours per student, trying to do this sort of thing. They might say they are paying you for teaching, but they really are paying you for research. If you spend that amount of time with students, you are never going to get to first base. I think the system was calculated to keep me from doing the kind of job I felt was appropriate to do as a teacher. But I still have a lot of opportunity to do teaching in industry.

Teaching in Industry

MP: *Do you get to do that kind of individualized teaching at Bell Labs?*

Pollak: The big difference between industry and university is not whether or not you teach. The difference is that in industry you typically have more students who want to learn. You've got people who are coming to you to discuss mathematics. They need help on problems and want to get ideas on how to solve them. They really want to do it. So often (in academe) you get students who don't want to learn mathematics. They are there because they have to be. At Bell Labs we offer an enormous number of internal education courses, which provide me with teaching opportunities. Another big difference is that, at a higher level, I've got no graduate students—I've got no slave labor. If there's a problem which I'm pretty sure can be done, and quite sure it's worth doing, but it's not as imporant as some other problem that I want to work on instead, I haven't got anybody whom I can force to do it. We have excellent, new people who come in, but they are full of their own good ideas. Why should they work on mine? If I had graduate students, they would have to do that problem, but that's the major difference—how to get something done that you don't want to do yourself.

Missed Opportunities in Continuing Education

MP: *Now we know why you don't want to teach in a university. Why are you so involved with mathematics education?*

Pollak: I simply am interested in education. I made the quite unconscious decision more than 20 years ago that I was going to spend much of my discretionary time at Bell Labs on mathematics education. I have done that and it has led into a lot of things in that area. It ended up with my being President of the MAA at one time, and Program Chairman of the Fourth International Congress on Mathematical Education at Berkeley, Chairman of the Advisory Board of the School Mathematics Study Group and many other things. And I say it was unconscious.

MP: *How important is continuing education to Bell Laboratories?*

Pollak: It's very important! Over the years, the educational pattern in Bell Laboratories has changed a good deal. One thing that a high-technology industry like ours understands is that you cannot expect that the education which employees bring with them will be enough for a lifetime career. We have available to everyone a very large internal continuing education program. Of couse, in recent years, it has been more computer science than anything else.

Now outside Bell Laboratories, in the larger national context, continuing education in this country is in some sense a crime, how little and unsystematic it is. Opportunities are being missed. If you want to make comparisons, you might look at the Open University of England—a national system of opportunity via television, first-rate courses with first-rate people. People watch courses at home on television and go periodically to centers where they can meet with instructors for help and homework. It's a tremendous system.

This country is full of industries that would be very happy to get help with their continuing education programs. You may have to go on their own premises. Engineers and other people who have been out of school for 10 or 15 years may not want to go back to the campus and compete with the kids there, but you can put on the programs and the courses where they work or somewhere else.

Model Building in the Schools

MP: *What are your current thoughts on the state of the undergraduate curriculum in mathematics?*

Pollak: My own feelings about the undergraduate curriculum are pretty heretical. It seems to me that our basic obligation is to provide people with the fundamentals in mathematical sciences, that is, the basics of the various fields that are going to come up in whatever careers they undertake as a result of their undergraduate education in mathematics. They certainly will have to teach these things if they end up as college teachers. It seems to me that the fundamentals are the calculus, linear algebra, probability, statistics (as a separate discipline with a separate ethic from probability), basic computer science, and some experience in modeling. I should probably include as a seventh one in this list some discrete mathematics such as combinatorics or algorithms.

MP: *What do you mean by model building and applied mathematics?*

Pollak: Applied problems—problems that use words from other disciplines and pretend to be applied mathematics in the textbooks—are usually pure frauds. The idea that seems to pervade much calculus teaching is that as long as you can learn how to say "moment of inertia," "center of mass," and "pendulum" with a straight face, that's all you have to do. I often have the impression that the meaning of these things and how in the world you get from the physical idea to the mathematical formulation of it, seem quite unnecessary to the teacher. My favorite caricature of the teacher of mathematical physics is: "Okay folks, today we're going to study the Coriolis effect. Consider the following partial differential equation." The student probably has absolutely no idea as to what that means and where you got it, what the variables are, what you kept and what you threw away in making this model.

So, what is applied mathematics? It really is starting out with a situation in some other field you're trying to understand. Then you try to formulate a mathematical problem that will help to shed some light on the situation you are trying to understand. If you succeed in formulating a mathematical problem, you then go ahead and solve that problem if you can, and, of course, in all honesty, you go back and forth many times, from the question and its formulation to the question and the solution.

Then you see what you have learned. What does the solution say about the original situation in the outside world? With a very high probability, the first time around you will find that the interpretation in the original situation is garbage; it's wrong, it doesn't make sense. So you say, well, we must have formulated it wrong, and we go through this loop many, many times. Now, that's what really happens, and I don't think you capture that by mouthing words from some other discipline. I think students should have model-building experience of this sort.

MP: *How do you go about implementing model building in schools?*

Pollak: If you are talking about changing curriculum and putting in some new things, the easiest place to do it is perhaps the elementary school. The next easiest is the university, and the most difficult is the high school.

MP: *Why is the high-school level the most difficult?*

Pollak: In elementary school, generally speaking, there is nothing whatever expected in the way of science education. There are no elementary science requirements in most states that I know of. So there is a certain amount of time there with which you can do anything you want. And if you've got some ideas about trying math and science and social science together you can get it in there by calling it elementary science.

At the college level, I think a lot of people can be persuaded that experiences in applied mathematics are important. It's a part of the total experience in mathematics that a student should have. You can argue it either way—philosophically or practically. You can argue that a part of becoming literate and knowledgeable in mathematics is to know how mathematics is used and how it is connected with the rest of the world. That's one end of the argument. The other end is to grab a student by the scruff of the neck and say: "Look, if you want a job, you have to know something about how mathematics is used." I'm not proud, I'll take either argument.

The trouble with high schools is that you've got entrenched departments and entrenched requirements. If you were to try to put in an applied mathematics or model building course, everybody would say: "Whose time is it going to come out of? Not mine!" It's got to be done jointly by science, social

Teacher Pollak with students at a National Science Foundation Institute in Atlanta, 1974.

science, and mathematics people, and I don't think we'll get to first base. The structure of the high school is harder to crack than either the structure of the elementary school or the structure of the college.

MP: *Wasn't that structure cracked in a big way during the Sputnik years?*

Pollak: Not the structure—the content. The fact that there is a chemistry course, a physics course, a biology course, and a math course every year—that structure didn't change at all. It was decided at the beginning of SMSG* that in order to have any chance at all of reforming the curriculum, we had to use the existing structure and the existing kinds of allocations of time, and just do it better. So, the first SMSG materials produced were a first-year algebra course, a geometry course, a second-year algebra course, and a half-year course in elementary functions. The one structural change that was made was to incorporate the solid geometry with the plane geometry. For nearly everything else, including the junior high materials, there was no major attempt to shuffle the time and the years and to reorder the materials in a major way because of the difficulty of getting that kind of thing accepted.

SMSG and the New Math:
Reflections by One of the Pioneers

MP: *You were deeply involved with the School Mathematics Study Group (SMSG) from its inception. What was it about the SMSG experience that was so enriching for you? Beyond that, what has been its long-term impact on mathematics education in this country?*

Pollak: That's a very big can of worms you are opening here, and I suspect they are going to be slithering out for the next half hour. As you may recall, Ed Begle was the first director of SMSG. I was an undergraduate at Yale in the mid 40's and I took Ed Begle's year-long point-set topology course. I started at Bell Labs in 1951 and occasionally went down to some of Al Tucker's combinatorics seminars at Princeton. When Ed Begle, with the advice of Al Tucker, planned the beginning of SMSG and the list of whom to invite, they invited me. Our first meeting was a four-week session at Yale. Our instructions were to start from the work of the College Entrance Examination Board and try to start outlining and gradually writing curricula. Initially, I was very dubious about the whole project. The people at Bell Labs said it was perfectly okay that I should go off to that first SMSG writing session, but I remember leaving a message with my secretary to call me after a few days and tell me that there was an important crisis or something in case I needed an excuse to go home. However, I enjoyed the first days of it very much. The crisis didn't turn out to be as bad as I had expected, and I stayed for the full four weeks of that initial writing session.

One of the people at that meeting who had a great influence on all of us was Martha Hildebrandt. She was head of the department at Proviso High School at Maywood, Illinois, and had been President of NCTM. The first thing she said to me was: "Now remember, Pollak, you can't teach anything after April." She was letting me in on one of the truths of what actually happens in the classroom. Spring fever comes, and the kids just don't listen anymore. I learned a tremendous amount from her. I found, as did many other mathematicians who came to that meeting, that the questions of what to teach, how to teach it, in what order, how to say it, and how to combine skills and understanding were tremendously interesting, difficult issues.

The problem of putting together a high-school course was not a triviality; there was a very real intellectual challenge to it. We were faced with questions such as these: What do you really mean by variables? What do you really mean when you say you're going to solve an equation? What do all those "simplify" exercises in the textbook really mean? How do you know when you are done? When is one thing simpler than another? How in the world are you going to teach word problems? All of this had lots of meat to it. That's how I got involved in it.

MP: *What is your evaluation of the SMSG experience?*

Pollak: One of the things that I was by no means clear about in those early days, and that I'm not happy about, is that I didn't think hard enough about the place of applications and about the way that applications of mathematics ought to be interacting with the rest of the curriculum. That wasn't really

---

*During the late 50's, the School Mathematics Study Group (SMSG) was formed as part of a national effort to upgrade school mathematics.

faced squarely until the second round of SMSG, which began in the late 1960's and never really got done because the funding started to run out.

The Second Round of SMSG

There were some very beautiful things written in the second round of 7th and 8th grade materials, which hardly anyone has ever seen, in which applications were much more intelligently interspersed and worked in and became a fundamental principle. But in the first round, we didn't do that very well. We put in a great deal of interesting mathematics which could and can still be beautifully justified and worked up on the basis of applications, but we didn't do it, and we've suffered because of it. That is 20-20 hindsight.

The second-round materials of the SMSG contained beginning work on computing, applications, and probability. Those materials were written somewhat closer to the English style of teaching in that different parts of mathematics were intermingled rather than presented as a solid year of algebra or a solid year of geometry. In the first round, the spiral approach, which had been preached for everything else, was used within each individual course, but not the intermixing of the different parts of mathematics. In the second round, we wrote totally different seventh and eighth grade books. Then we came to the conclusion that it wasn't possible to finish a total rewrite of high school mathematics, and if I remember correctly, there was a bridging volume written to take students from these new seventh and eighth grade materials back to a more traditional curriculum in the senior high school.

Failure of SMSG at the Elementary Level

MP: *How successful was SMSG at the elementary level?*

Pollak: The elementary effort was probably not the success it could have been. The number of secondary teachers of mathematics in the United States is about one hundred thousand, and an effort was made to reach every one of them with a summer institute or something like that. The total fraction of teachers that was reached by the summer institutes was around two-thirds. While the institutes covered a lot of different things and possessed varying degrees of quality, they did provide a sense of vitality and of thinking about the problems, opening up to new arrangements and new pedagogy.

The number of elementary teachers of mathematics is about one million, rather than one hundred thousand, and the people who made the decision threw up their hands at that and said: "Look, it can't be done; we'll try to write good teacher's commentaries and guides, but you can't possibly try to reach one million elementary teachers."

Now the problem with the new curricula was that the pedagogy was meant to be different from what was sometimes done. There was supposed to be an open-endedness; there was supposed to be a spiral approach, in which one did not teach for mastery of things the first time around; ideas would come back many times. It was difficult because of the way in which elementary teachers had been trained themselves and the experiences they had had in many cases. The result is that they taught things for mastery that no one ever intended to teach for mastery, like how to do division in base 8, to caricature it.

There was a failure to reach the elementary teachers. It could have been done; one could have insisted on at least a paid week's work in the summer with the elementary teachers so that they might get an idea of what people were trying to do and why. That was not done, and we paid the price of putting the whole process in disrepute. One reason for thinking about that so hard right now is that there is a certain amount of danger that the same kind of thing will happen again.

Facts versus Opinions on Mathematical Education

We have been talking a lot about my opinions, impressions, and prejudices as far as mathematics education is concerned, and I do think there is something important to say here. It is something that you find particularly in the writings of Ed Begle, for example, in his posthumous book, *Critical Variables in Mathematics Education*, and something that you have always got to remember: Mathematicians have very strong opinions about mathematics education. They instinctively think they know what's right and what's wrong, and they don't hesitate to tell you with the highest powers of certainty at their command. Ed Begle, I think, particularly pointed out that they were not always right and perhaps were quite frequently wrong when you put all of these things to experimental test. I think the many opinions that mathematicians have about teaching, which we don't mind pontificating, need to be examined experimentally. We have instinctive ideas about innumerable aspects of the teaching of mathematics, both in those levels that we teach and the levels that we don't. I think those ideas deserve to be listened to, but they also deserve to be tested.

MP: *It's very difficult to test opinions on the teaching of mathematics, isn't it?*

Pollak: The answer to that is yes and no; that's complicated. The bulk of research in mathematics education has traditionally been done in doctoral dissertations in mathematics education. Doctoral dissertations in mathematics education are typically, by their very nature, studies done in a relatively short time with a relatively small number of students. I think you are very lucky if you get somebody to spend a year or two examining a particular question with as many as a few hundred people. Many doctoral dissertations don't do that well. The statistical results of doing something like that are almost always suggestive rather than definitive. If you look at Ed Begle's book, *Critical Variables in Mathematics Education*, the typical kind of result you will find, and I don't think I am going to be caricaturing this very unfairly, is this: On a particular point there have been 110 studies, and 60 found that *A* worked better than *B*, 35 found that *B* worked better than *A*, and 15 found no difference between the two. Evidence in mathematics education does not typically come from a single study designed so carefully, massively, and long enough in time, that it has a chance of really settling a problem. Our evidence comes from a large number of studies on a particular point, each of which is itself only suggestive and not at all definitive. Therefore, in that sense, you are right. How are you ever going to get at all close to assurance on these matters?

On the other hand, it is also possible to do something like NLSMA, the National Longitudinal Study on Mathematical Abilities, that was done in the 60's and which followed 50,000 kids for five years. What was typically found in that study was that in fact many cause-and-effect relationships that mathematicians would naturally have suspected and then asserted as fact were not true. At least, they certainly were not proved. NLSMA did find that the textbook made a difference in the gains by the students between pretest and posttest. It did show how to test for understanding and transfer using multiple choice tests.

Let me mention one other problem with the picture of research in mathematics education. We do have a large amount of research going on, much of it in small studies, typically at the elementary and secondary level. One of the major interests of NCTM is to support research in mathematics education, to publish it, and to devote sections of their meetings to it. I know of practically no research in mathematics education at the college level. It is something that the MAA has not been willing to get interested in in any serious way, and I don't think that NCTM has either. In the college area, you are absolutely right; we know just about nothing.

Do Mathematicians Suffer from Reality Anxiety?

MP: *In one of your papers, you close by saying that the only thing that would prevent a curriculum redesign that incorporates applications would be control by those people who went into mathematics teaching because they wanted to avoid the real world.*

Pollak: I believe that mathematics education at all levels has more people in it than we care to admit who went into mathematics teaching because it was a way of making a living and avoiding the real world. This is a serious factor in the difficulty of getting applications of mathematics into our schools and colleges. There are not many teachers of mathematics who say explicitly: "Showing students how mathematics is used is bunk. The purpose of teaching mathematics is simply to get the mathematics across, and nothing else matters." Very few teachers say that openly. So you have meetings and everyone will agree that you ought to have applications. You look again a year later when the smoke has cleared away, and nothing has happened; and you do have to keep asking yourself why. If everybody agrees that connecting mathematics with the real world is a good thing, why doesn't it happen in the classroom?

International Activities

MP: *You have been extremely active in mathematics education, not only at the national level, but also at the international level. How did you get into mathematics education beyond the borders of the United States?*

Pollak greeting guests at the Fourth International Congress on Mathematical Education, Berkeley, 1980.

Pollak: I think it happened in the same way in which I got into things in the U.S. I had been a student of Ed Begle's and thus was in on the beginning of SMSG, and through that also on the beginning of CUPM. So I became known as a noisy fellow, and I was invited to go to several meetings in foreign countries. Remember that curriculum reform in the United States really came before that sort of activity in most other countries, and they were very eager to hear about our reform efforts.

MP: *So they were drawing on the experiences of SMSG and CUPM to some extent and hoping to learn from what had been happening in the United States. Did they learn?*

Pollak: Yes. There was quite a bit that happened in Europe. They certainly didn't copy the United States. That would have been very much the wrong thing to do. However, generally speaking, citizens of any country believe that its own students are likely to be smarter than those of any other country. When they hear what any other country is doing, they know they can do better! So they did get a lot of things started, and that's very valuable because in recent years, when there has been no large scale, organized support for curriculum work in the United States, we have been learning a great deal from the Europeans, and much of the major systematic activity in curriculum thinking is now in other countries. They started later and kept going after we stopped.

MP: *What's been able to sustain them?*

Pollak: I haven't seen any general pattern of what's been able to sustain them. They've got some very good people. I think that they certainly haven't developed any prejudice against national thinking. Remember that thinking about the curriculum nationally in this country and trying to do projects with federal funding were really quite new ideas, and fell into relative disfavor in the 70's. Education is basically a local affair in the United States. In European countries, it is no surprise to do work on education on a national scale. It's all done in the central Ministry of Education. If the central Ministry of Education says let's keep looking at this, it will happen. There have been a lot of exciting initiics and good ideas in other countries. For example, in the area of technology interacting with education there is more going on in a systematic fashion in England and France than in the United States.

MP: *You were in charge of the International Congress on Mathematical Education that was held in Berkeley in 1980. What did you learn from that Congress?*

Pollak: It was very clear at the Berkeley Congress that the United States had more to learn than to give at that point in time. That was not true of earlier congresses. I suspect I ended up as Program Chairman and Chairman of the Executive Committee for the Congress at Berkeley because I had been to all of them: Lyon in '69, Exeter in '72, and Karlsruhe in '76. There was a great deal of exporting of U.S. ideas still going on at the earlier congresses. It was disappearing by '76, and by '80 we were learning from everybody else.

MP: *Then you approve of international congresses devoted to mathematics education.*

Pollak: I think we have an awful lot to learn and we have a lot to give in various ways.

MP: *How can one learn more easily about developments in mathematics education elsewhere?*

Pollak: That's an interesting question. One of the things that doesn't exist and that should exist is an international clearing house for mathematics education. We don't even have a single library in the world which gets all curriculum materials and all work in research and development in mathematics education. There is such a library for the social sciences in Boulder, Colorado, but there is none in mathematics.

Bell Labs: Managing Mathematicians

MP: *Your job at Bell Labs is Head of the Mathematics and Statistics Research Organization. What does that entail?*

Pollak: Like all middle management at Bell Labs, I spend a lot of time doing committee work and managerial matters for the company—matters that cut across the whole organization. I also read every memo and paper from my department before it goes out. That takes a lot of time. The total number of papers runs around 200 a year. So it is pretty close to one a day that comes in. I read every one of them, make comments if I have any, make suggestions on exposition, pose questions on the mathematics or other aspects of the problem, and just plain proofread. It is a major way that I keep informed of what everybody is doing.

Part of the job of a middle manager at Bell Labs, which is quite different from the department chairman at a university, is the requirement that you must know about, be able to understand, and describe in some detail the work of every individual in your organization. That's vital if you are to evaluate people and determine who should get big raises and who should get small ones.

It is really a part of the responsibility of everybody in the organization as well as mine to look around to see if you're doing the most interesting things that you might be doing at the moment. At Bell Labs people do change what they do fairly often. The department heads within mathematics who report to me and with whom we do much of the planning also have this sort of responsibility.

MP: *Many people, particularly those in mathematics and the hard sciences, regard Bell Labs as a very special kind of place, a place of great excitement. What makes it special in your view?*

Pollak: Bell knows what sort of working conditions to provide for mathematicians and scientists. That's a much trickier problem than people realize. To caricature it, some technological organization that one hears about might say: "Oh, yeah, mathematicians are a good thing to have. Okay, let's go out and hire some and give them some office space, some blackboards and tables. We will waive the usual rules about having them punch a time clock. We will just have them sitting around doing mathematics because we have heard that if they do that then some valuable results will come of it." Typically, what happens (and remember this is all a caricature) particularly after a few years, is this: There is a slight recession, the company looks around at where it can save some money, for something that it's paying for that isn't contributing anything to the company, and discovers this nest of parasites sitting there. Out they go.

What happened was that you didn't provide for interaction between the mathematicians and the rest of the company. Mathematicians do not automatically do valuable things. They have to be motivated; they have to interact; they have to get good ideas from chatting with and living in the middle of lots of other people. So the next time you hire a mathematician, and you put him in a large room with 100 engineers, and tell them all to interact. The mathematician who tries hard and is honest about it, will interact a great deal and will probably leave at the first opportunity because there's no chance to sit back and think. There's overstimulation. What you need is an opportunity to interact with everybody else and to be stimulated by them. At the same time, you need to have a group of colleagues, enough to have critical mass, and the ability to pull back and think quietly without any unreasonable pressures. The proper balance between stimulation and protection is difficult to learn to achieve.

The great strength of Bell is that it has had a mathematics research organization since 1925 when Bell Laboratories was formed. There is an established understanding of how people work, and management's confidence that this succeeds because it has succeeded continuously for fifty years. You have the freedom to work on a great variety of problems and to work at your own pace and your own time. Since performance is measured over very long time periods, this freedom requires enormous responsibility, to think all the time about what you ought to be doing and what's the most interesting thing to do next—including the responsibility to get out of an area when it has passed its peak. But you can do this with assurance and a feeling of security because the place has learned over fifty or sixty years that this works.

Grading at Bell Labs

MP: *What does it take to be a success at Bell Labs? How does one decide if someone has been productive?*

Pollak: Well, the thing you look for basically—and it is easier to say in words than to carry out quantitatively—is how much of a difference that person's existence makes to research, to the company, to the country. There are many different ways in which one can make a difference. By opening up interesting new directions in some part of mathematics, by giving good consulting advice to people, or by running a good computer center, to give just a few examples. But basically, you see how wide a swath a person cuts. How much would things change if that person were not there? Once a year, the department heads within mathematics and I get together for several days to argue about performance, to look at each individual and try to understand about how much difference it makes that this person exists and does what he or she does.

Then, I get together with directors of several other centers or laboratories consisting of people working in economics, in computer science, in psychology, human factors, and acoustics. We argue about the merits across this whole large organization, which is the Division. This, in turn, means that I know something of what these other people do, and they know something of what the mathematicians and statisticians do. Finally our boss does the same thing with the Vice President of Research. The success of this sort of pattern depends on a maximum advance and continuing flow of information, on the opportunity to know as much as possible about what everybody else is doing.

Advice to Academicians on Keeping People Happy

MP: *Bell Labs seems to be a place that contains a lot of exciting, very different people with strong personalities. How do you keep all of them happy?*

Pollak: I think the fundamental issue in keeping people happy is to understand and appreciate their work. The critical thing is to keep open your understanding of what value is. Again, my prejudice is that mathematics departments would be very well served by working hard to maintain a broad view of what constitutes achievement in the mathematical sciences.

Take the problem of trying to house pure and applied mathematics in the same department. Consider an individual who takes a situation in some other field that has never before been successfully analyzed mathematically. By "success in analyzing" I mean that the model turns out to have interpretive understanding and maybe even predictive power. Let's assume this person takes a situation in engineering, or history, or economics, or physics, and he manages to make a mathematical model that is complicated enough to say something about the real world but simple enough to do something with the model—and he does this successfully from the point of view of the applied discipline. That, to me, is a successful job of applied mathematics. Furthermore, it is a success in applied mathematics regardless of whether the mathematics itself that was done in the middle of this job was new or old. If the problem turns out to be one in combinatorics or complex variables or linear algebra or topology that has been done before—great. The only trouble is that many mathematics departments will not say: "Great." They'll say: "You haven't done anything." What can happen if you insist on the novelty of the mathematics itself is that the person will then complicate the model, unnecessarily from the point of view of the application, but necessarily in order to do some new mathematics in the middle of it. You insert some stuff you really don't need in order to make it look mathematically new. Now maybe your pure mathematics colleagues will feel better about it, but you have loused up the job. So it's necessary to keep a pretty broad appreciation of what is success. I myself think it is a shame that there are so few departments in which pure mathematics (whatever that is) and applied mathematics, statistics, and computer science can all coexist.

Airplane Problems

MP: *You are regarded as an excellent problem solver. You have described some problems as being airplane problems, meaning problems upon which significant progress can be made during a lengthy airplane flight. Do you have a favorite airplane problem or two for us?*

Pollak: In fact, I have an airplane problem about airplanes. It started about two years ago when I flew from New York to Columbus for Bell Labs, to Raleigh-Durham for the State of North Carolina, to Tampa to visit my wife who was taking care of her mother, and to Washington, D.C. for the National Science Foundation. Bell Labs was willing to pay for a round-trip to Columbus, the State of North Carolina was willing to pay for a round-trip to North Carolina, I was willing to pay for the round-trip to Tampa, and the National Science Foundation was willing to pay for a round-trip to Washington. If I made them all pay for a round-trip, that would give me much more money than the cost of the trip. The question is: "What is a fair way to divide up the cost of the trip?" I didn't see how to do it, and I sat down to chat about it with Peter Fishburn, who had worked in areas of this sort. We made various models showing that if you put all sorts of conditions on it, various contradictions arise. This problem is going to result in a paper or two.

MP: *How about another airplane problem?*

Pollak: A number of years ago I was in charge of the UNICEF Collection for my church, and so I ended up late in November with, I think, around 8,000 pennies. It was part of my job—with considerable help from my daughter—to count them. So I decided that we would not only count them, but look at their dates because of my curiosity about the decay rate of pennies. That is, it is clear that you don't see many 1909 Lincoln head pennies around or even many pennies from the 20's or 30's. So we took the data, counting how many there were from each year. Then Ed Gilbert, one of the mathematicians in Bell Labs, took the results and worked up a model. He found two interesting things: First, the exponential decay rate fits the existence of pennies very well. The half-life is on the order of six years. The second thing that we wondered about was the degree of mixing. In the East where I am, you would expect to see mostly Philadelphia coins, and you wouldn't expect to see as many coins from Denver and San Francisco. How many years does it take, if at all, before the proportion of coins you see from Denver and San Francisco is not lower than you would expect, by decay rate and amount that was minted, than for those from Philadelphia? The answer to that was three years. Fitting the exponential to the data was a very good fit; it fit beautifully, except for one point which was off by a factor of five. It was that much lower than the amount you would have expected. Can you guess what year that would have been? Well, those are the 1943 steel pennies. This occurred because the banks at one point, about 1965 or so, were trying very hard to get rid of them and get them out of circulation. Their number really fell below what you would have expected. That was kind of a nice problem, and Ed wrote a nice paper on it.

Is Applied Math Bad?

MP: *You once said that in the case of John Horton Conway, although he is a pure mathematician, applications caught up with him. Do you think they will ever catch up with someone like Paul Halmos, who in a recent interview said the following about applied mathematics: "There is a sense in which it's just* *plain bad mathematics. It's a good contribution on the one hand; it serves humanity; it solves problems about waterways, sloping beaches, airplane flights, atomic bombs, and refrigerators; but just the same, much too often it is bad, ugly, badly arranged, sloppy, untrue, undigested, unorganized, and unarchitectured mathematics."*

Pollak: Although I consider mathematicians to be crazy, they are not suicidal. So I'm certainly not going to attack Paul. My serious reaction to his comment is that it may be fundamentally derived from the feeling that applied mathematics consists of mathematical physics. His initial examples from which he obviously drew both his instincts and adjectives were physical examples. It is perfectly possible for people not particularly to care for mathematical physics and not to care for what looks like a great mess and the complication that comes out of it. The first thing in thinking about what Paul has said is to remember that applied mathematics consists of a lot more than classical analysis applied to classical physics in many different ways. It consists of a great many other fields of mathematics applied to a great many other fields of human endeavor. To make a gingerly counterattack, I would comment that I recently chatted with Paul Halmos about the problem of division of costs of an airplane trip and that Paul asked if this could be submitted to the *Monthly*. So this problem at least wasn't as bad as some of the other things that he was talking about. Quite often it is true in applications of mathematics that there is a problem that needs work done on it, and you need some numbers and you need some answers. And often they are very sloppy, and dirty, and you work very hard, and you get very little aesthetic satisfaction. But the job needs to be done, and it gets done.

Pollak giving the Michigan Mathematics Contest Awards Lecture, 1963.

If you are doing applied mathematics on a slightly more fundamental level, and living in an environment like Bell Labs where lots of exciting things are going on, and ask yourself why this, and how come that, and so on, then you can get much satisfaction from formulating the problem as distinct from having to slug through what somebody else has formulated. You are much more likely to find a good-looking structure. We need to remember that many of the beautiful structures that mathematicians like Halmos work on so successfully have their roots not only in the generalization and understanding of other simpler mathematical structures, but also in the behavior of the physical, economic, and psychological world. Nice mathematical structures do come out of the real world. In fact, it often happens that the axiom system you get driven to *naturally* and the kind of questions you get driven to *naturally* in the real world are pretty nice, and are more successful mathematical structures than some of those you dream up sitting at your desk. I think Paul may owe more to the questions that come from outside of mathematics in the formulation of the structures that he works on than those he mentioned in that particular quotation.

Mathematics: Invented or Discovered?

MP: *You said that there are some problems that give you a certain amount of aesthetic joy as well as the joy of having completed a particular problem that has a need for an immediate answer. In your own career, are there some problems that you can describe in fairly general terms that stand out as being especially nice from both standpoints?*

Pollak: These are the ones that you are likely to talk about when you get a chance to talk to students or at a mathematics meeting. When mathematicians in industry (or any other mathematicians) talk about mathematical work, they use their own selection bias. They pick the things that are nice to talk about. The same is true in mathematics itself. If you start working on a problem, and at the end of a year you have 200 methods that will not solve the problem, you don't get a Ph.D. for that, and you don't talk about it very much. Now in the same sense, the problems that turn out to be particularly pretty are the ones that you are much more likely to talk about.

There are three problems from my own research that stand out in my memory: First is band limiting, the work on concentration of signals, which I did with Dave Slepian and Henry Landau. (It has become a pretty big theory and has had lots of interactions into other fields of applications and into mathematics itself. That turned out to be a very exciting structure.)

Mathematicians often argue whether mathematics is discovered or invented. I certainly had the feelings in that particular case that I was discovering it and not inventing it. After struggling for years, the insights eventually came to me that made it all fall into place. It all hung together in an incredible way—every loose end had its own natural location. When we looked at the end result, we realized that we had been luckier than we had any right to expect.

We couldn't have invented all that. We had discovered a structure that must have been there. At least, that's the feeling I had; it hung together too well. If you invent something and you put in n features, then if you're lucky, you've got a right to expect $n + 2$ or $n + 3$ nice properties to come out, but not n^2. The feeling is similar to that in the development section of a good sonata movement, where you have had your first theme and a couple of minor parts of that, and the second theme, and then comes the development, and all of a sudden these things dissect each other, and get played against each other, and interact with each other, and you realize how it all fits together. That's the feeling you get in a major mathematical development. As I say, I get the feeling that I didn't invent all that.

Next is the loop switching problem that Ron Graham and I worked on. It started when John Pierce, my boss, walked into my office and said: "I've invented a new system for data transmission, and everything is clear except how the messages are going to figure out where to go. See if you can figure something out." That's how that particular problem started. It turned out to be a very exciting graph-theory development. We had to consider the properties of the distance matrix of a graph. People had only studied the adjacency matrices of graphs before. The distance matrices of graphs had fantastically interesting properties that were a solution to this addressing problem that John was interested in. It has led me to some very nice mathematics. The paper that Graham and Lovacs wrote, going beyond what Ron and I had done, was one of the most beautiful mathematical papers I have ever seen—the complete analysis of distance matrices of trees and the characteristic polynomials of distance matrices of trees.

The third area that comes to mind that I have had particular fun with is the area of shortest networks, minimal trees in various matrices, the Steiner problem in various topologies, and so on. That started with us as a financial problem, as a problem of accounting and pricing that came from Long

Lines, and we have been running with that problem for twenty-five years and through a half dozen changes of what the problem was. We have learned a tremendous amount of mathematics. That for me is a good problem.

MP: *What branches of mathematics attract you most?*

Pollak: I started out as an analyst. My background is in complex variables, and I probably like complex variables better than anything else in some sense, although I have done a lot of other things. My experience is that if you're going to work on a problem in complex variables, you need large chunks of concentrated, intense time. For me, analysis requires nearly continuous work for months in order to be successful. You've got a much better chance of attacking combinatorial problems on an airplane. You don't have to keep as many fundamentals and as many angles in mind at once in combinatorics as you do in analysis.

MP: *Is that a function of the newness of combinatorics?*

Pollak: I am simply giving you an observation of my own way of working. Some people who dislike combinatorial analysis have the impression that it's an isolated bunch of tricks without a general structure. That's no longer true. But maybe the extent to which it once was true has something to do with what I just said.

MP: *You've done many things. You've worked in mathematics education to an extent which is most unusual. You've done some exciting mathematics of your own for many years. You've worked with some very interesting people. What have you gotten the most fun out of?*

Pollak: It's hard to say. I get a great kick out of whatever I am doing at the moment. When I'm involved in a mathematical discussion of a particular problem, I love it; and when I'm involved in some work in running the department at Bell Laboratories, that's fine; and when I'm involved in some work in mathematics education, that's fine. I just plunge into it and do the best I can. I'm managing to be the old fogy by now both in the Labs and in mathematics education. "I did this so long ago I have forgotten why it is wrong" is what we say about mathematicians. This is probably true of me by now. It's hard to picture yourself getting old in that way, but I got a Ph.D. very young at the age of twenty-three. By the time I was thirty, I had begun to be involved, in addition to research, in both the administration at Bell Laboratories and in mathematical education in the United States. So I really have been doing all of these things now, all together for twenty-five years, and that's a very long time to keep remembering things which for their own reasons worked or didn't work at a particular time, and that can be an awful wet blanket on other people. It's no good having somebody around who remembers that long. Dick Hamming has said: "The best thing that ever happened in the world is the burning of the Library in Alexandria, because it removed a millstone from around people's necks." They didn't have to keep going back to check to see if it had been done before. It's a great nuisance to have somebody around who knows what was done before.

MP: *Which of the things that you have done has given you the greatest sense of satisfaction?*

Pollak: In the fall of 1968, Gail Young and I were sent by the State Department to Africa to take a look at the effects of the Entebbe Project in English-speaking Africa. I remember being in a large room in King Menelik II High School in Addis Ababa in Ethiopia. That day they were showing a locally made film that was part of the local adaptation of the Entebbe Project, and by gum there on the screen was something that I had put into the SMSG Algebra course ten years earlier. In the middle of Africa, surrounded by junior high students learning English at the same time as mathematics! That was a strange experience. It represents one of the major reasons why I went into industry in the first place. I wanted to be able to see something done with my mathematics. I wanted to be able to say: Here is something done better because of some mathematics that I had found, and that has happened at the Labs in my research, but it has also happened in that funny way in that assembly hall in King Menelik II High School.

GEORGE PÓLYA

GEORGE PÓLYA

Interviewed by G. L. Alexanderson

On December 13, 1977, George Pólya, Professor Emeritus of Mathematics at Stanford University, celebrated his ninetieth birthday. To mark the event, the Department of Mathematics at Stanford gave a dinner at the Stanford Faculty Club to honor him. Many of Professor Pólya's friends and former students attended and tributes to his work were given by Dean Halsey Royden of Stanford, a former student of Pólya's; Professor Donald Knuth of the Computer Science Department at Stanford; Professor Peter Lax, Director of the Courant Institute at New York University and president-elect of the American Mathematical Society; Jerzy Neyman, Professor of Statistics at Berkeley; and Felix Bloch, Professor Emeritus of Physics at Stanford, Nobel laureate, and a former student of Professor Pólya's at the Swiss Federal Institute of Technology. Professor Gabor Szegö, his long-time co-author and colleague at Stanford, though in poor health, was also able to attend.

The audience's diversity was indicative of the wide influence of Professor Pólya on mathematics and education: research mathematicians who know Professor Pólya as a colleague and collaborator, two-year and four-year college teachers who know him for his mathematics, his work in mathematics education and from his long association with the MAA, and high-school teachers whom he taught in many teacher institutes at Stanford. Cited on this occasion were his many mathematical discoveries which as Dean Royden pointed out, will be studied by graduate students for many years to come. These discoveries span an impressive range of mathematical fields: real and complex analysis, probability, combinatorics, number theory, and geometry, among others. For these contributions he has been made a member of several national academies, including the prestigious Académie des Sciences, Paris.

Professor Neyman pointed out that Pólya's *How to Solve It* has been translated into fifteen languages and his *Mathematical Discovery* into eight. Teachers associate Pólya's name with these and his many other writings on problem solving and teaching using mathematical discovery, so that "in the Pólya style," "the Pólya method" and such are well-defined phrases to mathematics teachers everywhere. His writings have a clarity and elegance seen all too seldom in the profession, making his books and papers a great joy to read. His choice of topics in mathematics and his choice of problems show rare good taste.

A few days after the birthday dinner in his honor he was interviewed. The following are excerpts from that interview.

MP: *Looking over your mathematical work one is struck by the wide range of mathematical questions you have investigated and the many fields in which you have made significant contributions. This is not the usual record for many mathematicians, no matter how able. How did it happen? How did you find so many good problems in so many fields?*

Pólya: I was partly influenced by my teachers and by the mathematical fashion of that time. Later I was influenced by my interest in discovery. I looked at a few questions just to find out how you handle this kind of question. I was influenced also—this is farther away—because I did not come straight to mathematics. I was influenced by the tortuous way I came to mathematics.

Mathematics Is between Philosophy and Physics

MP: *Through philosophy and physics?*

Pólya: No, even more complicated. I started studying law, but this I could stand just for one semester. I couldn't stand more. Then I studied languages and literature for two years. After two years I passed an examination with the result I have a teaching certificate for Latin and Hungarian for the lower classes of the gymnasium, for kids from ten to fourteen. I never made use of this teaching certificate. And then I came to philosophy, physics, and mathematics. In fact, I came to mathematics indirectly. I was really more interested in physics and philosophy and thought about those. It is a little shortened but not quite wrong to say: I thought I am not good enough for physics and I am too good for philosophy. Mathematics is in between.

MP: *I can think of theorems of yours in probability, combinatorics, geometry, algebra, number theory, and, of course, function theory. Are there any mathematical areas in which you did not have any inclination to work?*

Pólya: Well, you didn't mention that I worked on the fringe of mathematical physics. One of my books with Szegö is on *Isoperimetric Inequalities in Mathematical Physics.* This book reflects my old interest in mathematical physics. Oh, that is enough.

MP: *It seems to me that some classical problems are being examined again with renewed vigor. For example, there are young mathematicians actively pursuing the solution of problems like the Riemann hypothesis, a problem that received a good deal of attention at the Vancouver Congress a few years ago. Would you care to comment on the direction of mathematics and mathematical tastes in the 50's and 60's, and now in the 70's?*

Pólya: Well, my comments are not to be taken seriously. First of all, I haven't studied the latest mathematics and I am biased. I was always interested in nice applications and it seems to me that, though this is perhaps not true of those who really advance mathematics, several of their followers at least are obsessed by the idea of generalization. Everything should be generalized and their ideal seems to be a mathematical theorem of perfect generality, of such perfect generality that no particular consequence of it can be derived.

Hermann Weyl

MP: *Some years ago Hermann Weyl and you made a famous wager on questions raised by Brouwer's ideas, so we know you are a betting man. Would you care to quote any odds on the proving of the Riemann hypothesis over the next n years?*

Pólya: I am not a betting man. On the contrary, I am rather cautious, but if I have to make a bet I would say, no proof in the next ten years. I know several people, very good people, who work on the Riemann hypothesis, but I would still bet—no proof in the next ten years . . . if I have to make a bet.

The Pólya–Weyl Wager

Between G. Pólya and H. Weyl a bet is hereby made, according to the specifications below.

Concerning the following theorems of contemporary mathematics :

(1) Every bounded set of numbers has a precise upper bound .

(2) Every infinite set of numbers has a countable subset .

Weyl predicts:

A. Within 20 years (that is, by the end of 1937), Pólya himself, or a majority of the leading mathematicians, will admit that the concepts of number, set, and countability, which are involved in these theorems and upon which we today commonly depend, are completely vague; and that there is no more use in asking after the truth or falsity of these theorems in their currently accepted sense than there is in considering the truth of the main assertions of Hegel's natural philosophy.

B. It will be recognized by Pólya himself, or by a majority of the leading mathematicians, that, in any wording, theorems (1) and (2) are false, according to any reasonable possible clear interpretation (either distinct such interpretations will be under discussion, or agreement will already have been reached); or that if it comes to pass within the allotted time that a clear interpretation of these theorems is found such that at least one of them is true, then there will have been a creative achievement through which the foundation of mathematics will have taken a new and original turn, and the concepts of number and set will have acquired meanings which we today cannot anticipate.

Weyl wins if the prediction is fulfilled; otherwise, Pólya wins.

If at the end of the allotted time they cannot agree who has won, then the Professors of mathematics (excluding the bettors) at the E.T.H. and at the Universities of Zürich, Göttingen, and Berlin, will be called to sit in judgement; which judgement is to be reached by majority; and in case of a tie, the bet is to be regarded as undecided.

The losing party takes it upon himself to publish, in the Jahresberichten der Deutschen Mathematiker-Vereinigung at his own expense, the conditions of the bet, and the fact that he lost.

Zürich, February 9, 1918
H. Weyl and G. Pólya

George Pólya

Pólya in the classroom urging students to "Guess and Test" when doing mathematics. Pólya, through his writing and teaching, has been a powerful influence on three generations of teachers of mathematics. In his ninety-seventh year, he remains active and just published his 250th paper.

Mathematical Influences

MP: *Which mathematician's work influenced you most? Why? Among your teachers, who influenced you most? Why?*

Pólya: Well, if I have to, I have to name four names. I was greatly influenced by Fejér, as were all Hungarian mathematicians of my generation, and, in fact, once or twice in smaller matters I collaborated with Fejér. In one or two papers of his I have remarks and he made remarks in one or two papers of mine, but it was not really a very deep influence. I was much more deeply influenced by Hurwitz. In fact I went to Zürich in order to be near Hurwitz and we were in close touch for about six years, from my arrival in Zürich in 1914 to his passing in, I think, 1919. And we have one joint paper, but that is not the whole extent. I was very much impressed by him and edited his works. I was also impressed by his manuscripts. And then there were Hardy and Littlewood; that is laid down in our book. Hardy had a very great personal influence on me. But the longest and closest collaboration is with Szegö. I wrote together with him two books and a few papers, and it was a very close collaboration. Our interests were sufficiently similar and sufficiently different. We were interested in the same questions, but about some questions he knew more answers, about other questions I knew more answers. We completed each other, and that through books and papers and over many, many years.

MP: *What about earlier mathematicians, say seventeenth- or eighteenth-century mathematicians?*

Pólya: Yes, of course, among old mathematicians, I was most influenced by Euler and mostly because Euler did something that no other great mathematician of his stature did. He explained how he found his results and I was deeply interested in that. It has to do with my interest in problem solving. I don't know all the works of Euler. I know a very small fraction. His works fill around seventy volumes of which I read only a small fraction. I am not a good reader. A few of his works I studied very intensively.

The Art of Problem Solving

MP: *That brings up my next question. How did you become interested in heuristics and the art of problem solving? Did anyone or any event influence you on this?*

Pólya: Well, I think I wrote it somewhere. In one of my books it is mentioned. I came very late to mathematics. I had an interest in biology, literature and philosophy. And as I came to mathematics and learned something of it, I thought: Well, it is so, I see, the proof seems to be conclusive, but how can people find such results? My difficulty in understanding mathematics: How was it discovered? And then I was deeply influenced by some books. I wish to mention just two. One was the book of Ernst Mach on the history of mechanics. For me personally this was the most beautiful book I read. I read it at the right time because I knew a little physics, but just a little. I was just right for it. His main theme is: You cannot understand a theory unless you know how it was discovered. His best book and best-known book is on mechanics, but he wrote also other books, on the theory of heat and still others. But that was the main idea: In order to understand a theory really, you must know how it was discovered. So he came to heuristics. In fact, in some of his other books there are a few direct remarks on problem solving. Then I thought about it and I came across the *Regulae* of Descartes, which is really a book on problem solving. That is not mentioned in any history of philosophy, because those historians who wrote about him didn't know about problem solving. My interest in literature contributed a little. When I was interested in literature, I was most interested in books of Hippolyte Taine and he wrote about literature in a quasi-scientific way. How in such a vague subject you can bring in something that approximates science, that deeply impressed me. It also contributed to my interest in heuristics. It is essentially a vague question, and that you can introduce something which has something to do with science, that I think I learned from Taine. I was also impressed by his style.

MP: *How did you meet Gabor Szegö and how did the* Aufgaben und Lehrsätze* come to be?*

Pólya: Oh, I am seven or eight years older, you see. I had already my Ph.D. In fact I had a stipend to study abroad and then I met him when he was a student. We had the same interests. There was some special question: I had some conjecture about Fourier coefficients, you see, and he proved that.

MP: *When was that?*

Pólya: I cannot tell you exactly; that was around 1913.

MP: *And where?*

Pólya: In Budapest. I was most of the time abroad, but when I went back to Budapest I visited the University. He was still a student when I met him and his future wife.

MP: *What was the reaction to those books when they first appeared? They were different from the usual.*

Pólya: Yes, they were different. It was a good reaction. Perhaps it was not sufficiently emphasized that they were different, that is, a *novum*. There is something essentially new in them: the problems are put together, they are classified, not according to topics but according to the method of solution. This was not recognized right away. But there was a good reaction from various sides.

MP: *That came out in 1925?*

Pólya: Something like that.

---

* *Problems and Theorems in Analysis, I and II*, Springer-Verlag, 1972, 1976.

The young Pólya and Szegö (left) in Berlin delivering the manuscript of the famous _Problems and Theorems in Analysis_ to the publisher, Springer Verlag. Today (1984), more than sixty years later, it still is in print.

How to Solve It

MP: *With* How to Solve It *you explored problem solving in a general way but it was with* Mathematical Discovery *that you gave us a book with many examples of problems from elementary mathematics demonstrating various patterns and principles. Earlier books had dealt with more sophisticated topics.*

Pólya: No, *How to Solve It* makes the fewest demands on the knowledge of the reader. A reader who knows very little mathematics can read it with some interest. Perhaps he will miss a few points.

MP: *But certainly the* Aufgaben und Lehrsätze *and* Mathematics and Plausible Reasoning *are more sophisticated mathematically than* Mathematical Discovery.

Pólya: Oh, yes, sure. *Mathematical Discovery* is between the two others and *How to Solve It*, and I think for the high-school level this is the most useful book. You see, there are detailed examples the teacher can, almost without change, use in his or her class. I think that is the most useful at the high-school level.

MP: *What I would like to ask about that, then, is: When did you develop an interest in teaching at the earlier levels and was this in any way related to your work with secondary school teachers in the teacher institute programs.*

Pólya: No, I was interested in it fairly early. I have also a teaching certificate, as I told you, for the lower classes, in Latin and Hungarian. In fact I have a teaching certificate for all classes of the gymnasium in mathematics, physics, and even philosophy, and in order to get the certificate you must

Pólya in his study.

be a practice teacher. So I taught in the high-school level. They call it gymnasium, for kids between ten and eighteen, but it is different in several respects from a high school. I was practice teacher for a year. So I had experience. I must tell you, *How to Solve It* was written really twice. I wrote something, a draft, in German while I was still in Zurich. Then I came to America and in this respect, my coming to America was, I think, useful, because here, in this country, there is more interest in the "How to" books. And, by the way, Hardy predicted it to me. When I told him about the "How to" book, he said, "Oh, you must go to America." And then I rewrote it, it appeared in final form in English. It is considerably different from my original German version, and I think to its advantage.

MP: *Are there any directions you would like to see mathematics teaching take over the next few years?*

Pólya: Well, it is going already away from the "new math" and it should go farther away. A French author, I think René Thom, called new math "a philosophical and pedagogical error." I don't question the good will of the promoters of the "new math" but I think they are in error.

MP: *What have you enjoyed most? Teaching? Research? Writing? Speaking?*

Pólya: I don't know. I like all of them. Like you ask the little kid, "Whom do you like more, Mommy or Daddy?" He answers, "Both."

MP: *Are there books and papers you would like to have written and haven't gotten around to yet?*

Pólya: Well, I would like to write still an epilogue to my work—it would be along the lines of this interview—but I didn't find a sufficiently good plan for it.

MINA REES

MINA REES

Interviewed by Rosamond Dana & Peter J. Hilton

Mina Spiegel Rees was born in Cleveland, Ohio, on August 2, 1902. She went to New York City public grammar schools, Hunter High School, and Hunter College, where she began studying mathematics. In 1923 she graduated summa cum laude, having been editor of the yearbook, president of the Student Council, and a member of Pi Mu Epsilon and Phi Beta Kappa. She received an M.A. from Columbia University in 1925 and a Ph.D. from the University of Chicago in 1931, with a thesis written under the supervision of Leonard Eugene Dickson. She taught mathematics at Hunter College until 1943, when she was granted a leave of absence to serve as technical aide and executive assistant to the chief of the Applied Mathematics Panel, National Defense Research Committee, Office of Scientific Research and Development. For this work she was honored by both the British government (King's Medal for Service in the Cause of Freedom, 1948) and the U.S. government (President's Certificate of Merit, 1948). After the war she was invited by the U.S. Navy to establish the mathematics research program in the newly created Office of Naval Research. She served as head of the Mathematics Branch (1946 to 1949), director of the Mathematical Sciences Division (1949 to 1952), and deputy science director (1952 to 1953).

In 1953 Mina Rees returned to Hunter College as dean of the faculty. In 1961 she became dean of graduate studies of the newly created City University of New York, established the Graduate Center of CUNY, and became in succession its dean of graduate studies (1961 to 1968), provost (1968 to 1969), and president (1969 to 1972). Since her retirement in 1972 she has been active as a member of several boards concerned with application of research to social problems. She is a member of the visiting committees of two universities and works with various foundations concerned with improving the effectiveness of the educational establishment.

In 1962 Mina Rees was the first recipient of the Mathematical Association of America's Award for Distinguished Service to Mathematics. She was a member of the National Science Board (1964 to 1970) and since 1973 has been on the board of directors of the Society for Industrial and Applied Mathematics's Institute for Mathematics and Society (SIMS). She has been president (1971) and chairman of the board (1972) of the American Association for the Advancement of Science. She has been granted numerous honorary degrees and awards, including, in April 1983, the National Academy of Sciences Public Welfare Medal. The award, one of the most prestigious honors the academy can bestow, was made for her contributions to the scientific enterprise, especially in mathematics and computer sciences, since World War II.

In 1955 Mina Rees married Leopold Brahdy, a physician who died in 1977. She lives on the East Side of Manhattan and is actively interested in the arts, including dance, painting, music and literature. She is an accomplished painter herself (for a time she studied in Mexico every year and nowadays goes to Maine for the summer and paints at the shore).

A Woman in Mathematics

MP: *You are very much a pioneer as a woman in mathematics—in leading the way and becoming an eminent mathematician yourself and showing that there is absolutely no contradiction between being a woman and being a mathematician.*

Rees: I'm afraid I'm *not* an eminent mathematician!

MP: *But you are eminent!*

Rees: Maybe I could be called eminent, but I am certainly not an eminent *mathematician*! There are some women who are doing high-quality mathematical research, but I'm not.

MP: *There are women in mathematics today, but there is still a very, very widespread feeling that it is somehow or other strange and unusual for a woman to take up mathematics—that maybe a woman who takes up mathematics isn't quite as female as one would wish her to be. All these prejudices exist in the minds of young people.*

Rees: That's important, of course. Julia Robinson has been acknowledged to be the most significantly productive mathematician among the women. One reason I was glad she got the MacArthur award was that I thought that this might make the public realize that there are women who really are significant mathematicians. She has had all the stigmata of eminence now.

MP: *Did you encounter any difficulties because you were a woman as you studied mathematics?*

Rees: Yes, I did, but not in the beginning at all. When I was in college—and this is still true to a lesser extent—there was a group of colleges in the East that were colleges for women: they included Hunter, where I went, the "Seven Sisters" (Smith, Wellesley, Radcliffe, Mount Holyoke, Vassar, Bryn Mawr, and Barnard), and some others. Now, by definition, a college has a mathematics department, so at a college for women, you have to have women in mathematics! At those colleges, women were not discouraged from studying mathematics. In fact, it was one of the most popular majors at Hunter when I was there. Later, after I had my Ph.D. and was teaching, I knew virtually all the women Ph.D's on the East Coast, and they were teaching at these colleges; that's where women Ph.D.'s made their careers. So I didn't meet any discouragement at all when I was going into math. Indeed, I didn't know it was a peculiar thing to do. I did what everybody did: I picked the field that I found most interesting and decided to major in it. It never occurred to me that there was anything the matter with that.

After I had my bachelor's degree, I studied at Columbia. When I had taken four of their six-credit graduate courses in mathematics and was beginning to think about a thesis, the word was conveyed to me—no official ever told me this, but I learned—that the Columbia mathematics department was really not interested in having women candidates for Ph.D.'s. This was a very unpleasant shock. Of course, this is certainly not at all true of the mathematics department at Columbia now.

I decided to switch to Teacher's College and take the remaining courses necessary for an M.A. there. A few years later, after I'd saved enough money, I went to Chicago. That was the only episode that raised a question about the appropriateness of mathematics as a field for women before I had my Ph.D. It was a really traumatic affair for me. But, in fact, Columbia served me well because by that time I had studied algebra using Leonard Eugene Dickson's book, and I knew that I wanted to work with him. So the change to Chicago, where he was Distinguished Professor of Mathematics, was really a welcome one.

MP: *How did you get interested in mathematics?*

Rees: I had had a regular course at Hunter High School with no emphasis on anything in particular. I was a very good student, valedictorian of my class, and I greatly enjoyed math, which we studied for four years. When I went to Hunter College, I found that the mathematics department was where I wanted to be. It wasn't because of its practical uses at all; it was because it was such fun!

MP: *Did you have any special teachers in high school?*

Rees: No. The teachers were women who had been educated at good colleges, knew what they had learned originally, and continued to teach it. It was a classical girls' high school that didn't try to do any fancy business. But I always had such a good time in math class that that's what I wanted to do.

MP: *How did your family feel?*

Rees: They didn't have the slightest opinion on the subject. They wanted me to be happy in whatever I wanted to do. They never would have thought of intervening.

MP: *What about your peers at that time?*

Rees: A lot of them also went into mathematics. Don't forget, we were in a girls' high school and a girls' college. Here is an interesting note. When I was in the eighth grade in a public elementary school, I had marvelous teachers. One day one of my teachers, a man who was my current hero, said to me, "Mina, you know, I think you ought to take the admission test for Hunter High School tomorrow morning." I said, "Hunter High School?" I had never heard of Hunter High School, but if he thought I ought to take the test, I would go down and take the test. I passed the test and was admitted. It was a superior school for gifted girls, and of course there was a mathematics department. And all the girls studied mathematics.

MP: *Did you have brothers and sisters?*

Rees: I was the youngest of two girls and three boys. The youngest brother, who was closest to me, was a businessman and engineer and was gifted in that kind of thing, but he was a very practical-minded soul. The other brothers were not mathematicians, at all. We were a family who didn't intervene in one another's lives. I certainly don't have any impression that anyone in my family really was concerned with my choice of subjects.

MP: *Some people seem to hold the view—almost as a neurophysiological view—that there is some difference between the capacity of men and women to be mathematicians, quite distinct from any environmental influences. Do you have any views on that?*

Mina Rees and her husband, Dr. Leopold Bradhy, on the steps at Delphi in 1963.

Rees: I don't believe that is true. The answer probably requires a long wait and then a careful study. Until the environmental influences have changed, I think it will be impossible to get hold of it. I was recently talking to Cathleen Morawetz about Julia Robinson and my hope that Julia's being singled out as much as she has been recently would make her a kind of model for girls who are considering mathematics as a career. I said, "I hope that young girls will see that it is possible to be a serious mathematician—to be what a mathematician calls a mathematician." Cathleen said, "Mina, I don't think girls are prepared to give the kind of single-minded devotion that you need if you are going to be a great mathematician." That, I think, is an entirely different question from the one you asked, except that it does go back to social influences. The "eye on marriage" is the background thing. In my family, that wasn't present. There was absolutely no parental pressure for me to marry. Each of us did what he or she pleased. That was not typical of American society, I think.

Mina Rees

Mathematicians and Public Policy

MP: *You have been and continue to be involved to an impressive extent in many significant ways with policy making. It is my impression that in this respect you are most unusual. That is, mathematicians generally confine their activites much more closely to the world of mathematics. Have you found that being a mathematician has been useful to you as a person in public affairs?*

Rees: To a certain extent. I refer to activities in an academic environment, and public affairs does embrace that to a certain extent. In dealing with academics, it is absolutely superb to be able to say you're a mathematician! Nobody dares to say mathematics is not important or not significant. I have always found it was an advantage when I was dean or president of a college. No discipline surpasses mathematics in purely academic prestige. But in general being a mathematician is useful chiefly because it is highly correlated with a well-organized mind.

MP: *Can you say that the discipline of mathematics, the mathematical way of thinking, is in some way helpful and relevant when you're considering issues of policy.*

Rees: Unquestionably, but in the same sense that lawyers have certain advantages when they're considering issues. If your habit is to organize things in a certain way, the way a mathematician does, it is clear I think that you are apt to have an organization that is easier to present and explain. I find that there are occasions when that irritates the people I'm dealing with. I can remember someone once saying to me, "Don't push me against the wall!" I had to learn to be less aggressive in stating what seemed to me to be the logic of a situation.

MP: *You learned to ally your logical talents with diplomacy.*

Rees: That's it.

MP: *Would you like to see more mathematicians involved in policy issues the way you have involved yourself?*

Rees: I feel that my involvement in this kind of question has been so time consuming that I haven't had as much time as I would have liked to do mathematics. There was a period of time when I concentrated on mathematics and participated a little in policy making. Then I switched and spent a lot of time on policy making and didn't do very much mathematics. I think that has been a deprivation for me; I have lost touch with things I cared about in mathematics. So I don't know that it's something one should urge upon people, for there's a real choice that has to be made. I don't think you can do both very well. I suppose people who are more gifted in mathematics than I am might be able to do both.

I think it would be helpful if there were more mathematicians who did play a public role in order to secure the continuing support of mathematics. When I was in Washington, I felt that the most important thing I did was to see to it that mathematics got its share of support, chiefly by repeatedly demonstrating its achievements. I got the support only by being on hand all the time and being watchful. It doesn't come automatically.

MP: *Do you regret making the choice you made? What have you lost?*

Rees: A possible mathematical career. I'm not sure that I would have had it, but I might have. You can't do everything, though. I just lost track of things that I cared quite a lot about and finally gave up because I wasn't on top of things anymore.

MP: *You couldn't have switched back?*

Rees: I don't think it's possible once you have left mathematics to get back into it. You're just lost.

MP: *It's bad enough to take on a chairmanship of a department for three years. When, as in your case, you really devote yourself to academic administration and then to policy making in the larger context, it must be terribly difficult. The subject moves so fast that it gets away from you.*

Rees: You just really have to make that kind of choice.

MP: *Earlier, while talking about Julia Robinson, you mentioned the "stigmata of excellence." What did you mean by that in relation to her and in relation to your own life?*

Rees: I was referring to the things that happen in mathematics. You're elected president of the American Mathematical Society, you're elected to the National Academy of Sciences, you give colloquium lectures of the Society and so forth. There are certain things that happen to somebody when *he*—in all other cases—has been identified as one of the *real* mathematicians of the time. None of these are relevant to me. Julia is the only woman who has been elected in mathematics to the National Academy of Sciences, and only a small number of women have been elected at all.

MP: *Do you have any advice for young women mathematicians on how to deal with men with these strange attitudes?*

Rees: I want them to want to be mathematicians if they like mathematics! That's the thing that troubles me. I don't care whether or not there are a lot of women mathematicians, but if other girls have the same motivations I had—and I'm sure they do because there were so many of us at that time —I wish they wouldn't have irrelevant things interfering with their desires. It may well be that in our society a woman must sacrifice too much if she wants a career that's offbeat. It didn't seem like a sacrifice to me because I had plenty of men friends, and I wasn't prepared to get married; that wasn't what I wanted to do.

Mina Rees with Ed Begle, 1960.

MP: *When did you get married?*

Rees: Not until much later—when I came back from Washington.

MP: *You've mentioned two choices now. You chose the public policy over the mathematics, and earlier you chose the mathematics.*

Rees: All this depends, I think, so much on the social milieu. I know that in Jewish families, girls are expected to marry and are sort of driven to consider marriage their first obligation. I saw this happen so much when I was dean at Hunter.

Marriage was first on the agenda, and there really wasn't much option for those girls. Really, the only thoroughly successful women mathematicians that I know are married to men mathematicians. I used to tell students at Hunter that graduate scholarships and fellowships at universities were a fine idea because universities were great places to meet the right man. If you can meet somebody in the same discipline, you're apt to be well taken care of later on.

MP: *In another interview, you mentioned that when you were at Chicago you found a lot of women there had come from the South—where they were certainly steered in traditional directions of marriage and family—to study mathematics, and at Chicago they overcame that influence and blossomed into excellent mathematicians. Can you expand on that?*

Rees: Chicago was a mecca for Southerners—and not just in mathematics. I lived at the women's graduate dormitory, which had people in all disciplines, and about half of the women there were Southerners. At that point, there weren't many real good universities in the South; there were some pretty good ones, but nothing like Chicago. It was fairly easy to come up the Mississippi to Chicago. Not all of the women were there for doctor's degrees; most of them were there for master's degrees. Many of them came in the summer—almost all were teachers. Now these girls were teachers in high schools and in a few girls' colleges in the South. The typical summer visitors were from the high schools, and the ones who stayed for the winter were from the colleges.

There never was any discrimination that I was aware of; I don't know what the faculty said to one another over at the Quadrangle Club, but I know that you couldn't feel any discrimination in the classroom.

MP: *What about in terms of marriage? They got to Chicago and no longer were under family pressures.*

Rees: There were a number of marriages; I don't believe family pressure was needed. Virginia MacShane was a mathematics student there; she married Jimmy MacShane. She didn't become a mathematician, but she found her husband there! The girl who was the secretary of the mathematics department—Tony Killien—married Ralph Huston there. They got their degrees considerably later than I did, so I wasn't there for the marriage. Tony didn't use her degree until after Ralph died; then she took an appointment in the department at Rensselaer where he'd been teaching. It is hard to name a woman mathematician who isn't married to a man mathematician, isn't it? I know of a couple in Minnesota, and there's Olga Taussky-Todd and Jack Todd, the Lehmers, the Stones; Cathleen Morawetz is an exception. She is married to a chemist—and an artist who likes my watercolors!

MP: *What did your husband do?*

Rees: He was a physician. One night he gave me a passage from George Sarton's *A History of Science*: "The mathematical and the medical minds are, if not antagonistic, at least very different and sometimes poles asunder."

Awards and Honors

MP: *You have an enormously impressive list of awards. First, your most recent award is the Public Welfare Medal given by the National Academy of Sciences. Can you tell us something about it?*

Rees: When the award was established in 1909, George F. Becker, who proposed it, wrote:

"Patriotism and justice alike demand that certain important public services involving the applications of science should receive a conspicuous recognition for which in this country there is no provision . . . . Such services in the application of science might however be recognized by the Academy in a manner that would be keenly appreciated by the men[sic] honored and which would command the respectful consideration of the whole country . . . . To accomplish these ends I propose that the Academy confer a medal for eminence in the application of science to the public welfare."

The reward is intended to focus on the benefit to the country of the development of science and its application. The award brings an "honorary" membership in the National Academy of Sciences; it gives you all the privileges of the academy except electing members.

MP: *To go back to one of your earliest awards, you were the first person to receive the Award for Distinguished Service to Mathematics given by the Mathematical Association of America, in January 1962. Can you tell us about that award?*

Rees: Of course, I was not privy to the discussions that led to the award. Indeed, I didn't even know it was going to exist. I think that it was largely a recognition of my work with the Office of Naval Research and my continuing work after I left ONR with mathematical organizations and with the National Bureau of Standards. When I was with ONR, practically all the mathematicians—and they were fewer in number in those days—felt that ONR had done a fantastic job of reestablishing and strengthening their position in universities after the war, as well as raising the general level of mathematical activity in the nation. Both the American Mathematical Society and the Institute of Mathematical Statistics adopted resolutions of appreciation for my work in ONR when I left Washington.

I think I ought to tell you a story that is related to this. When I went to Washington after the war, I could get no one to put my name on a waiting list for an apartment. I was confronted with the very real problem of where to live. I shopped around Washington and discovered a hotel that had been one of the great hotels: the Lafayette near Lafayette Park. They would let me live there for two weeks; then I had to go away for a week, but after that they would take me back. And I could park my things there while I was away. In making a virtue of necessity, every time I had to leave for a week I made a trip and visited leading departments of mathematics. I saw virtually all the leaders of American mathematics—a great number of the people who were widely respected in American mathematics— and consulted with them about how ONR could be useful, what it could do that would really help mathematics. I felt that this was an extraordinarily happy accident because I never would have taken the time to make all those trips if I hadn't been thrown out of Washington. It really meant that when I finally came back to the Navy and said, "This is the program I want," it was one that I had discussed with most of the senior people who were going to be affected by it. I had been worried that mathematicians would be suspicious of a military organization like the Navy supporting mathematics, but virtually everyone had been won over to accepting a contract with the Navy to support his own research and that of his students. ONR's research program proved to be an extremely effective operation. In addition to supporting faculty research, it gave support to young students who were getting their doctorates. It gave secretarial help, it gave travel opportunities; it helped *Mathematical Reviews*; it did most of the things that mathematicians felt had to be done. I think that virtually all mathematicians felt that this was an extraordinarily helpful program that had really saved mathematics and advanced it and strengthened it. Of course, America became very strong in mathematics in that period.

Mathematicians during World War II

MP: *Since you referred to the Office of Naval Research in those days shortly after World War II, I'd like to ask you about your own knowledge and experience of mathematics during the war. I know this is a delicate subject. When we compare what you wrote in the* American Mathematical Monthly[1] *with what Barkley Rosser wrote,[2] we see that you are more respectful of the quality of mathematics that emerged from the war than he is. Can you elaborate on some of the sources of good mathematics during World War II?*

Rees: Rosser's report was largely concerned with the work of mathematicians who were working within the military establishment. Of course, there were many mathematicians doing that. It is much more likely that mathematicians in that environment will be asked to do mathematics to solve

---

[1] "The Mathematical Sciences and World War II," *American Mathematical Monthly*, 87, October 1980, 607–621.
[2] "Mathematics and Mathematicians in World War II," *Notices of the American Mathematical Society*, 29, October 1982, 509–515.

immediate problems and will be expected to stick pretty close to the problem to be solved. Much of the work of the Applied Mathematics Panel was also of the kind Rosser reported on. But often a substantial background in a less familiar subdiscipline, or the mathematical way of thinking you talked of earlier, would give our people a quick solution to what might have been a difficult problem. For example, in the article you refer to, I mention our use of an 1887 result of Lord Kelvin to determine the speed and turning characteristics of Japanese ships in World War II. Moreover, the AMP was organized at universities. Therefore, direction was always by a university head of the project, and there was the kind of atmosphere that a mathematician would feel it was important to have if he was to do his best work. For example, some work of extraordinary importance immediately after the war—and now—came out of sampling inspection. The origin of that whole project was a luncheon conversation between two academics who were very thoughtful economists expert in the application of statistics. One has since won the Nobel Prize for economics and the other is at the moment an Undersecretary of State for Economic Affairs. These two men were puzzled because a Navy captain with whom they had discussed the problem of destructive sampling of munitions had said that he didn't see why he had to destroy so much of the evidence—that there ought to be a way whereby, after a while, an experimenter, like a savvy captain, would know that this was a good batch or a bad one and stop sampling. The academics thought that made sense and did what a university person is accustomed to do: they asked themselves questions that were directed to the point the captain raised, and this led them to formulate a tentative solution. Since they were unable to get anywhere with it because it required very special mathematical skills, they took it to an outstanding mathematical statistician who was associated with them—Abraham Wald, an emigré from Hitler's Europe. He was intrigued by the problem and solved it by developing a new technique in statistics that is now one of great importance: sequential analysis. The essential character of the new technique was determined in a few days, and the method was actually put to use during World War II.

University people can act like that. It's very hard to act that way when you're sitting in a military establishment in wartime. I think that's a partial explanation of what happened. Another one is that some of the people who headed our groups believed it was important to develop new theory as well as to solve problems. One of these was Richard Courant. He and Kurt Friedrichs, who came from Göttingen together and often worked together, developed a part of the shock-wave manual during the war on the basis of war-related problems. This manual has become important in aerodynamic developments. They wrote the basic first book on it in wartime. That would probably not have been done in a military establishment. Courant always told me he couldn't just do problems; he had to develop theory. Of course, he did problems, too. That's the point: in the war you just worked day and night; you did everything.

MP: *In the field of cryptology, which is still so sensitive, of course, some very interesting mathematics was done, and perhaps one day you will be able to talk about it.*

Rees: We had an interesting experience in that field, too, as part of the work on ONR after the war. In the course of my stay with ONR in Washington, the naval security group that became part of the National Security Agency decided that they had some problems that obviously needed mathematical analysis beyond what they could do in the course of their regular work. They asked me if I would set up a summer research project with people cleared to work on segmented aspects of the work. They formulated the problem to be pretty far removed from the immediate uses. We had an extraordinary group at UCLA including Barkley Rosser and Adrian Albert.

It was very amusing to me that Adrian always attacked everything via matrices—and other people attacked everything by their favorite form of mathematics. Anyway, he got going on this problem, and we began studying an approach to its solution through finite projective planes. At the end of the summer we hadn't solved our problem. One night some months later, my husband and I were sitting here having dinner when the phone rang. It was Adrian, telling me that a young colleague of his had solved the problem. But it was a completely abstract problem that had no connection with NSA; it was just a mathematical problem that was very important in its field. In the cryptology field, lots was going on, at least in the parts I became acquainted with, much was nontrivial. We can't be too specific, except we can say that there was some very exciting mathematics, much of which could not be published in the form in which it was originally done because of the context but did lead to publications after the war in which the work was stripped of its security context.

Teaching versus Research

MP: *Morris Kline, in some of the books he has published, has very strongly argued that there is a fundamental antagonism between teaching and research. That is, the research mathematician is impatient with the restraints imposed by the necessity of giving undergraduate courses and therefore will tend to neglect his teaching duties or will only teach what he is particularly interested in as a mathematician. I believe the opposite: that, as a generalization, if you are going for the best teacher, the only way to do it is to go for the best mathematician, and that means the mathematician interested in research. From all your years of research, how do you feel about this very vexatious question, which has undoubtedly been discussed in so many mathematics departments?*

A Russian Cathedral near Nice as drawn by Mina Rees.

Rees: Essentially, my position is with you. I am impatient with the other point of view. On the other hand, there are unquestionably very good research mathematicians who simply cannot communicate at all. The only time they are really effective teachers is when they are talking about the things they're doing in their research. They are effective with students who want to work in the field that they're specialists in. There are such people, and it's silly to close your eyes to the fact: those people cannot teach typical freshman! There are also people, and I think this is true very extensively in mathematics, who do not have the gift of original insight—of really original thought—but who nonetheless are devoted to mathematics and work very hard and are up to date with what other people are doing. I think that those people often *can* teach effectively. To get the idea that you can learn mathematics and then teach what you have learned at the university for the rest of your life (which I think has happened in many instances) is the really dangerous component there. I think what Morris says has a tendency to lead people to think that that's what you do: you get a Ph.D., maybe, and then you never pay any attention to new mathematics; if you do this, you can't teach! It's a fraud on the students. I'm sure that those who have no research program must keep up with the work of others, at least in the fields they are teaching. There are lots of very effective teachers who really work very hard to keep in touch with the way mathematics is growing, but who could not help make it grow.

MP: *It's only fair to say that Morris Kline does refer to the scholar.*

Rees: That's essentially what I'm saying.

MP: *I entirely agree that if you can find those people, they are wonderful teachers, but how do you find them and how do you produce them?*

Rees: I think I've known some.

MP: *Oh, yes, they exist, but I don't know any form of education or training that is designed to produce the scholar as distinct from the research mathematician.*

Rees: We had a long period of time when we considered a Doctor of Arts degree, and we had argument after argument about that. We would decide to go that route and then decide not to. Some institutions tried it, but then we didn't need any teachers for a while. And it's all folded. That was an attempt in the direction of developing scholars. I always felt it was a bad decision. I think it will fail unless it does build in this desire to keep up with the growth of mathematics, and I don't think that was put into most of the programs.

Mathematics and Aesthetics

MP: *In a lovely article you wrote called "The Nature of Mathematics" (Science, October 5, 1962), you say several things, one of which is enormously appealing: "Mathematics is both inductive and deductive, needing, like poetry, persons who are creative and have a sense of the beautiful for its surest progress." Do you think that we should, in teaching mathematics, explicitly try to make our students sensitive to the beauty of the subject, not simply to its vast importance?*

Rees: Yes. It seems to be so essential to the nature of mathematics. The whole emphasis on elegant demonstration, which every mathematician feels, would be lost if one were not aware of that. That is the essense of the mathematical approach to understanding. I get so mad at people who talk about mathematicians as practical people!

MP: *Do you see any risk that the computer is going to play so strong a role in the lives of our students and in their motivations for studying mathematics that they may lose some of their sensitivity to that aesthetic aspect?*

Rees: I may be quite wrong about this, but my hunch is that the number of people caught up by computers will be vastly more than the number of people who become mathematicians. I think that most of the people who are driven toward mathematics are *driven* toward mathematics, and they are not going to be diverted by the practical aspects of computers. They may be absorbed by the mathematical problems in computing or by the use of computers in mathematics—that's quite different. But there will be many, many more people using computers for practical purposes than would ever have gone into mathematics. The nature of mathematics requires this devotion, this essential quality, in my judgment, so I don't believe that's going to be changed. But the problems that are worked on may incorporate some of the things that are in computing. As we know, there are many important problems now that are deeply mathematical and must be handled by computers. Of course, it was that aspect that I didn't see clearly years ago. Although there are many aspects that I saw that I think were not obvious, that made me feel that we had to have a solid mathematical development while the construction of computers was going on. In an article I wrote for the *Annals of the History of Computing* on the computing program of ONR (Volume 4, Number 2, April 1982, pp. 102–120), the important section was the part on mathematics, but most people thought I was talking only about computers!

MP: *What are some of the less obvious aspects that you saw?*

Rees: The most important point was the need to revive among mathematicians active research interest in numerical analysis that had long been dormant. This need was critical because as soon as computers became operational, a whole new set of questions would be faced. Familiar problems like the solution of simultaneous linear equations would require a machine-oriented approach. Even Gaussian elimination was not well understood from this point of view. The Office of Naval Research established an extensive program to begin to meet this need.

There were other questions of a quite different nature. In the *Annals* article, I mentioned the airplane-reservation system, for instance. We saw all these vast uses of computers that lots of professionals pooh-poohed. It seemed pretty obvious that this was something computers could do. Also, the problems of banks and insurance companies—we had those problems identified as appropriate for computers long before most of the potential users were interested. Again, we were just looking at the types of things you had to do—not any details. In some problems, like those involved in military uses, some of the engineering problems turned out to be extremely difficult, of course.

MP: *Many people think that mathematicians are practical, dull, single-minded researchers. But your life certainly contradicts that. You are interested in music, dance, art. Here we sit surrounded by paintings, including some you've done yourself. Have you always painted?*

Rees: I always had so many other things to do that I didn't get to it. But during World War II, we had no vacations at all, and after the European victory, a friend and I decided to take a two-week vacation. We went to the Museum of Modern Art and enrolled in a course called "Understanding Painting through Painting," with a textbook called *Get in There and Paint*. During the first class I painted New York Harbor. Then we went up to Maine, complete with paintboxes, and my friend and I painted in oils for two weeks. It was just wonderful, and I've done it ever since.

A recent drawing by Mina Rees of a Maine vacation spot.

CONSTANCE REID

CONSTANCE REID

Interviewed by G. L. Alexanderson

Constance Reid was not trained as a mathematician, but she is well known for her books on mathematics and mathematicians. These include *From Zero to Infinity* and *A Long Way From Euclid* (published by Thomas Y. Crowell) and *Hilbert, Courant in Göttingen and New York*, and *Neyman—From Life* (published by Springer-Verlag).

She is, however, a member of a mathematical family. Her brother-in-law, R. M. Robinson, is professor emeritus of mathematics at Berkeley; and her sister, Julia Robinson, also a professor of mathematics at Berkeley, is the only woman mathematician to have been elected to the National Academy of Sciences.

Mrs. Reid majored in English at San Diego State University, received a master's degree at the University of California at Berkeley, and taught in the San Diego schools until her marriage in 1950.

Since the publication of her book on David Hilbert, she has been invited to speak at a number of mathematical meetings.

The Question That Everyone Asks

MP: *I would like to begin, if you don't mind, by asking the question that everyone asks: How did a nonmathematician come to write the life of the great mathematician David Hilbert?*

Reid: Maybe I should first tell you how I came to write about mathematics at all. I had given up teaching when I got married and had begun to do free-lance writing. One day my sister told me about a number theory program that my brother-in-law had run on a computer. At that time—it was 1951 or 1952—computers were very, very new. I was fascinated by the idea that one of them was being used to answer a question about numbers that the Greeks had posed—a question that even I could understand. I wrote an article and sent it to the *Scientific American*, which accepted it just after my first child was born in October 1952.

MP: *The article has quite a dramatic heading (reading):* "*Perfect Numbers. Six is such a number: it is the sum of all numbers that divide it except itself. In 2000 years only twelve perfect numbers were found; now a computer has discovered five more.*"

Reid: Yes. It was dramatic enough to be picked up by *Quick*, a popular little get-through-the-week-at-a-glance magazine. Robert Crowell, the president of the Thomas Y. Crowell Company, saw the item, read the article in *Scientific American*, and wrote me suggesting that I write "a little book on numbers" for Crowell.

MP: *You must have been surprised.*

Reid: At first I thought it was quite a joke, since I had never studied any mathematics beyond elementary algebra and plane geometry in high school. But then—all writers want to be published—I got the idea of a little book with a chapter written around each one of the digits. The chapter on "6," the first perfect number, would be the article I had already published in *Scientific American*. The Robinsons agreed to educate me. I think they were rather intrigued. So I was on my way.

MP: *That was "From Zero to Infinity"?*

Reid: Yes. I wanted to call it *What Makes Numbers Interesting*, because that was really the theme of the book. But the sales department vetoed it.

MP: *I think you mentioned once that there were readers who objected to your article in "Scientific American."*

Reid: Not to the article. To me as the author. The article was mathematically correct. After all, the Robinsons had read it. But the readers (maybe, just one reader, I have forgotten now) objected that articles in *Scientific American* should be written by authorities in their fields and not by housewives!

MP: *Having written "From Zero to Infinity," you just went on writing about mathematics?*

Reid: Well, Crowell and Routledge & Kegan Paul, the English publishers, kept asking for more mathematical books by Constance Reid. It was dreadful. I felt like a hoax.

MP: *I remember you said in one of your talks that you even considered studying more mathematics.*

Reid: Yes, but Raphael—my brother-in-law—discouraged me. He said I would have to learn too many uninteresting things before I got to subjects as interesting as those I had been writing about in my books.

The Switch to Biographies

MP: *How did it happen that, after you wrote "A Long Way from Euclid," you switched from expository writing about mathematics to biographies of mathematicians?*

Reid: Let's not call them "biographies"—that's really too formal a word for them. I think of them rather as "lives," in the Plutarchian sense. But to answer your question, the truth of the matter is that after I finished *A Long Way from Euclid*, which goes from Euclid to Hilbert and Gödel, I had lost my mathematical innocence. Things that were once new and exciting to me, like prime numbers, seemed like things that everybody must know. I felt I would be presuming to "explain" them as I had in *From Zero to Infinity*. I really couldn't write any more in the same way for the same audience.

MP: *I believe you said once, in connection with "From Zero to Infinity," that you had received a good deal of fan mail in response to the chapter on zero.*

Reid: Oh yes! A great deal on "0," not quite so much on "1," but quite a bit less on "2," increasingly less on "3" and "4." And I never got a single letter on the chapter on "9," which dealt with congruences. I think it's generally recognized that, even with the most popular books on mathematics, only the first few chapters are ever read. With *From Zero to Infinity*, because of its format, I had a pretty accurate measure. Also, even that book got progressively more difficult. By the end I was already losing the mathematical innocence I just spoke of.

MP: *To go back, we were talking about how you happened to switch to biographies.*

Reid: Well, while I was struggling with the question of what to write next (for I had already received an advance from Crowell for another mathematical book), my sister Julia suggested I do a volume of short biographies of early twentieth-century mathematicians whose names, like those of Volterra and Picard, are attached to things that college students find in their courses. She pointed out, I remember, that it was very difficult to learn anything about a particular mathematician's life because, at least at that time, *Mathematical Reviews* indexed obituaries by the author and not by the subject.

MP: *She had in mind a sort of "Men of Modern Mathematics"?*

Reid: Yes. But of course it would be much different. After all, I'm not E. T. Bell, who was a real mathematician.

The Story of Hilbert and Paradise Lost

MP: *Who were some of the mathematicians you were going to include in it?*

Reid: I wanted a good international mix. That was almost my first consideration. I remember Birkhoff and Veblen were the Americans. Of course I had Hilbert. Not Poincaré, because Bell had included Poincaré (although he wrote his last chapter on Cantor). I wrote short lives of a number of these people. Then I got to Hilbert, and I realized almost immediately that to treat his life in a comparable fashion I would need many, many more pages. I had a feeling for Hilbert that I didn't have for the others—a sense of his "story"—so I decided I would write a book just about him. When I told my publisher, I must admit, he was not too enthusiastic about the idea.

MP: *Had he heard of Hilbert?*

Reid: Probably not. But who has—other than mathematicians and physicists? I remember he told me that the only thing that would sell worse than the biography of a mathematician was a book about South America.

MP: *That is something that has certainly changed!*

Reid: Yes. There may even be hope for biographies of mathematicians! Anyway I was *determined* to write the life of Hilbert. I told Robert Crowell so and also told him that if he didn't want the book when it was finished I would be glad to return his advance. No hard feelings. So there's the answer to your question—how a nonmathematician came to write the life of Hilbert.

MP: *It seems to me that Hilbert would be a very hard person to write about, because, in Lady Bracknell's words, his life was not exactly "crowded with incident."*

Reid: Ah, but it was such a good story! If you think of mathematics as *a world*, which it is, then Hilbert was a world conqueror. In fact, I compare him to Alexander several times in the course of the book. He even cut a Gordian knot, very much in the style of Alexander, when he abandoned the struggle for a constructive solution to Gordan's Problem in favor of a proof of the existence of a solution.

David Hilbert

MP: *And Gordan said: "Das ist nicht Mathematik. Das ist Theologie."*

Reid: But then later he said, "I have convinced myself that theology also has its merits." A very graceful concession. I have always admired Gordan for that. But to go back to Hilbert, the reason I could write about a mathematician like Hilbert—in other words, write in a nontechnical way—is because he was someone whose personality and lifestyle were a very real part of his contribution to mathematics—*beyond* his mathematics. So it seemed that he was a particularly good subject for me and for the public I was writing for. At that time, you see, I still thought that the book about Hilbert would be published by Crowell and read by the same people who had read *From Zero to Infinity* and *A Long Way from Euclid*.

Constance Reid

MP: *When did you first hear about Hilbert?*

Reid: The first time I ever *heard* about Hilbert was when Julia told me about the Paris problems. Imagine a mathematician saying in 1900, "These are the problems we should work on during the coming century"—and of course being so remarkably right! I thought at the time that it was such a dramatic thing that everybody should know about it. But that was long before I ever thought of writing *Hilbert*. Well, of course, Hilbert came into *From Zero to Infinity* in connection with his solution of Waring's Problems. That was treated in the chapter on "8," the first cube. Also his work on the foundations of geometry and his efforts to establish the consistency of arithmetic were treated in *A Long Way from Euclid*. So, in a sense, my earlier writing "prepared" me for writing about him.

The Story of Courant, or Paradise Regained

MP: *You say in the first chapter of your "Courant" book that you came to write about Courant at Friedrichs's suggestion.*

Reid: I don't think I would have thought of writing about Courant if Friedrichs hadn't made that suggestion, and yet *Courant* was exactly the book I wanted to write next, the natural sequel to *Hilbert*. The scientific paradise that Hilbert had created in Göttingen had been destroyed, literally in a day, by Hitler; and Courant consciously tried to re-create it, different but with the same spirit, in the United States at NYU.

MP: *The two books almost fit together as a continuous narrative.*

Reid: Paradise lost and paradise regained, I sometimes thought. Someday I hope they will be boxed together as a paperback pair. I would really like that. I have always wanted them to be easily available to students. Right now they are just too fantastically expensive. But of course they are in libraries.

MP: *Nevertheless, in its way, "Courant" is a quite different book from "Hilbert."*

Reid: Oh yes. They were quite different men and quite different mathematicians.

MP: *Courant seems to fit the more conventional view of a man about whom a book could be written.*

Reid: Lady Bracknell would not have approved—his life was certainly "crowded with incident."

MP: *He was poor and became rich; he fought in a war; he wrote and edited influential books; he was involved with government, with the military, with wealthy men in business and industry.*

Reid: I think Courant, unlike Hilbert, could have been successful in any number of different occupations, but, like Hilbert, he was truly dedicated to mathematics and to the development of young mathematicians. There was, however, a difficulty in writing about Courant that I didn't have in writing about Hilbert.

MP: *What was that?*

Reid: Hilbert was almost universally admired, while Courant was much loved but also much hated. I am still amazed at the hostility that, even now, more than seven years after his death, is expressed against Courant. In fact, one of the things that I really regret is the fact that I let the knowledge of the hostility toward Courant keep me from writing many things favorable to him. After the book was published I heard that Lipman Bers said, "It is an honest book. *But will they know how charming he was?*" I'm afraid not. That was my failure.

MP: *You have said Courant and Hilbert were very different. I'm wondering if you had any stereotype of a mathematician in mind before you began to write about mathematicians?*

Reid: I don't believe so. I had met a lot of mathematicians through my sister and brother-in-law, so I knew they were not all alike. Still, I don't think I knew quite how varied they could be. I did have one preconception that may interest you. Like most people, I thought that mathematicians were very "quick"—certainly quick about anything that had to do with mathematics. I was surprised and intrigued, I remember, when Hans Lewy told me that Hilbert was not at all quick but "slow to understand."

Courant in his student days.

MP: *In your book you quote Hilbert's remark to Harald Bohr that he "always found mathematics so difficult" and when he read or heard something it nearly always seemed "so difficult and practically impossible to understand."*

Reid: Yes, and then the next part I love.

MP: *I'm afraid I don't remember the next part.*

Reid: I think I can quote it the way I wrote it: " 'And,' he added, with his still childlike smile, 'on several occasions it has turned out that it really was more simple!' " It was characteristic of Hilbert that he always started out in a new field by going back to the beginning and working the thing through for himself. He *had* to understand and, as Courant once said to me, "to get to the bottom of things." Courant, on the other hand, seems often to have been stimulated by the thought that he could make an awkward piece of mathematics more aesthetically pleasing.

MP: *It makes one wonder though. Would Hilbert have been an Olympiad or a Putnam winner?*

Reid: Certainly not if he had had a von Neumann competing against him! All the mathematicians I have talked to have said that von Neumann had the quickest mind they ever knew. He and Hilbert worked together, for a period, when Hilbert was old and von Neumann was young. I would like to have sat in on those sessions.

The Research on Hilbert

MP: *How did you go about researching the life of Hilbert? There certainly doesn't seem to be much material available on the lives of mathematicians, except for their mathematical works.*

Reid: The first thing my sister did was to give me the collected works of Hilbert. Then when people

Constance Reid

asked her about my qualifications for writing his life she could tell them that I had studied his works! Seriously though, I had several problems when I started. In addition to not being a mathematician, I did not know German; so I had to learn that language well enough to make my way through German material. I had been to Göttingen; however, I did not know the Göttingen milieu and that resulted in a faux pas or two on my part. But one person led me to another. Julia told me Hans Lewy had been in Göttingen and had known Hilbert, so I talked to him. He sent me on to Pólya, who had been in Göttingen earlier and had a recording of Hilbert's voice—the famous last lines of the Königsberg speech: *Wir müssen wissen, wir werden wissen.* "We must know, we shall know." Pólya sent me to Szegö, who, with Reidemeister, had helped to arrange the Königsberg speech and the recording. Szegö gave me the name of a mathematician who knew the married name of one of Hermann Minkowski's daughters who lived near Boston. That's how I met Lily Rüdenberg—now a very good friend of mine —and she generously gave me access to the complete collection of Minkowski's letters to Hilbert, dating from their university days in Königsberg to Minkowski's premature death in Göttingen in 1909. Incidentally, I later found out that another daughter of Minkowski's lived here in San Francisco, in a house that I passed almost every day. The Minkowski letters were really the making of the book, because they gave me fresh, live materials from a period of Hilbert's life that was really lost, from which no one survived.

MP: *Were you ever able to find the other half of the correspondence—Hilbert's letters to Minkowski?*

Reid: No. But I am still following up every clue. However, Courant always said that, if I could have only one side of the correspondence, the Minkowski side was the better one to have.

MP: *Do you think the Hilbert letters to Minkowski are still in existence?*

Reid: My theory is that some mathematician living in the Hilbert house during the war took them to preserve them after Mrs. Hilbert's death. I don't think anyone in Göttingen would have destroyed them.

MP: *You have some interesting letters in your Courant book, too. I was surprised that you had letters from Courant to his first wife, who later died in a concentration camp. How did you get those?*

Reid: They were among Courant's papers. Apparently he and his wife returned their letters to each other at the time of their divorce in the middle of the First World War. Courant always intended to write about his experiences in Göttingen, so he saved the letters in which he described these.

MP: *That is a wonderful letter you quote in which he describes his first Vortrag in the Hilbert-Minkowski seminar! But, tell me, what has been the reaction to your biographies—"lives"—of German mathematicians in Germany?*

Reid: That depends on which one you're talking about.

MP: *Well, the Hilbert book.*

Reid: There I think German mathematicians were rather miffed that an American lady who had not known Hilbert and who was not a mathematician had presumed to write about their great man. However, my doing so precipitated some long overdue action on their part. A memorial volume—the *Hilbert Gedenkenband*, edited by Reidemeister— was published the year after *Hilbert* was published. Also the Hilbert papers, which were just crammed into miscellaneous boxes at the Mathematics Institute when I examined them, were later moved to the University Library and finally properly organized.

MP: *I gather that the reaction to your Courant book has been different?*

Reid: Quite different. The Germans are very interested in what happened to Courant under Hitler and in his subsequent success in America. They are also very appreciative of his going back to Germany after the war, in spite of the way he had been treated, and doing what he could, in many different ways, to help German science get back on its feet. *Der Spiegel,* which you might say is the *Time* of Germany, devoted a full page—and a picture—to a review of the English edition of the book. Can you imagine *Time* doing anything similar? Since then, the book has been translated into German. A very good translation, I understand. There has been another review in *Der Spiegel* and another picture. I must admit, though, that Courant is identified as a physicist rather than as a mathematician.

MP: *I have always been surprised that your Hilbert book has not been translated into German.*

Reid: The interest in *Hilbert* is much less general. Also, the Hilbert book explains a lot of things about German academic life that would have to be taken out or rewritten for a German audience.

MP: *You know, I suddenly realized that I have not asked you how it happened that Springer published the Hilbert book instead of Crowell. Did Crowell refuse it?*

Reid: No. Not at all. But when Courant read the manuscript of *Hilbert* he insisted that it must be published by Springer. That was because Ferdinand Springer had been very close to Hilbert and the whole Göttingen group after the First World War. Courant was very loyal. He always gave Springer a great deal of credit for the remarkable post-war resurgence of German science. Max Born was also glad, for sentimental reasons, that the book was going to be published by Springer. And I think Robert Crowell was very relieved!

Hilbert and Courant as Teachers

MP: *From your books about Hilbert and Courant, I gather that both men were effective teachers. In fact, the list of their students is pretty impressive. Do you yourself have any ideas about the teaching of mathematics?*

Reid: I have had very little experience. The biographical material for that old *Scientific American* article refers to my teaching "a very elementary form of arithmetic to the more backward sailors at the U.S. Naval Training Station in San Diego during World War II." Frankly, I had completely forgotten that! But, to go back to your question, I have observed in my research that great mathematician-teachers seem to fall into two classes—those like Klein (and at a later period in Göttingen, Hecke and Siegel), who present the subject complete and finished, a perfect jewel; and those like Hilbert and Courant, who struggle with the subject in front of their students in the classroom. You remember Courant's remark that in Hilbert's lectures you could "feel his intellectual muscle." Some students are stimulated by one type; and some, by the other. Arnold Sommerfeld and Max Born used to argue over the relative superiority of Klein and Hilbert. Sommerfeld preferred Klein; Born, Hilbert.

Richard Courant

Hilbert giving a lecture.

MP: *I am glad you mentioned the names of a couple of physicists, because there is a question I have always wanted to ask you. I remember you said, in one of your talks, that physicists were much more responsive than mathematicians in answering questions. Would you expand on that remark?*

Reid: First, I have to remind you that, when I said that, Edward Teller was sitting in the front row. But, in fact, that has been my experience. You ask a mathematician, "What was so-and-so like?" and he squirms a little and then he tells you an anecdote. The anecdote, like a symbol, stands for something much greater and more complicated that he senses about the person but cannot put into words. Physicists, it seems to me, enjoy analyzing character—in words. I think of them as more literary.

MP: *It is true that there have been many more autobiographical memoirs by physicists than by mathematicians.*

Reid: And the mathematicians who do write autobiographically generally have a physical orientation—except, of course, Hardy.

MP: *To go back, however, to the teaching of mathematics, what do you, or what did you, think of the "new math"?*

Reid: I do not want to be drawn into that controversy. I will make one comment from my own experience. I believe that my children, both of whom had the "new math" in school, acquired a much better understanding of what they were doing mathematically than I had at the same age.

MP: *How did it happen that you stopped taking mathematics in high school? Were you bored? Discouraged by your teachers?*

Reid: No. I really loved algebra. I did not care much for geometry. I never felt I was proving a theorem but simply remembering how to prove it. I can't blame the teacher, because Julia had the same teacher and she went on. I think I was just more interested in boys and baseball—major league baseball—than in mathematics.

MP: *Are you still interested in baseball?*

Reid: Heavens, no!

Women in Mathematics

MP: *Hilbert seems to have been ahead of his time in his attitude toward women in mathematics. The German university was apparently full of what we call "male chauvinists." There were two quotes in your book that stood out for me. One was the oath for the Ph.D.—"that you will defend in a manly way true science". And then someone said in regard to the "Habilitation" of Emmy Noether: "What will our soldiers think when they return to the university and find that they are expected to learn at the feet of a woman?" That's a really incredible statement!*

Reid: The Göttingen mathematicians, however, have a pretty remarkable record in regard to their attitude toward women mathematicians. Sonja Kowalewski got her doctor's degree from Göttingen, albeit *in absentia*. Klein, as well as Hilbert, supported Emmy Noether's *Habilitation*. Courant very definitely encouraged women to continue their mathematical studies. An outstanding instance is the case of Cathleen Morawetz. An article about her career and the role Courant played in it appeared recently in *Science*.

MP: *Now that you have written "Hilbert" and "Courant," do you ever feel drawn farther back in the Göttingen story, writing about Riemann or Klein, say, or going ahead to Weyl and Noether?*

Reid: No. Riemann is too far in the past for me. I like what Henry James called "the visitable past." Klein, however, has always interested me very much. He has the quality that appealed to me in Hilbert and in Courant—dedication to mathematics and the creation of a stimulating mathematical center. He was a great figure. I think Courant had a sense of the drama and the tragedy of Klein's life, which he expresses very well in the obituary that he wrote for *Naturwissenschaften* in 1925.

MP: *You may have written Klein's story already.*

Reid: I think so. Klein is a strong, living character in *Courant* as well as in *Hilbert*. So, to a lesser extent, is Emmy Noether. I have been approached several times about writing her life; but what really interests me about her is the developmental period—her relation to her father and to Gordan, who was her teacher—and I don't think there is much material about that available. As fas as Weyl is concerned, he wrote marvelously well about himself and his life and his feelings about mathematics. There is really no need to write more.

Jerzy Neyman

MP: *But now you have written another "life".*

Reid: Yes, the life of Jerzy Neyman, a really great mathematical statistician and the man who built up the Department of Statistics at Berkeley—the best in the country. He was eighty-four when I started my research, but he was still very active, both as a teacher and as the director of the Berkeley Statistical Laboratory. Getting to know him was a great pleasure for me. Unfortunately he died before the book was published; nevertheless I called it *Neyman—From Life*. I had really completed the manuscript at the time of his death, having decided to conclude with the celebration of his eighty-fifth birthday.

Jerzy Neyman (right) with Harald Cramér in Stockholm, 1964.

MP: *How did you happen to choose Neyman as a subject?*

Reid: Very much the same way I "chose" Courant. As I told you earlier, Friedrichs suggested that I write about Courant after he read *Hilbert*, and Erich Lehmann suggested that I write about Neyman after *he* read *Courant*. He saw a similarity in the lives of the two men. Both were foreign-born, and both built up great scientific centers in this country.

MP: *What were the advantages (and disadvantages) of writing about someone who was alive at the time you did most of your research?*

Constance Reid

Reid: The most important advantage was the opportunity to get personally acquainted with my subject; next was the opportunity to ask him all the questions I wanted to ask. Of course, he couldn't always answer them. In the end he insisted that I knew more about his life than he did.

MP: *And the disadvantages?*

Reid: Well, you can't help being affected by the fact that you expect what you write to be read by your subject. Also you have to make a particular effort to see *beyond* the old man who is right there in front of you. That was a problem with Courant as well as with Neyman.

MP: *Nevertheless, it seems to me that in both* Courant *and* Neyman *you have given a vivid picture of what they were like when they were young.*

Reid: That's because I was very lucky in turning up firsthand material from the early years. In Courant's case, besides the letters which we have already talked about, I found in his attic three soul-searching journals which he had kept at various times from 1912 to 1918. In Neyman's case, I was able to obtain from Egon Pearson copies of the letters which Neyman wrote to him during the period of their famous collaboration on the testing of statistical hypotheses. Fortunately for me the collaboration was carried on almost completely by mail because Neyman was in Poland and Pearson in England.

MP: *Neyman and Courant (and even to some extent Hilbert) were, in a sense, organizers and administrators as well as first-class mathematicians. Are they easier to write about than those who solely do mathematics?*

Reid: I would say so. Certainly for me. You see, I feel that a very important part of my subjects' contribution to mathematics lies in their personal influence and the scientifically stimulating environments which they created around themselves. Their scientific work is always there for anyone who wants to consult it, but this other is very transitory—it will last only as long as there are people who personally remember it. That's why I began to write about Hilbert when I did—even though it was very presumptuous of me. But nobody else was doing it.

MP: *Neyman was a statistician. Are they different from mathematicians?*

Reid: I think so. Statisticians are not so much yes/no people as mathematicians. Even statisticians who like Neyman are basically mathematical in their approach know that they are dealing with things that cannot be fully "mathematicized."

MP: *What do you think it was about mathematics and mathematicians that originally excited you?*

Reid: That's easy. Mathematics is a world created by the mind of man, and mathematicians are people who devote their lives to what seems to me a wonderful kind of play!

HERBERT ROBBINS

HERBERT ROBBINS

Interviewed by Warren Page

Herbert Robbins is best known to most mathematicians for his collaboration with Richard Courant in writing the classic *What is Mathematics?*, but in the following forty years, Robbins' work has earned him the reputation as one of the world's leading statisticians. Jerzy Neyman once indicated how rare it was for a professor to place on a high pedestal one of his colleagues—of necessity, one of his "competitors." And yet, in Neyman's two survey papers on major advances in statistics during the second half of this century, the main breakthroughs cited by Neyman were all obtained by Robbins.

Robbins is the Higgins Professor of Mathematical Statistics at Columbia University, and is known among friends for combining a cynical humor with a deep involvement in humanitarian causes. Among his well-known sayings are:

> "No good deed shall go unpunished,"

and (when asked by a university administrator if there might be a risk in recommending a junior faculty member for tenure):

> "If he washes out in his research, we can always make him a dean."

In 1978 Jack Kiefer of Cornell wrote that "Robbins (then 63 years old) does not seem to have slowed down at all, and he is as lively and original now as in the past." That's still true. In 1982, I interviewed him at his home in East Setauket, New York, and it was immediately clear that he loves people and conversation. His broad interests, his many friendships, and meaningful involvements —like good vintage wine—seem to grow fuller and richer with time. All this, of course, made our meeting rewarding and memorable. It also made my job very difficult: trying to confine him to a few specific issues was like trying to pin down an enthusiastic, animated, intellectual octopus. The following is only a sample of what was retrieved from the occasion.

MP: *You first became nationally known in 1941 as the co-author, with Richard Courant, of "What is Mathematics?". So let's begin there. "What is Mathematics?" has been translated into several languages, and more than 100,000 copies have been sold thus far. What has made this such a mathematical best seller?*

Robbins: A Russian translation of *What Is Mathematics?* was published shortly after World War II, and it provoked the Russians to produce their own version—some of them call it the Anti-Courant & Robbins—because aspects of Courant and Robbins are not in harmony with certain Soviet mathematical tendencies. This Russian work presumably reflected the correct line on mathematics—its history, the contributions of Russian mathematicians, its current state, its importance, and its Marxist foundations. When I looked at the English translation, for the first time I realized what a good book *What Is Mathematics?* really is. It wasn't written by a committee, and it seemed to have no great practical importance to governments or their defense departments. It was written by two people who collaborated in an intense manner on a subject that concerned them very deeply all their lives. When I started working on *What Is Mathematics?* I was twenty-four, one year out from my Ph.D. It represented what I had learned about mathematics, what I hoped it would become, and where I wanted to find my place. For Courant, it represented a summation of what he had already done during his rich life as a mathematician and, in his later years, as an administrator and promoter of mathematical institutions. Courant and Robbins spoke for mathematics at a particular time (1939–1941) that can never be repeated. It would be impossible to write that book now; mathematics has changed so much, and the nature of the mathematical enterprise is so different. I would classify *What Is Mathematics?* as more a literary than a scientific work. It belongs to the tradition that started in French intellectual circles when Newton and Leibnitz discovered the calculus. People without formal scientific education wanted to understand this new notion. Salons were held, and philosophers gave lectures to crowned heads and rich bourgeois about the calculus. *What Is Mathematics?* belongs to that tradition of high vulgarization, as the French call it.

MP: *How did a young topologist like yourself come to work with Courant?*

Robbins: (*Laughter*) That's like asking, "What's a nice girl like you doing in a place like this?" Well, I had just earned my Ph.D. and I was beginning a one-year appointment at the Institute for Advanced Study in Princeton as Marston Morse's assistant. I needed a permanent job. When Courant came by looking for someone to work at New York University, Morse suggested me and Courant offered me a job. He had the book in mind, but I didn't know that at the time. When I began teaching at NYU in 1939, I was supporting my mother and young sister. For me, money was *a sine qua non*. My salary as an instructor at NYU remained fixed at $2,500 a year during 1939–1942. That was my sole support; there were no NSF grants then. Sometime during the beginning of my first year at NYU, Courant said to me, "I've been given a little money to work up some old course material into a book on mathematics for the general public. Would you like to help me with it? I can pay you $700–$800 for your assistance." I was in no position to turn down extra money for a legitimate enterprise, and the idea of communicating my ideas about mathematics to the educated layman appealed to me.

Confrontation with Courant

MP: *What was it like working with Courant?*

Robbins: I thought of him as an accomplished and worldly person who, in the fullness of his career, decided to devote some time to explaining to the world what mathematics was, rather than to write another research paper. I felt that I could help him in this. As work on the book progressed, however, the amount of time I had to devote to it became larger and larger, and I soon came to feel that it was interfering with any future career I might have in research. I used to commute regularly to his house in New Rochelle and, in a way, I became a member of the family for a while. In fact, I actually lived nearby for some time so that I could work with him when he wasn't busy. For about two years we worked very closely together exchanging drafts of chapters. But, as you probably know, the whole thing came to an abrupt and grinding halt in a rather dramatic confrontation described in Constance Reid's book *Courant in Göttingen and New York*.

MP: *Did your confrontation with Courant come right after the book had been written?*

Robbins: That's right. As Reid indicated, Courant felt that my collaboration was so helpful that he came to me early in our arrangement to propose joint authorship. He wasn't going to pay me any more, however, because as the joint author I'd probably want to spend even more time on the book. I agreed, since I had already become engrossed in writing the book. My first indication of what was really going on came when I went to the printers to go over the final page proofs, and the last page I saw was the title page '*What Is Mathematics?* by *Richard Courant*.' This was like being doused with a bucket of ice water. "My God," I thought, "What's going on here? The man's a crook!" By then the book had been written, except that Courant never showed me the preface in which he thanked me for my collaboration. The dedication page to his children was also written without my collaboration.

You mentioned earlier that more than 100,000 copies of *What Is Mathematics?* had been sold. That may be, but when I recently asked Oxford University Press how many copies have been sold they told me I had no right to know. Courant copyrighted the book in his own name without my knowledge. He had a wealthy friend who paid for having the plates made, and he got Oxford to agree to distribute the book. It was a unique arrangement in which he retained the copyright and received a much larger portion of royalties. After this had been done, Courant informed me that he completely controlled the book and he would remit to me, from time to time, a portion of his royalties. And so every year, for a number of years, I used to get a note from Courant saying, "Dear Herbert, Enclosed is a check for

Courant (right) and Robbins.

such and such an amount representing your share of royalties from *What Is Mathematics?*" I never knew how many copies were sold or how much he got, and I still don't. This arrangement continued up to the time of his death a few years ago, when his son Ernest became his legatee.

MP: *What happened then?*

Robbins: Three or four years ago, *What Is Mathematics?* appeared in paperback. Just prior to that, in order to simplify matters, Ernest offered to buy out my share, and I agreed to renounce all further claims on the book if we could set some reasonable figure for my doing so. But he never went through with this, although the sum had been agreed on, and when the book came out in paperback, I stopped getting anything at all. In fact, on the jacket of the hard cover edition (Robbins taking out his edition: copyright renewed, 16th printing, 1977) here you see something about the late Richard Courant and here's something about the present Herbert Robbins. But on the paperback edition, one finds that the mysterious Herbert Robbins appears only on the title page as co-author; on the back cover it looks as though it's entirely Courant's book. So, even after his death, there has been an intensification of the campaign not merely to deny my financial rights in the book, but even to conceal the fact that I was its co-author.

MP: *Were you ever given an explanation why Courant didn't treat you as might have been expected?*

Robbins: Some of Courant's friends came to me and said, "You see, in Europe, it's quite customary for a younger man to do the work while the older man is credited with being the formal author. This has happened before with many people and, in particular, with Courant. Don't be upset, etc." As a non-European, not acquainted with this tradition, I refused. "It wasn't fair! I had taken his word; I wouldn't have put the effort I did into this unless it was going to be a joint book." The drama continued, with more visits by Courant's emissaries, including some distinguished European mathematicians whom he had brought here. But I was adamant and wouldn't agree to be quiet. I threatened to make a fuss if Courant didn't include my name on the title page. Courant finally agreed to do so.

Important Influences

MP: *Were there any mathematicians who gave you guidance and encouragement during critical periods of your professional development?*

Herbert Robbins

285

Robbins: No. What they gave me was something perhaps more important. The leading mathematicians I encountered made me want to tell them: "You son-of-a-bitch, you think that you're smart and I'm dumb. I'll show you that I can do it too!" It was like being the new kid in the neighborhood. You go out into the street and the first guy you meet walks up to you and knocks you down. Well, that's not exactly guidance or encouragement. But it has an effect.

MP: *Who have been the most impressive mathematicians you've known?*

Robbins: The first mathematician who impressed me was William Fogg Osgood, author of *Funktionentheorie*, because he had a beautiful white beard. I was a freshman at Harvard and, being from a little town, I had never seen anyone like him. I was also impressed by Julian Lowell Coolidge because he spoke with a lisp that sounded very upper class. There are many ways to be impressive. These people impressed me as personae; I thought it must have taken several generations to produce people like them. The first mathematician I met who impressed me as a mathematician was Marston Morse, and I regard him as one of the two or three most powerful mathematicians America has produced. Morse was not a wide-ranging mathematician of the Hilbert type, but he created the theory of the calculus of variations in the large, and whatever he needed he learned, borrowed, or created for himself.

MP: *How did you come to meet Morse?*

Robbins: In the 1931 Harvard–Army football game, Harvard was losing at half-time. During the intermission, Harvard's President A. Lawrence Lowell said to the cadets' commandant: "Your boys may be able to beat us in football, but I'll bet we can beat you in mathematics." The commandant accepted the challenge, and it was agreed that Army and Harvard would have a mathematics competition the following year. Since cadets had only two years of mathematics at West Point, Harvard limited its team membership to sophomores. Lowell's relative, William Lowell Putnam, agreed to put up a prize—the forerunner of today's Putnam Prize in mathematics. In 1931, I was taking freshman calculus. Having just entered Harvard with practically no high school mathematics, I knew calculus would be useful if I ever wanted to study any of the sciences. At the end of my freshman year, much to my surprise I was asked by the mathematics department to join the Harvard math team. Marston Morse was our coach. We met with him on several occasions to prepare for the competition, and that's how I first met Marston. As it happened, incidentally, Army won that mathematics competition.

Marston Morse (right) with Robbins.

MP: *How and when did you make the decision to become a mathematician?*

Robbins: Morse, G. D. Birkhoff, and Whitney were the three mathematicians who most influenced me because I got to know them quite well for short periods of time and, in very different ways, they formed my image of what a mathematician was. Meeting these three early in my education turned my thoughts to mathematics as a possible career.

MP: *I'm sure you also had great teachers in other subjects. What, in particular, did these three convey to you about mathematics?*

Robbins: One of my professors at Harvard, a famous literary critic, used to walk in with a briefcase full of books and lecture on the Romance Poets. He'd take out a book, read a poem, and then comment on it. Now this represented real scholarship that left me totally cold. To my mind, this wasn't being creative. He was talking about what others had done. I would rather have done these things. He talked about Coleridge; I would like to have written "The Rime of the Ancient Mariner." On the other hand, Marston Morse impressed me deeply. Even though what he was talking about meant nothing to me—I didn't know the first thing about the Betti numbers of a complex and the number of critical points of a function defined on it—I could see that he was on fire with creation. There was something going on in his mind of a totally different nature from anything I'd seen before. That's what appealed to me.

MP: *Morse seems to have played a pretty prominent role in your life.*

Robbins: At that time, Marston's life was pretty much at low ebb. His wife had left him, and he was living as a bachelor at Harvard. I pitied him almost . . . but in a way, I didn't pity him; I was scared stiff of him intellectually. I was an undergraduate and, although I had never taken a course with him, I got to know him since he was living at the college. One day he said to me, "I'm leaving Harvard and going to the Institute for Advanced Study. You stay here and when you get your Ph.D. in mathematics, come to the Institute to be my assistant." Six years later—I hadn't seen him since—I sent him a telegram: "HAVE PH.D. IN MATHEMATICS." He immediately wired back: "YOU ARE MY ASSISTANT STARTING SEPTEMBER 1." Marston was, in a way, the type of person I would like to have been. He was a father figure to me—my own father died when I was thirteen. Marston and I were about as different as two people could be; we disagreed on practically everything. And yet, there was something that attracted me to Marston that transcended anything I knew. I suppose it was his creative, driving impulse—this feeling that your house could be on fire, but if there was something you had to complete, then you had to keep at it no matter what.

MP: *What was it that originally attracted you to topology?*

Robbins: My affair with topology was rather accidental. Hassler Whitney had come back from a topology conference in Moscow around 1936, and in a talk at Harvard on some of the topics discussed at the conference, he mentioned an unsolved problem that seemed to be important. Since I was then a graduate student looking for a special field to work in—not particularly topology, since I hadn't even taken a course in the subject—I asked Whitney to let me work on it. That's how I got started. I had set myself a time limit from the beginning: if I didn't get my Ph.D. within three years after starting graduate work, I would leave the field of mathematics. Midway through my third year, when they asked me whether I wanted to continue my fellowship for another year, I told them that I wouldn't be coming back next year. Although I did manage to complete my thesis that year, I didn't feel that I had become a topologist; I thought I had become a Ph.D.—a kind of generalized mathematician.

Becoming a Statistician

MP: *How did you become a statistician?*

Robbins: My first contact with statistics came when I was teaching at NYU. Courant had invited Willy Feller to give a course in probability and statistics, but at the last minute Feller couldn't come. The course had been advertised, but now there was no one at NYU with any interest in probability or statistics. As the youngest and most defenseless person in the department, I was assigned to teach the course. It must have been a pretty terrible course because I knew nothing about either subject. This was just before I joined the Navy in World War II, not as a mathematician but as a reasonably able-bodied person.

Statisticians at Chapel Hill, North Carolina, 1946. Front row, left to right, W. Hoeffding, Robbins, R. C. Bose, H. Hotelling, S. N. Roy. Back row, graduate students.

It was in the Navy, in a rather strange way, that my future career in statistics originated. I was reading in a room, close to two naval officers who were discussing the problem of bombing accuracy. In no way could I keep from overhearing their conversation: "We're dropping lots of bombs on an airstrip in order to knock it out, but the bomb impacts overlap in a random manner, and it doesn't do any good to obliterate the same area seventeen times. Once is enough." They were trying to decide how many bombs were necessary to knock out maybe 90% of an area, taking into account the randomness of impact patterns. The two officers suspected that some research groups working on the problem were probably dropping poker chips on the floor in order to trace them out and measure the total area they covered. Anyway, I finally stopped trying to read and asked myself what really does happen when you do that? Having scribbled something on a piece of paper, I walked over to the officers and offered them a suggestion for attacking the problem. Since I wasn't engaged in war research, they were not empowered to discuss it with me. So I wrote up a short note and sent it off to one of the two officers. In due course, it came to the attention of some mathematical research group working on the problem. However, I had no clearance to discuss classified matters, so there was a real communications problem: how were they going to find out my ideas without telling me something I shouldn't know? (What I shouldn't know was, in fact, the Normandy invasion plans.) Well, in some mysterious way, what I had done came to the attention of Marston Morse, and he saw to it that my note reached the right people. Shortly afterward, S. S. Wilks, then editor of the *Annals of Mathematical Statistics*, asked me to referee a paper by Jerzy Neyman and Jacob Bronowski (author of *The Ascent of Man*) on this very same problem. I recommended rejecting their paper as: "a rather unsuccessful attempt at solving a problem that is easily solved if it's done the right way, and here's how to do it." Wilks wrote back that he had to publish the paper because Neyman was one of the authors. But he also wanted me to publish a paper on what I'd written to him. So, after the war in Europe ended, there's an issue of the *Annals* containing the paper by Neyman and Bronowski, followed immediately by my paper which, so to speak, says "Please disregard the preceding paper. Here's the solution to the problem that they can't solve." That was my first publication in the field of statistics. But even then I had no idea that I would become a statistician. What I had been doing was not statistics, but some rather elementary probability theory.

MP: *What did you do after four years in the Navy?*

Robbins: I had a career crisis. My pre-war job had been as an instructor at NYU, and I had already burned my bridges there. Jobs were scarce, so with my back pay from the Navy I bought a farm in Vermont. I went there with my wife—I had gotten married during the war—to figure out what to do next. I thought I was going to leave mathematics and the academic profession completely. Then fate struck again with a telephone call from Harold Hotelling of Columbia University's Economics Department. Hotelling's primary interest was in mathematical statistics. Since Columbia had not allowed him to create a department of mathematical statistics, Hotelling had just accepted an offer to do so at Chapel Hill. The idea of such a department was being promoted at the University of North Carolina by a very energetic statistician, Gertrude Cox. Hotelling offered me an associate professorship in this newly created department. I thought he'd telephoned the wrong Robbins, and I offered to get out my AMS directory to find the Robbins he'd intended to call. Hotelling insisted that there was no mistake, even though I told him that I knew nothing about statistics. He didn't need me as a statistician; he wanted me to teach measure theory, probability, analytic methods, etc. to the department's graduate students. Having read my paper in the *Annals of Mathematical Statistics*, Hotelling felt that I was just the sort of person he was looking for. "Don't question it any further," he insisted. "The salary will be $5,000 a year." That was in 1946, and it was double my salary at NYU four years earlier. It was a very good salary at that time. So, with some trepidation, I agreed. At Chapel Hill, I attended seminars and got to know several very eminent statisticians. Soon I began to get some idea about what was going on in that subject and finally, at age thirty-two, I became really interested in statistics.

The Creative Process

MP: *Herman Chernoff characterizes your innovations as having been based mainly on extra-mathematical insight, and Mark Kac describes your contributions as marked by power, great originality, and equally great elegance. Is there anything you can share with us about the creative process—your feelings and experiences—during the germination of some new insight or breakthrough?*

Robbins: I'm always pleased to hear my work praised, but the things I've been associated with that are really important have not been done by me at all. I've merely been the vehicle by which something has done them. When something significant is happening, I have a feeling of being used—my fingers are writing, but there's a lot of noise and it's hard for me to get the message. Most of the time I'm just sitting there, in an almost detached manner, thinking: "Well, here's another day's wastebasket full of paper. Nothing's come through. Maybe another day. Maybe I should stay up tonight and try some more." I stay up nights when my wife and children have gone to sleep. Over and over again I keep working at it, trying to understand something which after months or even years turns out to be so simple that I should have seen it in the first ten minutes. Why does it take so long? Why haven't I done ten times as much as I have? Why do I bother over and over again trying the wrong way when the right way was staring me in the face all the time? I don't know.

MP: *How do you feel after having made a discovery?*

Robbins: I feel like someone who has climbed a little mountain the wrong way. Once I get the message, so to speak, I try to write it up as clearly as possible. Then I want to get away from it. I didn't do it. I don't want to see it again; I had enough trouble with it. I want to push this onto the rest of the world: "Look, there's lots more that has to be done, but don't expect me to do it. I've done my duty. I've contributed to the Community Chest. Now let somebody else carry on."

MP: *The fact is that you reached the summit, you made the discoveries.*

Robbins: Yes, I take some pride in that. Had I not lived, certain things would not have been done. But a world consisting of lots of me's would be intolerable. One is enough.

I always look for something terribly simple, because very simple things are often overlooked. In a way, my strengths are due to my weaknesses. Others are technically much better than I, but it never occurs to them to do the dumb kinds of things that occur to me. A good example is stochastic approximation. Lots of people said: "My God, we can generalize that; we can do it under much weaker restrictions; etc." And I thought, "Yes, that's true. But why didn't somebody do it sixty years ago?"

MP: *Let's stay with the creative process. What do you do when you're blocked or stymied?*

Robbins: There's nothing I can do except try not to get panicky about it. If I live long enough, one of these days I will stop and never get another idea. I have no idea when that will come, but it will come . . . or maybe it has already.

MP: *Why? I want an existence proof.*

Robbins: Look at Einstein. He seems not to have had an inspiration in physics during the last thirty or so years of his life—at least none that could be compared with the great ones he had from 1905 to 1920. Here was a man who had perhaps the greatest intellect that God ever created and, in the last years of his life, nothing much came of it. It wasn't because Einstein was frivolous or dissipating his energy; he had done what he could and there came an end. As I watched him at the Institute one year, he never complained about it and no one mentioned it, but everyone knew that he was essentially finished as a scientist. And Newton? The same thing. From about age thirty on, he did absolutely nothing in science. He had a career as Master of the Mint, he carried on a great deal of activity with friends—controversies over who invented the calculus, and so on—but the last half of his life was totally sterile from a scientific viewpoint.

MP: *Isn't this fear of "drying up" something every researcher and, in fact, every creative person has?*

Robbins: Yes, but it doesn't end here. Take the guy who cracked the genetic code, for instance. "What's he done recently?" In this country, the question is always: "What have you done recently?" . . . "Oh yes, you did such–and–such, but how about last year?" It's not so much that others are asking this question, we're taught here to ask it of ourselves. I constantly find myself asking what have I done during the last year or so, and how does it compare with what I did thirty years ago?

MP: *That sounds like sequential doom. Are you saying that we've come to expect an ever-increasing sequence of better and better encores of ourselves?*

Robbins: Right. And this creates a lot of frustration and anxiety. If I were a promising young tennis player, I'd hope to get better and better, and finally to win at Wimbledon. And then I'd become a teacher—one can't go on forever playing competitively with twenty-year-olds. No one would think less of me if I didn't enter Wimbledon at age 65. But what I'm doing is not tennis. As a mathematician, I'm using my brain, and there's no reason why it shouldn't be as good, if not better, than it was thirty years ago. "So why is it not?" I ask myself. I've been in statistics now for thirty-five years and I would like to try something else. Why don't I try going into molecular biology? Or sociology, or economics? Am I incapable of the mental effort, or am I just too weary? These complicated questions, raised by increasing expectations, I can't answer.

Competitiveness in Mathematics

MP: *Although physical prowess—say reaction time—is crucial in sports, it makes no significant difference if one's insight into a mathematics problem takes a second or a year. Perhaps we should consider the mathematician's personal drive to succeed and the price he's willing to pay for success.*

Robbins: Younger mathematicians have a greater desire to become known and make a reputation. This weakens with age, either through frustration if they don't succeed or through satiation if they do. And even if one does make it to the top, was it worthwhile? Is it worth continuing to strive for more? The really successful mathematician, if he's honest, must assess his life in terms of having foregone meaningful relations with others—wife, children, colleagues, friends, etc. As one becomes older, he becomes less likely to want to pay the price for new successes. The theorems that I've proved aren't going to be much good or as comforting to me as would be close friends when I'm old and perhaps infirm.

MP: *Having been both a topologist and a statistician, have you perceived any difference between those involved in these two fields or, more generally, between those in pure mathematics and those in applied areas?*

Robbins: I don't think I can distinguish any behavioral or personality differences. However, I recall that when I started out, applied mathematicians were looked down on by pure mathematicians. If you got a Ph.D. in mathematics and your professors thought you weren't really very good, then they'd suggest that you would do well to get an actuarial job in an insurance company, or an applied job with an industrial firm. If you weren't a pure mathematician, you weren't top drawer.

I remember a well-known mathematician, alive today, who started out in pure mathematics and then became interested in probability and statistics. While talking with him one day—I was quite young at the time—I asked what he thought was the most important work I might do during the next few years in the field of probability and statistics. To my amazement, he turned red with emotion and almost pleaded: "Robbins, the most important thing you can do is to show mathematicians that probability theory and mathematical statistics are really part of mathematics." I was absolutely dumbfounded. Evidently his former colleagues had made him feel that he was no longer a member of the elite when he became involved with probability and statistics. He had never been able to survive the blow to his ego that his defection from the realm of pure mathematics had caused.

MP: *Are there feelings of jealousy and/or competitiveness among individuals working on the frontiers of developments in mathematics?*

Robbins: Competitiveness and jealousy seem to belong more to my generation than to the current one. When I was young, there was a great deal of it. Young people now don't have as much. Although there's a reasonably well-defined pecking order, I don't see them motivated by the same burning desire to be Number One and to cast discredit on all their competitors. That was quite common when I was in my 20s and 30s. Maybe because it's easier now. Mathematics is a way of making a living, like selling insurance. When I started out, to be mathematician was a rare choice: there weren't many jobs, and one had to be prepared to give up certain things for the enjoyment of doing mathematics. Today, of course, mathematicians work everywhere.

MP: *Is a little bit of competitiveness healthy for those engaged in research?*

Robbins: Competitiveness, as far as I'm concerned, has an ambiguous quality. Sometimes I think that I'm the best in the world, since I'm the only one who looks at things the way I do. So, in this sense, I'm beyond competition. The other feeling is one of total ineptness. There are many fields of mathematics in which I don't even know the elements. I've tried to learn them, but I can't remember things from one day to the next. These are fields in which I just fall on my face every time I try. I can't help being anxious about not really knowing what others are talking about. I should know these things because my students have to. Thank God I'm not being examined!

Mathematical Reflections and Projections

MP: *In what directions is the field of statistics evolving?*

Robbins: Let's take just the field, called biostatistics, that deals with the application of statistical methods to human health and disease. The demand for trained biostatisticians is enormous, but there's absolutely no supply. If I were given ten million dollars to spend for advancing science, I could spend it trying to produce one or two good biostatisticians. Statistical methods that are currently being used were mostly developed in England for analyzing such things as agricultural experiments and industrial processes. Many of these techniques are being blindly applied to situations for which they are not adapted. The methodology for handling important problems in biostatistics does not exist. It's just beginning now; its Newton or Einstein has yet to appear.

MP: *How and where can one become trained as a biostatistician?*

Robbins: A mathematically capable student who wants to become directly involved with problems of human welfare, should be doing biostatistics. Unfortunately, there are very little encouragement to do this, and there are very few places now to learn biostatistics. A mathematics department would never think of advising anyone to study it. I would like to see a distinguished mathematics department in this country tell its students: "You are very capable and you could have a career in algebraic geometry, or whatever, but we would like to encourage you to go into biostatistics."

MP: *Since it's so demanding just to keep abreast of one's own field of specialization, is it possible to stay mathematically literate in general?*

Robbins: It's harder and harder. I am not totally illiterate in mathematics. If you're in a university, about the best you can do now is to go to as many seminars and listen to as many one-hour lectures as you can, and just hope that some of it will sink in. But to really keep up with the literature now is impossible. Even when I was a graduate student, in the 1930s, it was just barely possible to have a fairly good idea of most of what was going on. Maybe then somebody could still have said, "Anything that's going on in mathematics is of interest to me and with a little effort, if necessary, I'll read the latest paper on it." No more, it's not possible.

MP: *Is it better to be a mathematical specialist or a generalist, and how difficult is it to be either in a meaningful manner?*

Robbins: That's like asking if it's better to be a decathlon athlete or a high jumper. You do what's best for you. You do what God has given you the wherewithal to do it with. If you are pretty good in a lot of things without being world class in any one of them, you'll find some field or activity which requires exactly that, and nobody else will do it as well. No high jumper could win the decathlon. The person finds the problem. You can't decide what kind of mathematician to be.

Getting Known in Mathematics

MP: *How do today's mathematicians compare with those of earlier generations?*

Robbins: I once enunciated a law of human development: *The total amount of intelligence remains constant while the population increases exponentially*. If you ask who the great mathematicians of the present day are, and how they compare with those of fifty years ago, people will tell you that we've got so many bright people now who can do things which nobody could do fifty years ago. I take that with a grain of salt. I don't believe we've got all these greatly gifted mathematicians and all these young geniuses. Hilbert is Hilbert, and there won't be another one like him for some time.

MP: *Is there anything society can do to help produce future Hilberts?*

Robbins: There's not much difference between creativity in music and in mathematics. We have not seen, nor been able to create, a modern music that compares with the Baroque, even though millions of dollars are spent annually on music instruction in the high schools, and seventy-eight Americans have won the international competition in this, that, and the other thing. What comes out is pedestrian and not of much interest. During the Sputnik era, the country became concerned with its technical capabilities and we decided to strengthen our scientific establishment. Mathematics became a national priority. Everyone was running around reforming mathematics instruction, creating the new math, rewriting textbooks. More student scholarships and faculty research grants were awarded. There's no doubt that the effect of all that was to produce more mathematics, but I don't know that anything significant came out of it. It may have produced a lot of utility-grade mathematicians who have written lots of mediocre stuff. We viewed the problem in the same sense as our annual output of steel. Maybe we're only number three in the world in annual steel output. Would it do us any good to be number one? Who cares? The point is that nobody knows how to produce a Bach or a Newton.

MP: *There seems to be a greater publish-or-perish pressure today than ever before. Has this resulted in an increased tendency for academicians to jump on new mathematical bandwagons in order to take advantage of greater opportunities for publication?*

Robbins: Journals have proliferated to the extent that there's no real problem in getting published somewhere, although there's still a distinction between publishing in refereed and nonrefereed journals. Many published papers are of no real interest or value. I've often thought that when I become enormously wealthy, I'll establish the very prestigious Herbert Robbins Prize in Mathematics. It would have one condition: the recipient shall never publish another paper. As to mathematicians jumping onto new bandwagons, I believe that most people place self-aggrandizement and obvious rewards ahead of duty to the truth, so to speak. I like to think that when I was young, one did something because that was what one wanted to do, regardless of whether anyone paid for it or listened to it. But this is probably an illusion of age and selective recall.

MP: *Is it easier to become better known in some fields of mathematics than in others?*

Robbins: In number theory, there are a number of classical conjectures—Goldbach's, Fermat's, etc. Anyone who makes a contribution to them gets instant fame because these are such famous problems

—even though they seem to be somewhat outside the general domain of mathematics. An affirmative solution of Goldbach's conjecture would have no obvious consequences in any other domain of mathematics, or even in number theory itself. It's just a glorified champion Rubik's Cube puzzle. One could, of course, say: "I proved Goldbach's conjecture and that makes me the greatest mathematician of our time." That can be justified in the sense that some very powerful mathematicians have tried and failed. It's like saying that you're the first person to climb Mt. Everest. There are many conjectures hanging around—problems that nobody's been able to prove or disprove—and they represent standing challenges for young mathematicians to try their muscles on.

Jerzy Neyman (left) with Robbins.

MP: *But what about new breakthroughs or discoveries not rooted in historical precedent?*

Robbins: I wasn't able to prove Goldbach's conjecture, but I did invent empirical Bayes, stochastic approximation, and tests of power one. That's something like saying, "I failed to win Wimbledon, but I invented a new game called clinker ball, and I was the local club champion when there were only a few others who played it." Personally, I would rather have done some of the things I've done than some of those other things which I haven't. In a sense, I'm simply saying that I love my wife, and I'd rather have married her than somebody else who might be more famous. To have proved that π and e are transcendental were great accomplishments. These were outstanding problems, and everybody knew that anyone who could solve them would be famous. But, in a sense, nothing much came of it; nothing was created that wasn't there before. To have created some important new fields that didn't exist before doesn't make me a great mathematician, but it does contribute significantly to the general progress of the mathematical sciences.

MP: *So, in mathematics, there seem to be two different types of activities?*

Robbins: Yes. One is to find the answers to problems that have been raised earlier by others, and the other is to create techniques which will then find the problems to which they can be applied.

MP: *Let's broaden our focus. How are mathematicians regarded by society at large, and how does this compare with the public's view of physical and social scientists?*

Robbins: The public has a terrible fear of mathematics. I think it's quite real, and it's not going to be overcome by restructuring the curriculum or anything else—say, like painless dentistry. The ability and the desire to think abstractly and rigorously is not generally fostered in our society. Most people haven't the faintest idea of what mathematicians do, how they think, or what they contribute to society. Mathematicians are regarded with a sort of awe that attaches to any scientist—although we're not really scientists—because we are engaged in a very elusive form of activity.

What Does It Feel Like to Be a Mathematician?

MP: *How do you feel about being a mathematician?*

Robbins: Let me answer your question this way. Most people acquire a certain expertise, and they work in fields where their expertise can be used. I don't have any expertise. If I were a Picasso, I could wake up in the morning and say: "Well, I think I'll paint a Picasso today." And by the end of the day I would have painted a real, genuine Picasso. Although it may not be one of my best, it would be another Picasso and it would be discussed by art critics and sold to collectors, and so on. Another day, another painting. Now if I get up in the morning and say, "I think I'll do something in mathematical statistics," at the end of the day I've got a wastebasket full of paper and nothing to show for it. And likewise the next day, and the next. I cannot do something by willing myself to do it, and what I finally produce is usually complete junk. I've probably wasted more paper than any mathematician in the world. I have no idea whether I'll ever do anything worth talking about for the rest of my life. I'm not even like a dentist who comes home and can tell his wife: "Today I did three fillings and two root-canals, and I saved several people from serious tooth decay. Now let's have dinner." What did I do today? I talked to a few people. I tried to think about something and it came to nothing. Finally, I found that I was just repeating what some other researcher had already done. The day's been a total loss.

MP: *Doesn't this place a pretty severe burden on one's self esteem and character?*

Robbins: Most mathematicians are unable to cope with this. I see so many who have stopped working, or are just repeating themselves and basking in former glory. There are so many ways this emotional deprivation can get to you—the fact that you're just looking at the interior of your skull as though you were inside an egg, and there's no world except what you see inside. In most cases, there's no real contact with humanity, history, or culture in general.

Teaching and Learning Mathematics

MP: *Statements have been made to the effect that the good researcher who is a good teacher (undergraduate, or even graduate) is the exception. What has been your experience: are good researchers usually poor teachers?*

Robbins: Good researchers are often poor teachers; bad researchers are almost always poor teachers. The reason that you have poor teachers is that you have poor persons: undeveloped, ignorant, intellectually poverty-stricken individuals who have nothing to offer their students except the subject matter itself. They have no joie de vivre, enthusiasm, or curiosity for learning. They'd be poor in any profession.

MP: *Do you enjoy teaching? What, in particular, do you like and/or dislike about teaching?*

Robbins: I like to think that I'm a teacher by profession; research is what I do for fun. I want to show people what I've seen, that no one else has seen, so that they can share it with me. My teaching is like a man struggling with a bear. You don't know how it's going to come out, the result is not preordained. But that can be very painful too. Teaching should be like a competition between two antagonists with the outcome really in doubt. And yet you don't want it to be a clumsy job. Things are never settled: every answer raises new questions and begins a new cycle in the subject.

MP: *Have students changed much during your forty-five year teaching career?*

Robbins: There seems to be a regression toward mediocrity: lots of fairly good students, but not as many really bright or as many really dumb students. I don't see many outstanding, dedicated, obsessed, self-motivated freaks. Right now everyone wants to get an MBA, or to get into medical school, or into computer science, or some other highly remunerative graduate field. When I went to college in the thirties, not that many people went on to graduate study to prepare for a job in some special field. If your parents could afford it, or if you were very bright and got a scholarship, you went to college to get an education. I don't want to set myself up as a critic of today's youth, comparing them to a utopia that I envisage in the past. Nevertheless, I have a feeling now of teaching in a trade school; I didn't have the feeling of being a student in a trade school.

MP: *People do learn to learn differently, and today's youth seem to be getting an increasing diet of television, videogames, computers, and other interactive modes of learning. Will future students find classroom lectures uninspiring, and textbook or informational reading unbearably dull?*

Robbins: I have three children who spend a lot of time watching television and damn little time reading books. I don't know if any of them is going to get into college, or what kind of college it will be if they do get in. All I know is that they're a lot different from what I was like. When I was their age, I used to go down to the public library after school, and come home with an armful of books. I'd read them all before the next day—I must have read every book in the library. I don't believe that expertise at computer programming and interactive this, that, and the other thing is any substitute for the written word and the human voice. I don't have a home computer myself, and I'm not anxious for children to learn programming at an early age. I'm still hoping they will learn to read, think, and interact with people rather than machines. Anyone who reads a newspaper will see that parents are now being told that computers are the secret of success. "Send your children to computer school on weekends so that they'll get that edge in the race for success." We could all be replaced by computers, I'm sure. This would be advantageous to the efficiency of computations, but it's not the kind of world that interests me.

MP: *Will any important subjects become much easier or much more difficult to teach in the future because of changing technology, student intellect, or societal values?*

Robbins: Roughly speaking, this is the same as asking what will be the world's record for the 100-meter dash in the year 2500. I think we've gone about as far as we can go (maybe someone can shave half a second or so from the record) unless we mutate into a strikingly different breed. There's just so much energy and so much time for training, and that's it. The subjects of mathematics? I do not see them, as a result of efforts by some future Bourbaki, becoming simple and within the grasp of young children, so to speak, without painstaking introduction, slow step-by-step increments, and historical approaches. I don't see any reason to believe that there's going to be any great simplification or greater accessibility of mathematical knowledge in the future, no matter what amount of technology, training, or machinery is used. Nobody is going to run 100 meters in five seconds, no matter how much is invested in training and machines. The same can be said about using the brain. The human mind is no different now from what it was five thousand years ago. And when it comes to mathematics, you must realize that this is the human mind at an extreme limit of its capacity.

MP: *What about rapidly expanding scholarly disciplines such as mathematics: will more education and graduate study or training be required of future students who want to begin a meaningful career in mathematics?*

Robbins: I think less is being demanded now than used to be. You don't have to know any foreign languages, for example. If someone can write a doctoral thesis, all other deficiencies will be forgiven, and he'll get his Ph.D. even if he's never taught and has never convinced anybody that he knows the difference between mathematics and computer programming. There's a job market out there eager to swallow up such novices. They don't have to earn Ph.D's; an M.A. is fine, even a B.A. in mathematics is fine.

Knowledge and Power

MP: *We know that knowledge is power. But power is also knowledge insofar as prevailing political systems mold and determine what knowledge is created and utilized. Should scientists promote the creation and development of all kinds of knowledge for the sake of knowledge and the enlightenment of mankind, or should there be limitations—external or self-imposed—to the quest for knowledge?*

Robbins: Well, let me mention a remark that J. Robert Oppenheimer once made. As you know, his attitude toward the H-bomb changed from being opposed to being in favor of it. Oppenheimer said that originally he was opposed to the H-bomb because it served no useful purpose. But once a really clever way of making it had been proposed, it was so "sweet"—from the point of view of physics—that it was impossible not to try it. My blood ran cold when I read that. What kind of enterprise were we engaged in when something can be so technologically attractive that, even though it may involve the death of millions of people, a scientist must do it because of its scientific sweetness? One of the things I'm happy about is that I didn't work on nuclear weapons. I know many mathematicians who contributed to producing fission and fusion weapons. I'm glad I didn't. But then, nobody asked me to.

MP: *What about those who teach mathematical techniques that may be used for destructive purposes?*

Robbins: I hope that nothing I do will be used for purposes I don't approve of, but I know perfectly well that it will. It's inevitable. There's nothing I can do about it. As a teacher, I have become increasingly alienated from teaching because it gives me so little opportunity to explain to students that technique is not what it's all about: the desire to prove theorems is not what made me go into mathematics; there's more to life than learning how to get a Ph.D. I'm trying to tell them that the world we live in is not what it should be, and that they should spend most of their time not directly learning techniques, but rather learning what the world ought to be like, and how they should act to help make it so. What worries me is that, in my own student days, I had the benefit of contact with a very small number of people whose lives—not written words—influenced me profoundly. My students know nothing about me outside the classroom. They have no idea of how I live, why I'm doing what I'm doing, or what I think about the world. I feel frustrated. I can't turn my classroom into a pulpit; I'm in the wrong profession for that. But I damn well don't want to teach arc-welding to a bunch of robots who'll go out and arc-weld everything in sight for whoever's paying for it. So, in a sense, I might just as well admit that I'm not all that different from those who worked on nuclear weapons, because I'm teaching techniques to young people without knowing what use they'll make of them, or whether they understand what it's really all about. And I'm afraid they don't in most cases. How can I tell them?

MP: *Do you feel that scientists have a responsibility to become involved in issues of social concern?*

Robbins: Nobody has the responsibility to extend himself into a field beyond his competence. But if you feel this is something that concerns you, it's your duty to become involved. Otherwise, you'll be frustrated and bitter, and the world will be all the poorer. Proving theorems should be permitted to anyone. But if you're not an idiot in the Greek sense of being a private person, then you'll want to talk to others—scientists and nonscientists—about issues that concern everybody. And you will do so. Mathematicians have a very poor track record in this respect; the one exception being their participation in issues of human rights. The proportion of mathematicians defending human rights probably exceeds their proportion in the sciences as a whole. Perhaps I should mention that the chairman of the National Academy of Science's Committee on Human Rights is Lipman Bers. In Russia, many mathematicians try to support the Helsinki Accords.

MP: *To what do you attribute this?*

Robbins: Part of the reason may be that their original concern for issues of human rights led them into careers as mathematicians. People who have a predilection for resisting abusive social policies will tend to prefer activities which are not directly useful to their government's enactment of these policies. In Russia, for example, if you wanted to work in the sciences and not be controlled by the Party apparatus, you'd choose a science as far removed from practicality as possible. Pure mathematics would be a good career choice—in which case, the State would be more likely to place you at an institute not directly concerned with military matters. It seems quite unlikely that algebraic geometry can be used for military or political purposes. The Soviet Union can afford to allow very capable mathematicians to do pure research since they'll bring credit to the State in an indirect manner. Most other scientists are directly involved with something that can be used by the State.

Statistics and the Law

MP: *Have you ever used your mathematical expertise in matters of civil importance?*

Robbins: During the last ten years, I've become interested in the applications of probability and statistics to legal proceedings. Recent developments now make statistical evidence not only admissible in court, but preponderant in certain cases. In the past, someone would file a discrimination suit as an individual, citing direct anecdotal evidence of being denied fair treatment. Today, however, one files a legal suit as a member of a class, and the evidence is the data on how people are being hired or rewarded as a class. Thus, the evidence is statistical: although no single individual can be said to have been maltreated, the class as a whole may have been found to be treated unfairly.

A really serious problem that emerges is due to the fact our purely scientific statistical apparatus is not really well adapted for legal proceedings: statistical tools created for quality control in the chemical industry must not be misapplied in deciding issues of discrimination. This problem is exacerbated by the use of computer programs, since one can feed numbers into a computer and then interpret the output any way one wishes. Calling something evidence of discrimination doesn't really make it such. Imagine how difficult it must be for a judge and jury to interpret this type of evidence. As a consultant in legal matters, I find over and over again that "evidence"—results based on putting numbers into formulas and computer programs—is being misrepresented and totally perverted by statistical "experts."

R. A. Fisher (right) with Robbins at Chapel Hill, North Carolina, 1950.

MP: *Is there anything you still want to accomplish?*

Robbins: I'd give up my next five papers to write a good string quartet, but I've never been able to. In fact, I don't seem to have any choice in the matter. Sometimes, maybe at 2 a.m., I'm awakened by a feeling of someone knocking: "Hey Herb, you've been to the movies, you've taken your kids to the beach, and you've socialized with the neighbors. Now let's get back to business. You never really figured out what went on in this problem, and you don't even remember where you left it. But I remember, so please get up. We've got some work to do." Finally, I get up and start working, feeling as if I've been Cinderella at the ball and, now that midnight has struck, I've got to go back to cleaning up my mathematical house. After all, that's what I'm for. The last few years of my career have been unusual—I've actually gone back to working on empirical Bayes and stochastic approximation after a thirty-year hiatus. I feel that I didn't do quite as much as I should have, and no one else has done them justice. Once I give them another push in the right direction, I'll be able to relax and not worry about them.

MP: *Albert Schweitzer once said that the great secret of success is to go through life "as a man who never gets used up." At age sixty-seven and still going strong (more than eighteen publications in the last five years), what's your secret for not getting used up?*

Robbins: The one thing I must express is my great fortune in having found some wonderful young people to collaborate with. They've helped prolong my mathematical work far beyond what it would have been in isolation. Of course I'm not doing pure, abstract, postulate-theorem-proof mathematics. I'm involved with the mathematics that relates to the mysterious and fascinating phenomena of chance that I see around me. I'm trying to create methods for looking at the real world, and I'm better able to do this now than when I was young because I know more about the world and its problems. I'm trying now to solve some mathematical problems that I've been thinking about for fifty years.

Perhaps the real difficulty in answering your question stems from the fact that you're interviewing a sixteen-year-old kid who happens to be inhabiting the body of a sixty-seven-year-old man. You're looking at the body, but I'm afraid you're listening to the kid.

RAYMOND SMULLYAN

RAYMOND SMULLYAN

by Ira Mothner

Remember "The Lady or the Tiger?", Frank Stockton's classic story of the prisoner forced to choose which of two doors he will open? Behind one waits a lovely lady. Behind the other crouches a hungry tiger. Pick the right door and he gets to marry the lady. Pick the wrong door and he gets eaten.

Now, Stockton's prisoner didn't have a clue to what he'd find behind the doors. But suppose he had. What if there were signs on the doors, and—to make matters trickier—he knew that only one of the signs was true? The other had to be false. The sign on Room One read: IN THIS ROOM THERE IS A LADY, AND IN THE OTHER ROOM THERE IS A TIGER. The sign on Room Two read: IN ONE OF THESE ROOMS THERE IS A LADY, AND IN ONE OF THESE ROOMS THERE IS A TIGER. With that information he chose the right door. Can you?

If you have figured it out, then you are on your way to being hooked by Raymond Smullyan, author of a book called *The Lady or the Tiger?* (just published by Alfred A. Knopf) that takes from the Stockton tale its title and double-door dilemma. Called by the publisher a collection of logic puzzles, paradoxes and other curiosities, the book is much more than that, and Smullyan is not just a clever puzzle maker.

Professor of mathematics and philosophy at the City University of New York (CUNY), Smullyan, 63, is the author of two highly regarded works on mathematical logic, and published his first popular book just five years ago. There have been five more since then.

A spare, angular man, he seems taller than his slightly more than six feet and, with his full beard and long white hair, looks a bit like an Old Testament prophet. But the Jeremiah image is demolished by a boyish grin, and his life is full of the paradoxes he enjoys so much. Smullyan, the college professor, was a high-school dropout. He is a concert-caliber pianist who hardly ever plays in public, and a basically shy man who adores appearing on the lecture platform or reading aloud from his own works (breaking himself up at almost every humorous line). He is a logician who has embraced the Tao, a talented astronomer who has set aside his telescope, and a magician with professional skills and patter who once earned a living working nightclubs in Chicago.

As the author of *The Lady or the Tiger?*, Smullyan is much like Alice's white rabbit (the one with waistcoat and watch who leads Lewis Carroll's heroine down the rabbit hole and into Wonderland). Smullyan lures readers deeper and deeper into mathematical logic with tricks and games, broad humor and a bizarre cast of borrowed and original characters. Popping in and out of *The Lady or the Tiger?* and his earlier book of logic puzzles (*What is the Name of This Book?*) are Carroll's Alice and Humpty Dumpty, Edgar Allan Poe's Doctor Tarr and Professor Fether, plus an assortment of werewolves, vampires, knights and knaves and Smullyan's all-purpose problem solver, Inspector Craig of Scotland Yard.

Just about all of Smullyan's puzzles are fun: many seem simple, but few are as easy as the first lady-and-tiger problem. (Some other fairly easy ones appear on these pages; the solutions are on page 307.) If, as I, you had trouble with that one, don't despair. Logic puzzles take some getting used to. Look at it again. Remember, one sign must be true and one false. Now, if the sign on Room One were true, then the sign on Room Two would also have to be true, and that simply cannot be. So, the sign on Room Two must be true, and the sign on Room One false. Since the false sign on Room One says the lady is in that room, she clearly is not. She must be in Room Two.

The Smullyan puzzles stand out brightly against the rest of today's recreational logic. Martin Gardner, who wrote *Scientific American's* "Mathematical Games" column for 24 years, marvels unabashedly at Smullyan's inventiveness. "His output of brand-new problems is absolutely fantastic. Aside from the old chestnuts he puts up front in his books, the problems are all original."

What makes the books more than just collections of increasingly complex brain teasers is the ground the puzzles cover, and Smullyan's great trick is how many readers are able to tag along. Because even his most daunting puzzles are simply put, and require so little knowledge of math or formal logic, Smullyan draws even dogged and defensive nonscientists further into his books and deep into the other culture, toward the foundations of modern mathematics. The puzzles move from propositional logic to concepts like set theory and formal systems. In both books, they reach Kurt Gödel's Incompleteness Theorem.

A landmark piece of mathematical logic, Gödel's Theorem blasted the age-old dream of a single, vast mathematical system, a kind of mathematical truth machine, that could prove any proposition true or false. It set mathematicians seeking instead for limitations on mathematical systems of reasoning and limitations on the computers that depend upon them. What Gödel showed was that any system of logic, except the very simplest, could be made consistent—free from contradictions—only by allowing the formulation of certain propositions that could not be proved either true or false. In less exotic terms, Gödel proved that one could never construct a system of thought that is complete—that does away with uncertainty and paradox—which is just what logicians of the time were trying to accomplish.

By basing his puzzles on Gödel's proof, the discoveries that followed it and other modern mathematical basics, Smullyan puts readers in touch with ideas that even technologically ignorant folk now feel some need to grapple with. The need grows out of anxiety, as we watch mathematics' most enterprising offspring, computer science, plugging into just about every aspect of our culture.

The Politician Puzzle

A certain convention numbered 100 politicians. Each politician was either crooked or honest. We are given the following two facts:
- At least one of the politicians was honest.
- Given any two of the politicians, at least one of the two was crooked.

Can it be determined from these two facts how many of the politicians were honest and how many of them were crooked?

(Answers to puzzles are on p. 307.)

The Case of the Smithsonian Clocks

Two friends, whom we will call Arthur and Robert, were curators at the Museum of American History. Both were born in the month of May, one in 1932 and the other a year later. Each was in charge of a beautiful antique clock. Both of the clocks worked pretty well, considering their ages, but one of them lost ten seconds an hour and the other gained ten seconds an hour. On one bright day in January, the two friends set both clocks right at exactly 12 noon. "You realize," said Arthur, "that the clocks will start drifting apart, and they won't be together again until—let's see—why, on the very day you will be 47 years old. Am I right?" Robert then made a short calculation. "That's right!" he said.

Who is older, Arthur or Robert?

For ignorant grapplers yearning to learn more, Smullyan makes a dandy guide. He teaches logic at CUNY Graduate Center and mathematics at the system's Lehman College. "Elegant" was the adjective most scholarly reviewers grabbed for first when his works, *Theory of Formal Systems* and *First Order Logic*, appeared. "What characterizes Ray's books and papers," says George Boolos, professor of philosophy at MIT, "is how astoundingly clear they are. He is able to strip away what is inessential and give you the core of an idea, undiluted and unvarnished. He is a great simplifier,"

In addition to his puzzle books, Smullyan has produced two other works of recreational logic, *The Chess Mysteries of Sherlock Holmes* and *The Chess Mysteries of the Arabian Knights*, both filled with nontraditional chess problems that challenge the reader to discover not what moves come next but what moves occurred last. Turning almost 180 degrees away from logic, he has taken a wry and thoroughly approving look at Eastern thought in *The Tao is Silent* and published a kind of catchall collection of paradoxes, personal essays and philosophical speculations in *This Book Needs No Title*. Another collection of logic puzzles, *Alice in Puzzleland*, is already in galleys with the more serious if whimsically titled *Philosophical Phantasies* set to appear early next year.

How does he do it all? "Consecutively," explains his friend and former student, Melvin Fitting, who also teaches mathematics at Lehman. "Some people do things simultaneously. He doesn't. Ray has always been episodic in his work. He will get interested in something and more or less abandon everything else. He wrote an essay at one point," Fitting recalls, "and then, for the next couple of years, there was this enormous stream of essays. Pretty much everything else stopped, and the house was filled with piles of paper.

"After the essay stage," Fitting goes on, "he started doing puzzles. For the next two or three years, everything was puzzles. They were finding their way into all of his work. Now he's gone back to math, but the puzzle element is still there. Often, when he is telling me a new theorem, he will tell it in the form of a puzzle and then give it to me in straight math."

Smullyan does most of his work at a large table in the living room of his home near Hunter Mountain in the Catskills, more than 100 miles and nearly three hours away from Manhattan. Twice a week, on Monday and Wednesday, he heads into the city to teach.

These days, the telescope he made (grinding his own six-inch mirror) is packed away, and he rarely touches either of the pianos or the clavichord in the house. Music is provided by Smullyan's Belgian-born wife, Blanche, also a pianist, who ran a Manhattan music school before the couple moved to their home in tiny Elka Park—a community so small that an issue of the *Journal of Magic History* addressed to Raymond What's-his-name actually reached Smullyan. In the two-story house, manuscripts, proofs, and journals are stacked on just about every horizontal surface, and shelves are filled with books of philosophy, religion and handsomely bound editions of Thackeray, Lamb, Defoe, Trollope, Eliot and others, all acquired at bargain prices by Smullyan, who does not read much fiction these days. He does prize, however, his Edgar Rice Burroughs first editions.

Far from ascetic, he is an unreconstructed smoker with a teen-ager's sweet tooth and a positive passion for packaged pound cake. (Blanche keeps stacks of his empty cake containers on the porch to use for growing her seedlings.)

At home, Smullyan spends almost all of his time working quietly, filling pads of lined white paper with his jagged scrawl, while Blanche makes the most of healthy outdoor living, gardening in summer and getting wood for the stoves in winter. Her husband, she notes with affectionate resignation, does neither. "He does no exercise at all," she claims, which is certainly living up to what he calls his "idealized self-image" as "a philosopher of leisure, idleness, and quietude" and altogether in keeping with the Taoist doctrine of *Wu Wei*, action through inaction.

In the classroom, Smullyan is anything but leisurely or quiet. Early this spring semester, I watched him teach a graduate-level logic course, as he lurched to the blackboard (where he writes in a serviceable hand and complete sentences) and paced about his desk, fidgeting and chuckling. He would break into a small sibilant laugh at problems that seemed to leave his students more confused than amused.

Before class began, he tried to warm up the group, tossing out some simple puzzles the way a baseball coach bats out grounders. "There are twin brothers," he told the dozen or so students in the crowded little room. "One lies and one tells the truth. One is named John. How can you discover which brother is John by asking one three-word question?

The answer Smullyan gave them, allowing that there might be others, was the three-word question, "Is John truthful?" As he explained it, if John were indeed truthful, then John would answer Yes to the question, and his brother, the liar, would answer No. If John were not truthful, then he would also answer Yes to the question, while his brother, the truth teller, would answer No. In either case, whether John were truthful or not, the brother who answered Yes would be John.

Five-Ace Merrill Rarely Appears

Smullyan, who loves to write and tell absentminded-professor stories, real and apocryphal, could easily pass as the subject of one himself as he ambles through the corridors of the Graduate Center in midtown Manhattan, seeming neither surprised nor distressed when he wanders off the elevator onto the wrong floor. It is hard to imagine him when he played as a suave nightclub magician, Five-Ace Merrill. But there seems to be an almost infinite number of different Smullyans and the magician hardly ever appears in class. Smullyan's current teaching style, according to Malgorzata Askanas, who earned her Ph.D. at the Center in 1975, derives from his puzzles. "He sort of strings you along, and then you find yourself in the heart of Gödel's Theorem. You don't notice that you are doing what some people would consider 'hard stuff'. It all seems very easy. You just don't realize you are learning something profound."

As an educator, he holds to the notion that children are eagerly waiting to be filled with the right kinds of knowledge. "If they are interested," he declares, "they will learn." "And if they are not interested?" I ask. "Leave them alone," he insists. Smullyan cheerfully admits that he doesn't believe in grades and considers reporting a student's failure as an utter betrayal. He doesn't believe in degrees either. "Let kids learn what they want to learn," he says. "Let employers test them when they compete for jobs."

Smullyan's unorthodox views on education spring naturally from his experiences in and out of school, for he was, he admits, "a perennial dropout" and is "mainly self-taught." Born in Far Rockaway, Long Island, then a pleasant seaside village at the ultimate reaches of Queens county, Smullyan moved to Manhattan with his family when he was thirteen and attended Theodore Roosevelt High School in the Bronx to take special music courses the school offered.

Music and science were his chief interests, and he saw no reason why he could not pursue careers in both fields. "I was a bit of a prodigy," he recalls; he won a gold medal in a citywide piano competition when he was twelve. The decision to quit in mid-high school was made he explains, "because no one there would teach me what I wanted to learn," mostly modern algebra and logic.

He learned them on his own and, after several years of truly independent studies, took the College Board exams and was accepted by Pacific University in Oregon, the first of five colleges he was to attend. Next was Reed College, followed by a year in San Francisco, where he studied piano, and a return to New York to work at math and logic, begin devising chess puzzles and start learning magic.

At twenty-four Smullyan gave formal education another shot, enrolling at the University of Wisconsin, where he stayed for a year before transferring to the University of Chicago. After a semester there, he dropped out again but kept on studying, and taught music at Chicago's Roosevelt College. Back in New York for two years, he began performing as a magician in Greenwich Village nightclubs and finally went back to the University of Chicago when he was thirty.

Smullyan was clearly in no rush to wind up his studies. He took courses at the University for the next several years and moved his magic act to downtown Chicago. In 1955, Rudolf Carnap, a celebrated philosopher of science under whom Smullyan studied, recommended him for a post in the

"I'm primarily interested in puzzles related to deep ideas in mathematics."—"My favorite puzzles are those in which it seems there's no possibility of a solution."

math department at Dartmouth. He got it, based on a paper he had written, although he lacked even a high-school diploma. After teaching at Dartmouth for a year, the University of Chicago awarded the then thirty-five-year-old Smullyan his B.A., giving him credit for a calculus course he was teaching but had never taken.

In 1957, Smullyan moved to Princeton University, where he earned his Ph.D. and taught until 1961, when he joined the mathematics department at Yeshiva University. Seven years later, he came to CUNY.

While at Princeton, Smullyan had shown one of his chess problems to a student. Sometime later, the problem appeared in Britain's *Manchester Guardian*. It seemed the student's father had wanted to show the *Guardian*'s editors the kind of material he hoped they would run. The paper took the hint, and Smullyan began to publish chess problems there regularly.

Sometime in the mid-1970s a few of his chess problems appeared in *Scientific American* with the note "author unknown." Smullyan contacted Martin Gardner, who had already run some of his logic puzzles and knew him as a fellow magician. "I told Gardner I had a whole bunch and planned a book. Gardner wrote back and wanted to know why I didn't stop shilly-shallying and get the book done."

Smullyan's chess problems are not the kind that chess enthusiasts are used to. "They belong," Smullyan explains, "to the field known as retrograde analysis. Unlike the more conventional type of chess problem (which is concerned with the number of moves in which white can win), these problems are concerned only with the *past* history of the game." They pose such questions as which way the pieces were moving or which piece has been promoted. They can ask for the location of a piece missing from the board, the identity of an unidentified piece whose position is known, or the square on which a particular piece was captured.

Exercises in Reasoning Backward

What the reader gets out of the puzzles, in addition to pure pleasure, is plenty of exercise in reasoning backward, and the case for this kind of deduction was best made by old Sherlock himself (not Smullyan's but Sir Arthur Conan Doyle's) in *A Study in Scarlet*: "In solving a problem of this sort, the grand thing is to be able to reason backward. That is a very useful accomplishment, and a very easy one, but people do not practice it much. In the everyday affairs of life it is more useful to reason forward, and so the other comes to be neglected. There are 50 who can reason synthetically for one who can reason analytically."

Smullyan is not planning to delight or distress chess players with more problems in retrograde analysis, at least not in the foreseeable future. His next book out, *Alice in Puzzleland*, introduces some new examples and new varieties of logic puzzles. What Martin Gardner seems to value most in Smullyan are "the very tricky and funny ways he has of getting you into really deep water. Nobody has done that before. Lewis Carroll invented all kinds of logic puzzles, and they were very amusing, but when you get down to what they are all about, they're all about syllogisms. Ray has started where Carroll left off. He's done what Carroll might have done if he were living today."

Inspector Craig Visits Transylvania

Inspector Craig of Scotland Yard was called to Transylvania to solve some cases of vampirism. Arriving there, he found the country inhabited both by vampires and humans. Vampires always lie and humans always tell the truth. However, half the inhabitants, both human and vampires, are insane and totally deluded in their beliefs: all true propositions they believe false, and all false propositions they believe true. The other half of the inhabitants are completely sane: all true statements they know to be true, and all false statements they know to be false. Thus sane humans and insane vampires make only true statements; insane humans and sane vampires make only false statements. Inspector Craig met two sisters, Lucy and Minna. He knew that one was a vampire and one was a human, but knew nothing about the sanity of either. Here is the investigation:

Craig (to Lucy): Tell me about yourselves.

Lucy: We are both insane.

Craig (to Minna): Is that true?

Minna: Of course not!

From this, Craig was able to prove which of the sisters was the vampire. Which one was it?

Raymond Smullyan

Another admirer of Smullyan's is Douglas R. Hofstadter, author of the Pulitzer Prize-winning book *Gödel, Escher, Bach: An Eternal Golden Braid*. They both taught at Indiana University. Hofstadter recognizes a variety of different Smullyans or a series of Smullyans, one inside the other. "I have never penetrated very far into that many-layered structure," he says. "I don't know much about the inner Smullyan." Not many do.

Dr. O. B. Hardison, Jr., director of the Folger Shakespeare Library in Washington is another fan who appreciates the many different aspects of Smullyan and tells this story about Smullyan the magician: "I was at a lovely dinner party with him in New York and he was performing tricks for our host's eleven-year-old daughter. It was dazzling, astounding, almost sinister. He was pulling coins out of her ear and out of her nose and carrying on this delightful, witty patter the whole time.

"Finally, he had her clench her fist and put it on the table. He touched the top of her hand, and when she opened her fist there was a crumpled dollar bill in it. That's when the girl shrieked and ran upstairs. It was about ten minutes before she could compose herself and come back down. How Smullyan did it, I can't possibly say."

The Island of Questioners

Somewhere in the vast reaches of the ocean, there is a very strange island known as the Island of Questioners. It derives its name from the fact that its inhabitants never make statements, they only ask questions. The inhabitants ask only questions answerable by Yes or No. Each inhabitant is one of two types, A and B. Those of type A ask only questions whose correct answer is Yes; those of type B ask only questions whose correct answer is No. For example, an inhabitant of type A could ask, "Does two plus two equal four?" But he could not ask whether two plus two equals five. An inhabitant of type B could not ask whether two plus two equals four, but he could ask whether two plus two equals five.

I once visited this island and met a couple named Ethan and Violet Russell. I heard Ethan ask someone, "Are Violet and I both of type B?" What type is Violet?

Logician Smullyan has been described by some as the "Lewis Carroll of our times."

PUZZLE SOLUTIONS

The Politician

A fairly common answer is "50 honest and 50 crooked." Another rather frequent one is "51 honest and 49 crooked." Both answers are wrong. Now let us see how to find the correct solution.

We are given the information that at least one person is honest. Let us pick out any one honest person, whose name, say, is Frank. Now pick any of the remaining 99; call him John. By the second given condition, at least one of the two men—Frank, John—is crooked. Since Frank is not crooked, it must be John. Since John arbitrarily represents any of the remaining 99 men, then each of those 99 men must be crooked. So the answer is that one is honest and 99 are crooked.

Another way of proving it is this: the statement that given any two, at least one is crooked, says nothing more or less than that given any two, they are not both honest; in other words, no two are honest. Also (by the first condition), at least one is honest. Hence exactly one is honest.

The Case of the Smithsonian Clocks

We must first determine the number of days it will take for the clocks to come together. Since one clock is losing time at the same rate as the other is gaining, then the next time they will be together is when the slow clock has lost six hours and the fast clock has gained six hours. (Both clocks will then read six o'clock; of course they will both be wrong, but together.)

Now, ten seconds an hour is four minutes (240 seconds) a day, which is one hour in 15 days, or six hours in 90 days. And so the clocks will come together exactly 90 days after the day in January on which they were set right. Also, they will come together on Robert's 47th birthday, which is in May.

How can 90 days be fitted between a day in January and a day in May? Consulting a calendar, we see that there are exactly 90 days between January 31 and May 1 *providing it is not a leap year*! In a leap year the shortest possible time between any day in January and any day in May is 91 days (since February 29 falls in between). This proves that Robert's 47th birthday cannot fall in a leap year, and therefore Robert could not have been born in 1933 (since 47 years after 1933 is 1980, a leap year). Therefore Robert must have been born in 1932, and it is Arthur who was born in 1933. And so Robert is the older (incidentally, the year of the story—the year in which the clocks were set straight—must be 1979).

Inspector Craig Visits Transylvania

Lucy's statement is either true or false. If it is true, then both sisters are really insane; hence Lucy is insane, and the only insane Transylvanian who can make a true statement is an insane vampire. So, if Lucy's statement is true, then Lucy is a vampire.

Suppose Lucy's statement is false. Then at least one of the sisters is sane. If Lucy is sane, then, since she has made a false statement, she must be a vampire (because sane humans make only true statements). Suppose Lucy is insane. Then it must be Minna who is sane. Also, Minna, by contradicting Lucy's false statement, has made a true statement. Therefore, Minna is sane and has made a true statement; so Minna is human and, again, Lucy must be the vampire.

This proves that regardless of whether Lucy's statement is true or false, Lucy is the vampire.

The Island of Questioners

We must first find out Ethan's type. Suppose Ethan is of type A. Then the correct answer to his question must be Yes (since Yes is the correct answer to questions asked by those of type A), which would mean that Ethan and Violet are both of type B, which would mean that Ethan is of type B, and we have a contradiction. Therefore, Ethan can't be of type A; he must be of type B. Since he is of type B, the correct answer to his question is No, so it is not the case that he and Violet are both of type B. This means Violet must be of Type A.

OLGA TAUSSKY-TODD

OLGA TAUSSKY-TODD

An Autobiographical Essay*

(The truth, nothing but the truth, but not all the truth.)

Childhood

I was born in Olmütz, then in the Austro-Hungarian Empire; it is now called Olomouc, in Czechoslovakia. My parents had a good marriage. My father was a very interesting man, very active, very creative. He was an industrial chemist, but also a journalist who wrote for newspapers. We were three children, with three years between us, myself the middle child. My father was most anxious that we should have a good education. We were expected to do well in school and to have a profound sense of duty all around. He wanted very much that we should seek careers connected with the arts, but we all took to science.

My mother was a country girl. She was rather bewildered about our studies and compared herself to a mother hen who had been made to hatch duck eggs and then felt terrified on seeing her offspring swimming in a pond. She was not an educated woman, but she was intelligent and practical. She had a mind of her own. She was able to manage the household with three little children during my father's many absences in faraway countries where he did consulting work (outside of World War I). She was a rather quiet lady. She was educated to be a housewife and she made a nice home for all of us. Some evenings when I did not fall asleep readily I heard my parents in the kitchen making a late supper for themselves and the relaxed tone of their voices made me feel good. In some ways she was less old-fashioned than my father. The idea of us children using our education later to earn our living seemed all right to her, but not to him.

Shortly before I reached the age of three, my parents moved to Vienna. There we stayed until about the middle of World War I, and then we moved to Linz in Upper Austria when my father accepted a position as director of a vinegar factory there. I had started school in Vienna, and as was customary in Europe then, I had no training whatsoever in reading, writing, or counting before I started school. However, the teachers were very experienced, and somehow, at the end of the first school year I could read as well as all children could at this age. Our grades were given as 1, 2, 3, 4, 5, with 1 the top grade. I do not think that I ever received anything lower than 2, but I was not a *Lauter Einser Kind*—a child who gets only grade 1—as my older sister was. It did not seem to worry me. My best subjects were grammar and essay writing. (German grammar is not easy!) Apparently I did not always make grade 1 in *Rechnen* (this really means computing; the word mathematics was reserved for much older pupils). I conclude this because I remember vividly our teacher in *Rechnen* mentioning to me in the corridor that she could now give me grade 1 in *Rechnen*. However, I did not recall any particular achievement, nor did it particularly excite me apart from the fact that I could point to it in my report card and please our father. This was still in Vienna. By the time we moved to Linz, it was established that I was doing well in *Rechnen*. In Vienna, I had started to compose on my own—like many children at that time, we had private music lessons. When we moved to Linz, I started—again entirely on my own—to write poems whenever some event stirred me very much. This habit comes back to me to this day.

*This personal memoir was written by Olga Taussky-Todd at the request of the Oral History Project of the Caltech Archives. We originally approached Dr. Taussky-Todd about doing an oral history interview to add to our series of interviews with Caltech professors. However, she preferred to write her story and set it down in installments over the course of the spring of 1979. I worked in close cooperation with her over the next year to edit and rework the first draft into its final form.—Mary Terrall, September, 1980.

(The publisher would like to acknowledge that part of this material is published by courtesy of the Archives, California Institute of Technology.)

Life in Linz was harsh. First of all, there was the war, and the climate is rougher than in Vienna. The people are somehow heavier than the Viennese—in any case they are different. They can, however, be very kind when they get to know you. Food during the war years was slightly more plentiful in Linz, though still very scarce. In Vienna we were often near starvation; another piece of bread was a serious problem. Lining up for small pieces of food kept the family busy day and night. Even my younger sister had to take part in it at times. There was no university in Linz at that time—in fact, there was none until quite recently. Because of the lack of a university, the schools tried very hard to create an intellectual atmosphere by having high-school teachers giving popular or semi-popular lectures on various subjects. I hardly missed any of these. In those days, in a country with only a few universities, most Ph.D.'s became high-school teachers. There they were safe from the danger of having to go in for the agonizing job of doing another piece of creative research, and could still lead truly fulfilling lives through the subject they had chosen. In the countries in Europe with which I am acquainted, a high-school teacher is a highly respected person, partially due to the fact that he or she has to go through an arduous training, sometimes exceeding the work for a Ph.D. However, Vienna had incomparably more intellectual events going on, and of course, there was the big university there. A town without a university just is not the same thing.

Coming from a big city like Vienna into a small one like Linz, where we lived for several years in the last house on the edge of town with open spaces behind us, was a big change for me. I enjoyed the novelty of this change. In winter we were only a few steps from the snow-covered fields and meadows. In spring we could roll in the grass and watch the wild flowers. The hills could be reached by an easy walk.

My growing ability in mathematics must have been observed by my father, for he seemed to select me for mathematically related chores. One of them arose out of his work in the vinegar factory. The workmen had to send quantities of vinegar to the local grocers at an acidity level conforming to the law. The vinegar produced in the factory exceeds this level and hence water can be added. Sometimes several types of vinegar are mixed in addition to the water. The workmen had a pretty good idea of the proportions in question, but my father challenged me to work this out. This leads to a diophantine equation to be solved in positive integers. I apparently managed this all right and produced a little table with colored pencil entries which was posted up in the relevant room.

Another chore I was given at a very early age was to rearrange my father's magazines after he had upset their chronological ordering when searching for back numbers. My method for doing this coincided with the routine of the relevant computer program. To this day I have the habit of ordering the periodicals in libraries if I find them out of order.

I, however, kept my main interest in grammar and essays for a bit longer. This was particularly encouraged by a rather older girl, the sister of one of my classmates. This girl, too, wrote poetry and had heard about my attempts. It meant a lot to me to have an older girl as my friend, particularly when I was slowly becoming more serious about my studies, my self-education and my responsibilities.

My older sister had, of course, at all times a great influence on me, though maybe not too much at that particular time. Three years of difference in age between sisters means nothing later, but a girl of fourteen is a lot younger than a girl of seventeen. By the time I reached the age of sixteen, my sister had become a student at the university of Vienna. She commuted between home and Vienna and I did not see too much of her, although she brought me a gift every time she came home—usually a book—and these books were very much to my liking.

When we were children, my older sister talked to me about many things while I talked rather little to either on the whole. While I did help my younger sister at school and even later, in my childhood I stuck more to the older one, which I suppose is natural since she was more interesting. I was told that the younger one wept when I started school. This seems to indicate that I had spent quite a bit of time with her. My older sister is an extremely capable person in almost anything she tries. She is also very strong. She is also very kind to people. Since I left home rather early, going to Zürich, Göttingen, later Bryn Mawr, Great Britain, and then got married, my contacts with the family got weaker, though they are very strong to this day. Since we are all females and grew up in long-ago days, we should be expected to have had a good training in housewifely duties. Well, this was not the case. My father did not really want this very much. However, we all had to do quite a lot of housework. After all, we were children during the harsh times of World War I and its aftermath in a poverty-stricken country. Housework, including cooking, comes hard to me. I am by nature very clumsy and not practical. But I have always done my share. Both my sisters were far more practical for it, particularly my younger sister. But as soon as they grew up they refused it with a stubbornness that cannot easily be matched.

Maybe they are wiser! My younger sister got a degree in pharmacy and later held a research position at the Cornell Medical School in New York Hospital. My older sister is an industrial chemist who took over my father's work. She is now a much-sought consultant to many factories and a pioneer in the exploitation of the Jojoba plant, gradually adding new and more modern ideas to it.

Another person who played an important part in my life in Linz was Resa. She was the wife of one of my father's bosses at the vinegar factory. The factory was actually more than a vinegar factory; they also made jams, soft drinks and related items there. It was owned by an elderly man and his four sons, one of whom was Resa's husband. Resa was an elegant lady who came from Vienna and was most unhappy about living in a provincial town. She was interested in literature. She would have loved to have had a high-school education before marriage, but this was not yet possible in her generation for a girl. She was somehow jealous of my older sister. But when she met me she took to me at once and made no secret of this, maybe subconsciously to make my sister jealous. It was decided between my family and her that I was to give her young daughter lessons in mathematics, an absolutely hopeless task. However, I went to Resa every Saturday afternoon and spent a considerable time there, maybe more than I could afford. Resa's husband was only too pleased to have found a friend for his wife, for she was very lonely, having lost her friends in Vienna. Nobody seemed to notice that the little daughter made little progress in mathematics. Although Resa's husband was very well-to-do I had strict orders from my father not to accept a penny of payment for my tuition there, an order I obeyed even after he had gone. However, Resa gave me very expensive books of high quality—books about which we had heard in school, but barely dreamt of possessing. So there developed quite a friendship between Resa's family and all of us which lasted for quite a long time. Resa took a great interest in my future. I had to tell her about my teachers and I was often amazed when she reminded me years, even decades, later of certain incidents. Of course, my contacts with her became less and less over the years.

At the age of fourteen, I transferred to the high school (called *Mittelschule*) and at the age of fifteen to the *Gymnasium*. There we learned Latin, a fact about which I was initially very proud. There was only one school to choose from for girls. Not much later, when my interests changed rather abruptly, I would have given a lot could I have interchanged eight hours per week of Latin for the same amount of any branch of science or mathematics. However, initially I was very pleased with the challenge of grammar in this language and made out exceedingly well in it.

The war had come to an end in the meantime, but the shortages remained. Furthermore, having lost the war, the previously prosperous country plunged into a state of poverty that was to remain until quite recently. I remember the head of my school asking the children to come to the gymnasium, the largest room, and he informed us of what had happened. He then handed over to each of us a small book. This contained among other items the story about a man whose house had burned down during the night. In the morning the neighbors observed him searching through the rubble. He told them he was searching for his axe to build a new house, saying the old one was not so good anyhow. This made quite an impression on me.

In time, the worst scars of the war faded away. Food supplies improved almost daily. My family became citizens of Czechoslovakia overnight. It seemed strange to belong to a country that had just been born, a country with great resources, beautiful mountains, beautiful old cities, and great ambitions, only I had hardly ever lived in it. I myself had changed from a skinny, rather miserable looking child into a taller and slightly heavier girl. In school I was now doing very well, but I spent a good bit of time on my self-education. While I tried to read any books that came my way, the realization that the greatest wisdom was not to be gained by reading books struck me suddenly. I felt that scientific experiments provided an almost unlimited insight, even if used in a limited way at a given time. Science to me meant almost any scientific subject, but above all astronomy. Mathematics, too, came to me at that time as an experimental subject, for I started to study the laws of the integers computationally. Gradually it became clear to me that the latter was to be my subject. However, I had no idea what that meant. First of all, I was fully conscious that the fact that I was doing well at school had nothing to do with it. The work at school was really not that difficult if one applied oneself to it, but it was so uninteresting that you could not wish to apply yourself. I felt there was another mathematics.

I later found that the yearning for and the satisfaction gained from mathematical insight brings the subject near to art. While talent is undoubtedly needed by itself, it does not always make a person a mathematician. Yet most people who go into mathematics do it because they know they are good at it. When their talent slowly declines they find themselves occasionally quite lost. This happens to some people at an early age. But what are they to do then? As G. H. Hardy said in "A Mathematician's

Apology," mathematicians are not good at anything else. So, many of these people take up administrative work, thinking this is the solution. This can work out well if the person in question uses the power of such a job to help others who feel they want to go on. R. Courant was such a person. He worshipped devoted mathematicians and helped them in any way he could. However, there is also the other kind. They are the people who work their frustrations out. In my long and complicated life I have been under the influence of both types and I know what I am saying.

When I say that mathematics is at times linked to the arts, I am not the only person to say this. Many poets have expressed tremendous admiration for mathematics. I suppose the best known example is Novalis. Anatole France said that poetry was more important than mathematics, but his saying that shows that he applied his mind to mathematics.

Like many other children we had private lessons in music. Our father also took us to many concerts. He also made us do a lot of drawings. I myself found the music lessons very arduous, but I was very ambitious about them. My teacher even suggested a career in music for me. This would have been very wrong! I did however compose; this came easily to me. I also wrote poems without anybody asking me to do it. It came naturally and the poems were not bad. However, as soon as my interest in science and in mathematics awakened I saw this alone as a career, as my profession. No dreams of receiving honors had anything to do with it. Absolutely none!

Careers for women before World War I were, as far as I can remember, primarily as teachers in girls' schools, secretaries, shop assistants, domestic service, nurses, dressmakers, and things of that sort. All this was changed greatly during the war and it never went back to the way it had been, though some of the positions acquired during the war years went back to men afterwards. I remember very well that in the buses and trams, the fares were collected by women and even the drivers were occasionally women. All secretaries were now women. They had a sort of uniform: skirt and well-ironed white blouse. Nurses were given very intense training, including university courses, and their profession became highly respected. Women teachers had to have a far greater training than was required before—even Ph.D. in the better high schools. The schools expected the girl pupils to be very serious; cosmetics and hairdos were practically forbidden.

I hoped that I would make it into a university job eventually. In any case, I had at all times considered myself a teacher, and I have in fact done a great deal of teaching in my life in one form or another, starting at the age of fifteen when I gave tuition in chemistry to a classmate and even earned my first pay that way. My father did not want me to be paid, but my teacher insisted, "No pay, no tuition." My father wanted me to help the girl merely out of friendliness. He was embarrassed at the thought that I should be paid. I myself welcomed the thought of the challlenge to do the work since I felt up to it, and furthermore, I was quite pleased about the thought of earning money at such an early age (fourteen and a half). Later I did quite a lot of tutoring and my father withdrew his objections; on the contrary, it gave him a good feeling that I should be able to earn so much.

I had my first experience with creative research while still at school. At that time the government ruled (the government regulates all teaching in Austria) that the pupils in their final year were to write an essay in any subject they wished, on a topic of their own choice. I immediately decided to declare that my subject was to be mathematics. But I had no idea what I would be writing on. I defied the advice of several older people who tried to tell me that in spite of my dreams, all I knew was homework questions. This was true; however, when the date for submitting the essay came, I had written an essay entitled "From the binomial to the polynomial theorem." It described Pascal pyramids of all dimensions, instead of the Pascal triangle, and other work connected with binomial coefficients. This gave me a great deal of confidence.

Although my life between the ages of fifteen and seventeen was very busy, it was rather pleasant. This was partly because my father was able to return to the life he had led before the war; namely, working as an industrial consultant to firms in many countries. The war had stopped travel to foreign countries entirely. He had developed a number of chemical processes, some of which are still being used, with his name known to the users. After the war, he changed his contract with the vinegar factory to allow him to go on these trips whenever he wished. This made him very happy. On returning from a consulting trip, he went back to the factory, and it may have been that he strained himself too much in this way, for without any apparent illness he seemed to go downhill in strength. Added to this was the fact that he suffered an injury in a train accident. During the last year of his life, my mother insisted that my older sister, already a student at the University of Vienna, accompany him on these trips. In this way she learned much from him and was, in fact, able to take over his work later, adding new ideas to it all times. It was remarkable that such a young person could, in those days, make such difficult trips and arduous assignments on her own. But she was always a very strong and most able personality.

We lost Father in the middle of my last year at school and felt exceedingly grieved and lost. Although my father had not left us without savings, we had no income whatsoever. With a feeling of anxiety, I increased my tutoring activities to the utmost, and, in addition, took on a contract with the vinegar factory. I had observed some of my father's activities there, not realizing that I would have to carry them out myself someday. I earned a good bit more that way than I had expected at that age. Worn out with grief and responsibilities during the last years, the top pupil in my class, with the final examinations—the so-called *Matura*—in front of me, all this was no small matter. However, my teachers were my friends, the people at the factory, particularly the workmen, were my friends, and this helped a good bit.

I worked all through the summer after school was over, but was burdened by the fact that my future in mathematics was at stake. Although my earnings were not trivial, they were not more than extended pocket money, and my tasks at the factory were not interesting. I surely deserved better. Further, my family thought that I would do better to study chemistry, and to join up with my sister. In any case, what was I to do with mathematics? There did not seem a prosperous future in it. And all this was true. I spent the whole summer worrying. One day I met a lady, a friend of my family, who had heard of my dreams. She was decades older and mentioned that she too had hoped to study mathematics. That was more than I could take. In a flash, I saw myself decades older saying exactly the same words to a young woman. It seemed unbearable. I cannot say that this created the final decision. However, when the summer was over it was decided to let me begin studies in mathematics at the University of Vienna, taking also a major in chemistry, a truly wonderful subject. Eventually, my sister did very well, making trips all over Europe, later to the U.S., and even to India and Egypt. I did not seem to be needed. So I dropped the chemistry—but my younger sister returned to the family subject later.

With all this past on my shoulders, I entered the university a very worn out, grief-stricken young person, hoping to prove that my decision had not been wrong. There could hardly be a more serious, hardworking, thrifty person.

Olga Taussky-Todd, 1939.

University Days

When I entered the university in the fall of 1925, I had no idea what it meant to study mathematics. I did, of course, plan to work hard. Although I did not expect to fail, I had no idea how I would compare to my colleagues. But that was my least worry. I had come to study and not to engage in a competition. Whenever a high-school student enters the university there are two problems waiting. Firstly, is one really as good as one was made to believe by one's high-school teachers? And secondly, (this concerned mainly the continent of Europe, and may not be quite so painful nowadays) how was one to manage with complete freedom from the supervision that had regulated one's education up to a short time ago? That was particularly difficult for me, coming from a little provincial town, with no friends or colleagues in Vienna.

In those days, when you entered the university you were given two little books; the smaller one was an identity card with your picture in it and a student had to carry it all the time. This was because students might take part in political demonstrations. The second one recorded the courses you had registered for. There was a minimum of hours you had to take. However, nobody bothered whether you attended these courses. Nevertheless, you had to obtain the signature from the teacher at the start and at the end of the semester. If you had registered for eight semesters you could present a thesis, and if this thesis was found worthy by two professors, you could ask to be examined by two mathematics teachers, one teacher in a minor subject, and two philosophy teachers. The latter included psychologists, since their subject was then counted as philosophy. In principle, nobody cared what you did during the eight semesters. No credits were required for the courses. However, if you wanted to reduce the tuition fees, you could make the teachers give you a voluntary examination on their courses, a heavy chore for these teachers! In practice very few students were able to produce a thesis without having worked with and received a problem from some teacher. On the other hand, the teachers were anxious to have thesis students. The professors also had the colossal task of examining candidates for positions in high schools, where a doctorate was not always requested. This examination is called the *Lehramtsprüfung* in Austria and it leads to what is called the *Diplom* in Germany. High-school teachers in Great Britain and the United States would shiver if they knew what was involved in these countries. One could not be trained solely in mathematics or physics; to be qualified to teach either of those subjects, one had to be examined in both subjects and had to take chemistry as a minor as well. For example, my high-school mathematics teacher was trained as a physicist—he did his thesis in physics, as well as writing an essay in mathematics. Students training to be teachers had to take courses in some other subjects, too. Credits for performance in seminars and labs were obligatory. Essays in the two major subjects had to be submitted, though in the case of a Ph.D. candidate, the thesis could be used for one of the subjects. After all these and some pedagogical credits, the student was given several written examinations, and if these were satisfactory, an oral one.

I was very fortunate on entering the university in knowing right away who my teacher would be. I knew what my main subject was, and for this there was a unique choice, Philip Furtwängler, a famous number theoretician from Germany. I suppose he was the top of the mathematics department. He had no academic training as a number theoretician. He had started off with geodesy* during World War I, and studied Hilbert's work in class field theory on his own. He proved—and in a few cases disproved —Hilbert's conjectures in this field, but he never met Hilbert personally. He was nearing the age of sixty and could walk only with a cane and the support of two people, though he had been an athlete when he was younger. He could not write at the blackboard, so somebody had to do this for him. In my later student years, I took on some of this work, which was quite a challenging task. His lectures had to be well prepared under these circumstances.

He ran a big Ph.D. school, finding problems on all levels for everybody. I suppose that his best students were O. Schreier, E. Hlawka, W. Groebner, H. Mann, and A. Scholz. The students did not always see much of Furtwängler. This was particularly bad during my first year of thesis work. Furtwängler traveled to the mathematical institute by taxi, was then guided to the lecture room by two people and stayed there for two or three hours. He also spent some time in his office and a long line would form outside of this office, mainly of thesis students, waiting to see him. When your turn came you were not given much time. Occasionally, Furtwängler saw students in his apartment in the suburbs of Vienna. After you had completed your studies he welcomed even unexpected visits, mainly because his ill health made him feel isolated. But he cut his teaching duties quite frequently, particularly when the streets were icy.

---

* At one time his determination of "g, the constant of gravitation" was the best.

He gave an introductory course in number theory in my first year, which I was very pleased with. In my second year he treated algebraic number theory in a two-hours-a-week seminar without homework, and even included some of his work in class field theory. At the end of that year I called on him and asked if I could write a thesis in number theory. He immediately decided it was to be in class field theory. This was a great honor. This decision had an enormous influence on my whole future, in a positive, but also in a negative way. From the positive angle, it helped my career, for there were only a very few people working in this still not fully understood subject. It was definitely a prestige subject. It led to my appointment in Göttingen as one of the editors of Hilbert's collected works in number theory and made me a known number theoretician at an early age. With my need to enter the job market as early as possible, that was very beneficial. On the other hand, I had a very tough time as a thesis student. I had no colleagues whatsoever and hardly saw my teacher, who for quite a while did not direct me towards a specific problem. He had had a girl student in class field theory previously, but she developed TB and spent several years in Switzerland. She finally returned to Vienna, asked Furtwängler for an easy subject, and wrote a thesis in almost no time.

While I was struggling by myself with the different literature, E. Artin had developed a most ingenious method for translating one of the then still unsolved major problems, the principal ideal theorem, into a statement on finite non-Abelian groups. Furtwängler did actually tell me a little about this, but without explanations, and made me almost desperate. In the meantime, he proved Artin's group theoretic statement to be true and hence solved the principal ideal theorem. This was a tremendous achievement, but the world of mathematics was not very grateful, and considered his proof as ugly. In fact, they had little admiration for his earlier pioneering work either. In spite of my grievances against him as a teacher, I feel that his work deserved better credit. I have now been a Ph.D. teacher for many years and none of my students is being treated the way Furtwängler made me suffer. Of course, usually students go through a period of frustration, but I do not let them go through what had happened to me.

However, after his success Furtwängler became a bit more humane and announced that he now had plenty of problems, by applying Artin's method to other questions. Unfortunately, that was not the case. This particular set of problems is still very tough. Artin called them hopeless problems. The problem that Furtwängler assigned to me then concerned odd prime numbers. He had already solved it for the prime number 2 but did not show this to me. He knew that it would be different for the odd ones. After some struggle, I did indeed solve it for 3. While trying to generalize it for prime numbers larger than 3, I made an unexpected discovery which helped me to pay my teacher back a little for his meanness. I found that every prime number p behaves differently, and actually it all depended on $p - 2$. Since this is 0 for 2, this number appears an exception, but can actually be fitted into the general picture. In any case, my results showed that the problem was not a very attractive one. Furtwängler left the whole subject from then on and devoted himself to geometry of numbers. I was left with achievements in finite group theory rather than in number theory, and they were p-groups, one of the toughest areas of group theory. In Germany there were great teachers, like Artin, Schur, Hasse. They also were more modern than poor sick Furtwängler, who was much more isolated. It would have been wonderful to be their student, but I am nevertheless grateful for what I did after all learn from Furtwängler.

I recall that at the height of my desperation over my thesis problem, one of the *Privatdozents*, Walter Mayer, asked me how I was getting on. I mentioned that I saw no progress. To this he replied, "Remember that you are not married to Furtwängler." I understood this. He was in search of thesis students and would have gladly given me a problem and helped me with it. But his subject was n-dimensional differential geometry—he became an assistant of Einstein later—and I was married to number theory.

Mayer was also one of the little mathematics crowd that met in the Café Herrenhof. Actually he was the one who introduced me there. He had private means; among other sources of income he owned a café near the mathematical institute. But he very much wanted to be a professor. He did not care very much for Professor Hans Hahn and even less for Hahn's young protégé, Karl Menger. He aired his feelings quite loudly in front of students and I did not quite approve of this. In order to gain some sympathy from the established professors, and also simply to be useful to the department, he offered himself as assistant to Hahn. Hahn handed over to him the preparation of student lectures in Hahn's seminar. Hahn insisted that these lectures were to be rehearsed by an assistant so that they were reasonably sound. In my third year, at the beginning of the semester, they had not yet found a victim for the first lecture in Hahn's seminar. Someone suggested me to Mayer for this. So Mayer worked with me and the lectures turned out well. So a slight friendship developed. Although his subject—*n*-dimensional differential geometry—was not in my line, I registered for his course, but would actually have liked to leave it. This was quite impossible since he would have been left with only two students. Before the end of the semester I did leave after all because I was overworked. He never forgave me that.

When I was still in Vienna we heard that Einstein was looking for a mathematical assistant to work on differential geometry. I think Mayer's name was suggested to Einstein by some of the former Viennese mathematicians settled in Berlin at that time. Mayer was ideally suited for this and made out very well. Einstein took Mayer with him to Princeton. I think after some time Mayer did not work there any longer as assistant, but on his own. When still in Vienna he had started to work on algebraic topology, in Princeton he worked on Lie groups, at least at the time when I met him there in 1934–35 during my visits from Bryn Mawr. After that, the next time I saw him was in 1947–48 when I worked in Washington, when we spent some time in Princeton working with von Neumann. He was already rather ill then, and died not very much later.

Olga Taussky-Todd with Kurt Gödel in the early thirties.

In my first year at the university I took courses in chemistry, as mentioned earlier, but also in astronomy, hoping to make it the subject of my minor. However, the two professors in astronomy, Oppenheimer and Hepperger, seemed so eager to catch students that I feared they would not let me treat the subject as a minor and so I withdrew from it in my second year. Another subject I included was philosophy of mathematics. I attended a course by Schlick, then one of his seminars, and even later the meetings of one of his private circles. He ran an even more esoteric circle to which I could later have been admitted, but by that time I had withdrawn from these studies in order to spend more time on my mathematical pursuits. Like Bertrand Russell, Schlick and his followers combined philosophy with science and mathematics. This is again very much in the limelight because of the achievements of Gödel, who was a student of Schlick and a member of the same seminars which I attended, though he had his academic home in the mathematics department. He had taken his Ph.D. in mathematics, and he became a *Privatdozent* in mathematics several years later. He knew a great deal of mathematics and you could talk to him about any branch of mathematics. If you asked him a definite question which required some mathematical manipulation, he would write it down in logic symbols. He was an enormously gifted scholar; discussing any subject with him was a rare intellectual pleasure. I feel flattered by the fact that he spoke to me frequently. I am not surprised that Einstein valued him.

When Gödel had arrived at his most famous result concerning undecidability, Hilbert, on one side, ignored it, and Zermelo on the other side, claimed that he had known this anyhow and that it was not very impressive. (At least this is what I heard.) Zermelo was apparently a rather frustrated man who had some justified grievances (the details of which I never found out about). He did not wish to meet

Gödel when these two scholars attended the same congress. Some people had planned a lunch in an inn on top of a small mountain and I was invited. Some friends of Zermelo were in the group and they thought he ought to meet Gödel. But Zermelo had mistaken somebody else for Gödel and replied he could not speak to somebody with such a stupid face. Well, the misunderstanding was explained to Zermelo. But then he said there would not be enough food if we also invited Gödel. Finally he said climbing the mountain would be too much for him. Finally, Gödel, who knew nothing of all this, was somehow introduced to Zermelo. And then a miracle happened almost instantaneously. Only seconds later the two scholars were engaged in deep contemplations and Zermelo walked up the mountain without even knowing that he did it!

Philosophy of science and mathematics is a subject much cultivated now and linked to many branches of pure mathematics, but this was not the case then. Wittgenstein was the idol of the Wiener Kreis, the name of this particular group in Vienna. I never saw him there, but his famous book, the *Tractatus*, was used to settle all disputes, as the last authority. The Wiener Kreis was a mini-association. I was probably the youngest in age there and I did not associate myself with it for the purpose of working in it, but in the expectation of using their ideas to further my mathematical work. This did not work out for me then. Hence I left it. It was only much later that logic was used to prove mathematical theorems, notably by Abraham Robinson. The Wiener Kreis was concerned, if I understand it correctly, with continuing the development of a language for science and mathematics. It differs from other movements of this kind by stressing science. It was a successor of the Mach Verein (Mach Club). Mach was a scientist and philosopher, but not a mathematician. I understand that Schlick was the founder of the Wiener Kreis and that, while Wittgenstein and his *Tractatus* seemed to me the idol of the group (it was a changing group), they really wanted to improve on him somehow.

Among my other teachers were Wirtinger, an expert on algebraic functions, and Hans Hahn, of Hahn–Banach fame. He himself considered his characterization of *Streckenbild* his best achievement. Unfortunately, hardly anybody knows nowadays what a *Streckenbild* is, but his name is not forgotten. Then there was Karl Menger, the son of the famous economist. He was Hahn's student and pride. He was very talented and had many original ideas. He wrote a book on *Dimensionstheorie*, and another on *Kurventheorie*. After these books were published, he applied himself to the study of abstract sets in which a distance is defined which satisfies certain axioms. He studied the embedding of such sets into n-dimensional Euclidean spaces and obtained interesting results. My own ideas fitted into some of this research and were incorporated into a monograph by L. M. Blumenthal who had come to work with Menger in 1934–35. Menger's student Wald, who later made a name for himself in statistics, started off in connection with these ideas and even contributed to some of mine. Another visitor to Menger's circle was G. T. Whyburn who with his wife Lucille was very active in topology. Later Whyburn became one of the pioneers in analytic topology. Menger's work on embedding n-dimensional spaces into Euclidean spaces was continued by his student Nobeling. The lowest possible dimension for such a space was given by van Kampen and by Flores. Another teacher at Vienna was Helly, whose courses in algebraic geometry and non-Euclidean geometry I attended. He was a very scholarly man whose pioneering work on sequence spaces and on convex sets has gone into history. He was also very interested in the flourishing of the department, and in particular the students, irrespective of the fact that he was only a *Privatdozent*—he had no salaried academic position and had to earn his living outside the university. He was a friend to all of us. Then there was W. Mayer whom I mentioned previously, and Vietoris, who made a name for himself in algebraic topology as a very young man. (Many, many years later I saw him at the University of Innsbrück in Austria and he said that he remembered me from my student days. But since he always mistook me for another girl student I do not know which of the two of us he meant.)

Among colleagues was Franz Alt whom I re-encountered many years later at the National Bureau of Standards, a Ph.D. student of Menger, a man helpful whenever help was needed. Henry Mann was still a student then, due to the fact that he had to earn money first. He began by rediscovering already known results, but later solved the famed $\alpha-\beta$ problem. My last semester at the university was spent in Zürich.* My thesis was completed by that time. I was even allowed to lecture about it on the weekly colloquium of the department in Zürich. I attended courses by Speiser, Fueter, M. Gut, Plancherel,

---

*Students had to be registered for four years to get a degree, but all four years did not have to be spent at the same university. My uncle lived in Zürich, and he invited me to live with him for a semester and finish my university requirement there.

and Pólya. The latter gave me some good advice concerning lecturing. He attended my colloquium lecture and did not approve of my style. I am grateful to him to this day. I still had my oral examinations in Vienna looming over me, but during my semester in Zürich, I somehow relaxed for the first time in many years. After receiving my doctoral degree I continued my studies, earning some money by tutoring, and doing unpaid work in the mathematics department, and continuing with research I had started during my student days. In order to enter the job market, I attended two meetings of the *Deutsche Mathematikervereinigung*, one in Königsberg, a second one in Bad Elster. That was extremely hard on me. It was not only that traveling to faraway places was a terrifying experience for me, but I was pushed by my teachers to give two papers at these meetings, in front of world-famous people, and the terror of this is hard to describe. However, I somehow survived it all. At one of these congresses, I met A. Scholz, who had been a student of Schur. We discussed our related results and started collaborating on some research. In this work I turned out to act more as a group theoretician than a number theoretician. Some of our joint results became very well known and stimulated work in group theory which in a way led to the resolution of the class field tower problem.

In spite of my difficulties at these mettings, I learned a lot and even captured a very prestigious temporary job at Göttingen helping with the edition of Hilbert's papers in number theory. An appointment like that would not have come about had I not presented papers. My lectures were appreciated. Professor Hahn, who attended one of the meetings, spoke to Courant, the boss in Göttingen, about me. This was an act of great kindness, since I was not really a student of Hahn in Vienna, though I had been active in some of his seminars.

Göttingen

Hence I entered the famous old town and university of Göttingen where Gauss is buried. Unfortunately, my duties were excessive and I had little time to profit from the vast amount of talent there. There was Landau, Weyl, Herglotz, and, of course, Hilbert. There was Emmy Noether with her crowd of students, and there were brilliant young students like Heilbronn and Deuring in number theory and algebra (a whole crowd in analysis, including Fritz John and Hans Lewy). But I had to work on the galleys and page proofs of the Hilbert Volume I, a deadline job. Hilbert had no interest any longer in number theory. He worked only in logic then and annoyed Gödel immensely by publishing a proof for *tertium non datur* which contradicted Gödel's achievement. However, Hilbert did say to me as explicitly as one can say anything that despite the fact that he had worked on many other things, number theory seemed to him the most important. My co-editors were Magnus and Ulm. It turned out that Hilbert's work was not free from errors of all magnitudes. There were conjectures that had in the meantime been shown to be incorrect (some by my teacher Furtwängler). We worked very hard on all this, but later even more errors emerged. However, these errors do not in the slightest take anything away from the mountainous achievements of Hilbert. He had an enormous influence on the development of mathematics to this day and an ability to create new problems and simultaneously solve the problems posed by others. (The Waring problem is an example.) The fact that a man like him, who could forecast the main facts in class field theory based on only relatively simple examples, guided by his enormous insight, could also make mistakes, and even make a few wrong conjectures, is amusing, and makes him a more human being rather than the monster that he occasionally appeared to be. For he was known to insult and tease people occasionally out of pure naiveness, not realizing what he was doing. Since he wrote a colossal amount, he was probably sometimes too busy to check his ideas. However, there are more serious criticisms against him, and they created serious troubles for me at the time when I was an editor for his work in number theory. I had to deal with resulting correspondence on my own and finally decided to ignore it, because the task would have been impossible. It concerned his so-called *Zahlbericht*, published in the journal run by the *Deutsche Mathematikervereinigung*, but used like a book. It was a sort of text on algebraic number theory, which was greatly needed in those days. My teacher had learned the subject from this book and he taught it in the same spirit, and so did many other people. Books can create a subject; they can benefit it and they can harm it. (I suppose abstract algebra and its worldwide acceptance owes an indescribable debt to the van der Waerden books—which actually were never criticized.) However, when word spread about the republishing of the *Zahlbericht*, letters came in criticizing the book and urging me to rewrite it instead. These critics were advocates of abstract treatments and generalizations of algebraic number theory. Strangely, the greatest champion of abstract treatments, Emmy Noether, who after all at that time worked in the same building where I was doing my editing, never said a word about that to me. She probably realized that these people were entirely unrealistic. Their remarks were utterances of feelings, not concrete suggestions. Some people sent me pages of small errors which I simply worked into the book. However, the incorrect conjectures were incorporated as editor's comments. Many years later, Emmy did lash out about the *Zahlbericht*. It was when I was supposed to lecture to a small group of novices, at her request, on the fundamental facts of algebraic number theory and used the Hilbert treatment. I had a tough time with her, for she was not good at explaining, and some of the time I did not know what she meant and how I was to make changes on the spot. However, I did profit from some of her criticism finally. With advancing insight and experience, I finally understood the shortcomings of the *Zahlbericht* in my own way. Emmy at that time quoted Artin as having said that the *Zahlbericht* had delayed the proper advance of the subject by decades. However Emmy did not write such a book and a book was due to be written. The book by Hecke was never criticized. It is still a beautiful book. Also, a number theory book ought to be written by a number theoretician; too many books on algebraic number theory which appear nowadays are written without numbers in them. That is what Emmy would have done.

There is another criticism being raised against Hilbert nowadays. We live in an era of cynicism against everything and on the other hand blind faith in many things. It is popular to look for the faults of people and to bare them mercilessly. Hence, people say that Hilbert at times robbed people of their ideas. I am inclined not to go along with that. People say that much of his class field theory was already in Weber, and that Minkowski had inspired him greatly. The first person who attacked Hilbert to me in this way was the Dutchman Brouwer, after I had mentioned my editorship to him. But he added that he was certain that Hilbert's proof of the Waring problem was truly Hilbert's own work for he had arrived at it when staying in Brouwer's house. One thing seems certain to me and that is that Hilbert had no need to rob anybody of his/her work or ideas in a cold-blooded way. The creative thought processes of a mathematical mind are not easily explained, and it is hard to know what does subconsciously stimulate them.

I will now turn to the people in Göttingen with whom I was in contact. It was rather a small number. Courant had given me the appointment, he worked me very hard, but he was somehow very proud of me as his "discovery." He made me his assistant for his differential equations course. Since I had hardly any training in this, it turned out a tremendous chore for me, added to my other assignment. However, he had planned to make me give up number theory and join his famous group. I had no intention of doing this, of course. However, many years later when I was working in a scientific war job in London, I was practically commanded to solve a difficult boundary-value problem for a hyperbolic differential equation and finally succeeded in it. How I wished at that time that I knew a little more of the subject, though it would not have helped much, because it was not a problem of the classical type.

Next there was Emmy Noether. At that time she had decided to study number theory and reprove some of the facts in these subjects by generalizing to more abstract concepts. She was in a very good mood because she had achieved the proof of the principal genus theorem. This means that she had generalized Gauss's characterization of the square classes in quadratic fields (actually Gauss stated it in the language of quadratic forms and squaring means duplication, a special case of composition). Emmy was the first to generalize to arbitrary normal fields, actually relative fields in a language of abstract algebra with tools which are nowadays expressed by cohomology. Squaring is replaced by taking $1 - S$ powers, where S runs through the Galois group. I myself have made a small contribution to this recently, applying facts concerning integral matrices. Her former student, Deuring, was still in Göttingen, and she had visits from Hasse and van der Waerden and felt that she had arrived. She was definitely popular with the students, but her colleagues either mistrusted her work or disliked her. She was not an easy person to get on with. Although she was very kind, she was also very naive and thoughtless in her treatment of people. However, I had the good fortune to gain her confidence through an act of concern for her that had seemed very natural to me, and we became good friends. I had been present when one of the top people of the department spoke rather harshly to her. I really did not like this. The next day I told him that this had upset me. I really had no right to interfere and it may have hurt my own future. Fortunately, he was not a mean man. He went to apologize to Miss Noether and told my colleagues that he had done so, assuming that I had told them all about it, though actually I had not.

She ran a seminar in class field theory because of my presence during that year in which I also lectured. She ran a course on representation theory, but I found it hard to follow and did not have the time to put more effort into it. Among the students was Witt.

Among the people of my own age group, there were my editor colleagues and Heesch, the assistant of Weyl. My main topic of conversation with Heesch was trying to understand the infinitesimal elements connected with a Lie group. He was not making out too well in the highpowered atmosphere of Göttingen, but in quite recent years I suddenly saw his name appear again, for his work played a decisive role in the proof of the four-color problem. The other editors were Magnus and Ulm, and the three of us were united in our despair over our main duty, to edit the first volume of Hilbert's works. Ulm had written a very important thesis under Toeplitz which led him to be cited in connection with the Ulm theorem on Abelian groups. Magnus was a very active person. He was a student of Dehn in Frankfurt and he too had already quite a name in the theory of infinite groups. He became interested in my thesis work on groups and later in my work with Scholz. The former interest helped him to develop a new proof for the principal ideal theorem in class field theory. The second one came out of a conjecture of mine based on my work with Scholz, concerning group towers associated with class field towers. Magnus used a correspondence between group elements and formal power series in these elements as a tool and made some progress on my problem by establishing arbitrarily long group towers. Next came progress by Noboru Ito, next the thesis of my student Hobby, based on a construction by Zassenhaus, and next came the achievement of Golod and Shafarevich showing that there are actual fields with infinite group towers, and not only infinite group towers. The last finishing touch on the group tower problem itself was achieved by Serre. Magnus became a member of Courant's group at NYU after the war, after a spell of work at Caltech working with Erdélyi. I still have much contact with him, professional, and even some personal, since we are both interested in crystals.

Back in Vienna

The next two years were spent back in Vienna, doing a lot of tutoring, but also being active in the mathematics department, at first on a voluntary basis, later with a fixed appointment, carrying a small stipend. Considering the fact that about that time the world started on a course of turmoil and suffering, but also on a course of increased scientific activity, one feels embarrassed to report on one's own life history. I was heavily engaged in my joint work with Scholz. This was done by correspondence and personal contact at conferences. It was years before we finished our paper. At that time few people bothered about it, but it is now a much appreciated work. Scholz was a very talented man, and also a very fine human being. He was one of the few young people in mathematics who harmed their careers by not making a secret of their dislike of the Nazi movement (he died during the war in Germany). He had visited Vienna during his student days and worked with Furtwängler on his thesis, for one semester, then returned to Berlin where he did his main thesis work under I. Schur. The thesis of Schreier written earlier under Furtwängler was connected with Scholz's thesis work. I had noticed Scholz in Vienna then, but we did not meet. It was only later, when I was working on my thesis, that I contacted him in connection with a numercial example for a result in my thesis. Scholz was a marvel at numerical work. He did much very notable work, particularly in connection with the so-called inverse problem of Galois theory. (Some of this work was done independently by H. Reichardt and T. Tannaka and continued by Shafarevich.) That was immediately appreciated, but his later work is now being studied for the first time, and the young mathematicians in Germany are particularly proud of him. He was very poor at explaining himself, both orally and in writing, so people could not easily appreciate his ideas. I had quite a case of hero worship for him and did not dare to bully him about his difficult style of writing. But one day I overheard him saying that he had hoped I would tidy up his work, but that apparently I was too timid to do so!

During this period in Vienna, I made my first (temporary) break from number theory work. I became interested in topological algebra, almost overnight. It was through a reprint of Pontrjagin's work on *Stetige Körper*. In this paper Pontrjagin gives a characterization of the real, complex, and quaternion fields via topological properties of a field which is at the same time a topological space, so that addition, multiplication and division are continuous functions. This paper impressed me very much. Though other people at that time were also fascinated by problems in this subject, they were mainly in Princeton and in Zürich, and I became somehow infected by this on my own. Nobody had so far explained how mathematical fashions can emerge in completely separated regions at the same time. Some of the problems that emerged there are still unsolved, but great progress was made, particularly by the school of H. Hopf in Zürich. During my year at Bryn Mawr, when I accompanied Emmy Noether to Princeton I discussed my own ideas in this subject with Professor Alexander, who brought me in contact with N. Jacobson who had also studied Pontrjagin's paper. We wrote a sequel to it, with Jacobson contributing more than I did because of his better training. I also take credit for another idea I found entirely on my own. It was later completely immersed into the achievements of the Hopf school, though I am happy to say that one of Hopf's thesis students, H. Samelson, gives me the credit due to me. It concerns a proof for Frobenius's theorem concerning associative division algebras over the reals, via n-dimensional spheres. By a theorem of E. Cartan the latter can be group spaces only for dimensions 0, 1, 3. I possess a flattering letter from the great E. Cartan concerning this observation of mine. Otherwise, very few people know about this small initial contribution to a much larger issue. This is why it is of particular pride to me. One of the strongest members of the Hopf school was E. Stiefel. He reproved another related achievement of mine concerning Laplace equations in n-dimensions which are derivable from generalized Cauchy-Riemann equations. I was able to show that this is possible only for dimensions 1, 2, 4, 8. Later in life I had many more work connections with Stiefel, in completely different subjects. In a letter received from him quite recently, only a few days before his sudden death, he remarked that although we meet so rarely we always seem to be working on related subjects.

Well, all this was started for me during these lonely two years in Vienna. While in the first of these years my earnings were mainly from tutoring, they received a certain boost and more distinguished source when Professors Hahn and Menger arranged for a small assistant position for me at Vienna University. Since they were very enterprising people they found a way to earn money with which to pay young people like myself for their hitherto voluntary work. They gave a series of lectures during evenings on mathematical subjects on a level to be understood by less trained people. These lectures took place in the large auditorium of one of the physics institutes of Vienna University and were rather elegant affairs. The lectures were even published soon afterwards. For me there was only one drawback about this. These two former teachers of mine did not work in my line! My own teacher Furtwängler never cared about helping people and did not worry about the fact that I was giving him much assistance completely unpaid. So I had these three bosses and I worked hard for them. My load was truly a multiple load. However, I learned a great deal in subjects which were quite new to me, like functional analysis. I prepared the thesis for one of Hahn's students for publication and took on the supervision of another one almost entirely. His problem was a sequel of the thesis of the previous one. These are instances of my duties. Hahn was a great expert and contributor to the subject of functional analysis. The theses were on sequence spaces, a subject in which Helly had made earlier contributions.

As mentioned earlier, at that time I was also working on abstract spaces on which a Euclidean-type distance was introduced. A final item into which I was initiated then and which has stayed with me to this day is sums of squares. This is a subject that enters into many branches of mathematics, linking them together in a most attractive way. In recent years I have written several survey articles concerning them. One of them—for the *American Mathematical Monthly*—earned me a Ford Prize. At that time I was stimulated by some problems posed by van der Waerden in the *Jahresbericht d. deutsch. Mathematikervereinigung*.

While I was in Vienna I applied for a Girton College (Cambridge) science fellowship, which I had seen advertised in the newsletter of the IFUW (International Federation of University Women).* After I sent in the application I received an invitation to Bryn Mawr College in Pennsylvania, which I accepted. I thought the chances for the Girton fellowship were extremely remote. But after I had agreed to go to Bryn Mawr, I got a letter from Girton. It looked like a form letter, and I almost threw it away without reading it. It turned out to be the notification that they were awarding me a three-year fellowship with a very generous stipend of £300 a year, and with great freedom to do what I liked. Although I had already accepted the Bryn Mawr offer, I was allowed to keep the Girton fellowship and to spend the first year at Bryn Mawr. In the meantime, Miss Noether had arrived there, which would make that year more fruitful for me.

Bryn Mawr

The invitation to Bryn Mawr came about because I had met O. Veblen in Göttingen, where he had an office next to mine. (In fact, I heard him rehearsing his lectures on relativity through the walls.) He had told Anna Pell Wheeler about me. She was chairman of the mathematics department at Bryn Mawr, but had previously lived in Princeton with her husband who had been a professor there. She was a very interesting person, very active, but plagued by ill health and accidents during all the time I

---

*The AAUW (American Association of University Women) is a branch of this organization. Both groups have been helpful to women in the academic profession.

knew her. She was dignified, warm-hearted, broad-minded. She was very interested in helping women and in the advancement of women in general. She predicted that women would become physically stronger in the future. She seemed to be the outstanding woman mathematician in the U.S. at that time. Her subject was functional analysis and she had given the colloquium lectures for the American Mathematical Society on this subject during one of their summer meetings. That was a great honor for a woman. One seems to be expected to write these lectures up for one of the series published by the American Mathematical Society, but she did not manage to achieve this. She gave a related course at Bryn Mawr during my stay there. My position there amounted only to a graduate scholarship and the financial side of it was very poor. But it was a wonderful opportunity to visit the U.S., a country that in those days was quite unknown to many Europeans, an opportunity to learn more mathematics, maybe even a stepping stone to a position in those days of great unemployment. The latter hope was certainly not realistic. The depression was at its height, and unemployment for young mathematicians was desperate. The Bryn Mawr girls of my age group were pleased that I would not have to compete for a position with them—I had still two years of my fellowship in Cambridge, England waiting for me. But I suppose that I might have accepted a position with tenure in the U.S. if one had offered itself.

When I left for Bryn Mawr, my speaking knowledge of English was extremely poor, but I took a few lessons from an English lady who claimed that she knew no German at all. She was not an educated lady, but she certainly was a gifted teacher, for she got me to speak English. I was busy with various duties almost to the last moment. I suffered the shock of Hahn's death, losing a real friend, and there were also terrible political upheavals going on in Austria then. I traveled to London with all my luggage, changing many trains and crossing the channel, and then got on the boat train for Liverpool. I had chosen this type of boat on purpose. I expected that most passengers would be English-speaking while the passengers from European countries would travel from a more southern port. This turned out to be correct. Most of the English that I speak now I acquired then. I never found time again to take lessons.

I had never before lived at a college like Bryn Mawr. In European universities, nobody bothers about a student outside of teaching. At the beginning I was delighted about it, but after some time I found the noise of the graduate hall in which I lived, the lack of privacy et cetera, almost unbearable. On the other hand, many things were provided for me: food, cleaning, bedsheets. I held the so-called foreign women scholarship that year, so I was registered as a graduate student and had to obey the rules of a student—register for classes and pay the college back most of the money that was allocated for me.

Emmy Noether

Emmy Noether, who had arrived there a year earlier on a Rockefeller fellowship, was very happy there; she liked the girls and they treated her well. She became interested in the life of girl students, probably for the first time in her life. She strongly disapproved of non-coeducational colleges. She taught a seminar for two hours on Mondays, and repeated it in Princeton on Tuesdays. This gave her a chance to rehearse it. She had not done a great deal of systematic teaching in her life and although she would have liked a more senior position in Göttingen, she actually would not have liked the hardship of greatly increased teaching duties. In fact, she once said to me that women should not try to work as hard as men. I do not know whether this is true, and furthermore, she was able to afford a smaller salary on account of her simple lifestyle and the inheritance she must have obtained as the only daughter of a very important university professor. (Her brother already held a senior university position at that time.) She also remarked that she, on the whole, only helped young men to obtain positions so that they could marry and start families. Since she was very naive, she somehow imagined that all women were supported.

Miss Noether had been quite happy at Bryn Mawr in her first year. She studied van der Waerden's first volume on algebra with students and staff members. But by the time I arrived there at the start of her second year, I did not find her in a very good mood. She had been back in Göttingen during the summer and had arranged for all her belongings to be shipped to the U.S. She had found everything very difficult in Göttingen, and she had not yet found a position for the next year in the U.S. Although she knew that her friends from the old days would not let her down, she did not know where she would be settled. She was only fifty-four years old, but that was considered quite old at that time. She was determined not to train herself for undergraduate teaching at Bryn Mawr—she was paid by a Rockefeller grant at that time. And unbeknown to all of us, she was ill! She tried to hide that fact and to visit Göttingen the next summer for some surgery. But when certain troubles bothered her increasingly, she confided in a doctor at Bryn Mawr and he persuaded her to undergo surgery immediately. A week later she died of heart trouble—at least this is what we were told. During that year she had tried to work with us on some seminar notes prepared by Hasse on class field theory. She had one of the graduate students write a thesis under her, on normal bases in fields. Another young woman there, as a fellow, wrote a paper on work suggested by her. She spent some time advising Deuring, her former student in Göttingen, on his Ergebnisse volume of algebras. Altogether, we were four women working with her.

On her weekly trips to Princeton I accompanied her frequently, though not every week because of the high train fare. She was very pleased that I went with her, and we had nice chats. Otherwise I often irritated her—she disliked my Austrian accent, my less abstract training, and she was almost frightened that I would obtain a position before she would. These trips to Princeton were the highlight of my year in Bryn Mawr. Since I travelled with Emmy I was invited to dinners in the evening together with her. The Institute for Advanced Study and the department of mathematics were in the same building; I could see people like Einstein walking in the corridor, and I was even introduced to him several times. There was also von Neumann, H. Weyl, R. Brauer, Lefschetz, Alexander. I worked with two younger people, Jacobson and Magnus, and I learned a great deal there. It was a dream place for me.

Emmy Noether was definitely appreciated as an important researcher in Europe by people who understood her line of work. There were people in Russia, Japan, and the U.S. who greatly admired her. But the tremendous admiration that she has earned in recent decades, also as a human being, exceeds what she would have expected (and certainly what she received) in her lifetime. An Austrian lady, Dr. Auguste Dick, wrote a Noether biography, published by Birkhäuser. Since she is not an algebraist—her thesis was in differential geometry, I think—I asked her what had made her undertake this task on which she spent much time and even money, for she travelled long distances in Europe to interview people. She said that she had nobody financing this enterprise; it was her hobby and she had always admired Emmy very much. I receive frequent inquiries from her about Emmy, from schoolgirls writing essays about her or potential authors, or people who want information on Emmy's work.

Although to me, since I came from a poverty-stricken country, much of life at Bryn Mawr seemed quite luxurious. I had attended a European university and the attention students received practically brought tears to my eyes. I remembered the tough time I had in my student days. Nevertheless, life at that time was far from carefree. The depression was in full swing. Actually I was supposed to have arrived a year earlier, but the college had to cancel the plans because of financial losses they had suffered. Somehow they managed it a year later, maybe by applying to their donors. Some of the students in other subjects did not appear to be rolling in money; they could not afford to buy oranges to add to what the college provided, and I remember helping them carry their luggage to the station to avoid taxi fares.

Girton College, Cambridge, England

I left the U.S. in June and checked in at Girton College, Cambridge. There I was a fellow, a so-called don, with all the many privileges of one, no longer a Ph.D. treated as a student, which for a whole year becomes a bit tiresome. Still, I carry quite a bit of gratitude and loyalty for Bryn Mawr. Life at Girton seemed great and being attached to a place like Cambridge University is a wonderful thing. But I had a number of difficulties, some of them of my own making. There again was nobody in my line. At Bryn Mawr I had had mathematical contact with Emmy, Mrs. Wheeler, "the girls," and my occasional trips to Princeton did much for me. Following the current fashion in Princeton, I had become deeply interested in topological algebra and nobody in Cambridge was working in this area at that time. (Soon after I left, there was quite a lot of interest in that subject.) There was quite a bit of number theory going on, but not in algebraic number theory; it was either analytic or elementary. Occasionally Mordell or Erdös visited there; there was the brilliant Heilbronn who was practically waiting to work with me, but not on the subjects I knew best. There was Davenport, who worked with Heilbronn; there was Hardy; there was the great group theory man P. Hall; but somehow they all seemed on different planes. In my first year there, I got quite nice work done, partially aided by discussions with B. H. Neumann. If my mind had not been so deeply anxious to continue on topological algebra, I might have been able to attach myself to one of the research groups that existed there. But it did not work out. In my second year there I spent an enormous amount of time applying for jobs, going to interviews, and supervising students—partly to gain experience teaching in English.

University of London

The next year saw me in London at one of the women's colleges of London University. I held a very junior position with extremely arduous duties—nine courses to teach each week, each of them one or two hours, with homework to assign and grade. I was grading practically all the time. The work was partially not interesting, and partially not well known to me, particularly the geometry, which the other teachers disliked and dumped entirely on my shoulders. It was one of the conditions of the job offer that I would accept this situation. Also, I still had difficulties with the language, and there were no helpful books that I could use for my courses. So I had to work hard. My only consolation was that the students were extremely kind and pleasant, but unfortunately also quite without scholarly ambitions. My boss, a lady who had given up scholarly ambitions herself long ago, was not sympathetic to me. I had actually been squeezed into the college by Hardy and by the Head of Girton College.

I was able to make friends with some of my colleagues, but not with all of them. They saw me as a foreigner. This had not been the case at Girton College where people had quite a liking for my foreign accent and other foreign facts about me. Girton also had a scholarly attitude in everything and the level of research achievement and general culture was definitely higher. But I like teaching on almost every level, and I made the best of it. Furthermore, in spite of my appreciation of the beauty of Cambridge, London is not a place to look down on. My college was almost in the suburbs, but easy to reach by underground and buses, and on the few evenings I had free from grading or lecture preparations I rushed into the center for a bite. Actually, in spite of my terrifying duties, I got quite a bit of research accomplished, and I was in touch with some colleagues at other colleges of the university. There were intercollegiate seminars at which I lectured, and at one of these I met John (Jack) Todd who held a position similar to mine in a different college, but worked in analysis. In spite of the difference of our subjects, we had definite scientific contacts and so had to confer frequently. Not much later we got married.

World War II

We were barely married for a year when the war broke out. Jack was given leave from his college to take up a scientific war job. This materialized only a year later, and in the meantime we moved to Belfast, the home of his family. There we both taught at his university, Queen's University, and I was quite active in research with a fellowship, still from Girton College. I became interested in two subjects in matrix theory which still form a large part of my active research, namely, generalizations of matrix commutativity and integral matrices, which are part of number theory.

When we returned to London a year later, the war had already taken on very threatening aspects. Jack was now assigned to work on scientific duties in the Admiralty, and I returned to my London college, which had been moved to Oxford for greater security. After some time, I returned to London to take on work in aerodynamics with the Ministry of Aircraft Production, at the National Physical Laboratory in Teddington, outside of London—actually near to where Eisenhower had his military headquarters. Needless to say, with war anxieties, air raids, food shortages, heavy work loads, homelessness—we moved eighteen times during the war—research did not always proceed with great speed, but nobody complained, particularly not in London. One of our landladies polished her brass utensils every morning, even when we had had heavy air raids the night before.

The duties in my aerodynamics job were very heavy. This time I really had to give up all my previous dreams. But there were some rewards. For the first time I realized the beauty of research on differential equations—something that my former boss, Professor Courant, had not been able to instill in me. Secondly, I learned a tremendous lot of matrix theory. My boss, R. A. Frazer, was an algebraist who with two other authors, Duncan and Collar, had written a very impressive book on matrix theory, a book of particular use for applied work. The matrix theory was used in flutter research, a very difficult subject which is not yet completely mastered, even with high-speed computing machines. Actually, I was not assigned to the matrix theory by Dr. Frazer, although he claimed me for his flutter group because I was an algebraist. But matrix theory was simply oozing out to me simply by my working in these surroundings. At this time I heard about the so-called Gersgorin circles attached to a matrix with complex numbers as entries. I heard about them from Aronszajn. I became immediately extremely interested in them and hoped to use them in the flutter work where one has to test a matrix for stability. I would not say that they are an ideal practical tool for this purpose, but they certainly have a great many uses, and I can say that I stimulated much research concerning them, while I myself did not pursue them further after some initial achievements.

There are other theoretical tests for the stability of a matrix. I became interested in them much later, when I was already working at Caltech. If my interest in criteria for stability had not been aroused during my years in aerodynamics I would not have taken to doing and stimulating research in these quite fascinating theorems. With high-speed computing possible, people are no longer very anxious to test a flutter matrix, they simply compute all of its characteristic roots, particularly since J. Wilkinson has devised such ingenious methods for finding them. The tests I am now talking about are expressed as Liapunov's theorem concerning matrix stability.

Olga Taussky-Todd, at the Bryn Mawr College Commencement, 1935.

USA: National Bureau of Standards, Princeton, Institute for Numerical Analysis

I left the Civil Service in 1946, rather exhausted, and worked with a research grant on my own for a year. In 1947 we went to the U.S.A., initially for a year, because Jack had been invited to help in the exploitation of high-speed computers. He was invited to do this work at the National Bureau of Standards at their new field station in UCLA. When he came to check in at the headquarters in Washington, he was told that they wanted me to join too in some capacity, and also that the UCLA quarters were not yet ready.

I then settled down to work quietly on matrix theory, more quietly than I had been able to do for a very long time. And I learned a great deal. The Bryn Mawr people had spread the fact that I was visiting Washington, and the mathematics group that met weekly in Philadelphia invited me to address their colloquium. That was initially quite an anxiety for me since I had not given a lecture for years. But then I welcomed the opportunity and lectured on my matrix research. It turned out well. Soon afterwards, the mathematics group at Johns Hopkins invited me too, and I repeated the same lecture. There was also van der Waerden in the audience then. Wintner asked a number of grilling questions and made many comments. But I accepted all this very well. In this same period, I was invited to a colloquium at MIT to lecture on my work on a boundary-value problem for a hyperbolic differential equation that had come out of my aerodynamics work. On our way back from Philadelphia to Washington we called in at Princeton. The Institute for Advanced Study now had its own building. I was rather sorry about this. When I had visited there before the war the Institute was joined with the Princeton mathematics department at Fine Hall. Still, it was wonderful to be in the Institute again. I had gone through a number of very harsh years, overworked and overstrained, removed from my previous mathematical interests. However, this little visit did much to restore me. Almost as soon as I entered the building I ran into Schafer, an expert on non-associative structures, and when he heard who I was, he reminded me immediately of all the items that I had published on this subject. Later I ran into S. Chowla and I. Reiner, who seemed to know about my work on integral matrices and independently asked me the same question in this subject. Luckily I was able to help them. Reiner later made great use of this information and embedded it into the book he coauthored with C. Curtis. Chowla urged me to work on a different treatment of the fact I had told them. This I did later that year in my paper "On a theorem of Latimer and MacDuffee," which became the first of a long string of papers which form a good deal of my research and keep me busy to this day. Hence, I feel deeply grateful to Chowla for bringing me back to my peacetime work. When we returned to Washington only a few days later, we found out that the facilities in Los Angeles where we were to do our work were still far from ready. We then inquired whether we could spend the remainder of our waiting period in Princeton working mainly in the von Neumann group, which was housed in a separate building. This request was granted and we became members of the Institute. We lived in their housing facility and had offices in von Neumann's building, but mingled freely with the other members and the large number of visitors. Von Neumann was only rarely there and very busy whenever he returned. Goldstine was in charge during his absence. The group were all very friendly to us and at a recent reunion they all recognized us. I still remember how Bigelow, a famous engineer who was in charge of building von Neumann's machine, sat on the floor with a hammer in his hand putting nails in a wooden box we were trying to send off to Los Angeles prior to our journey there—for our long-awaited trip to UCLA was now imminent. My relations with the members of the Institute were cordial, but there was much nostalgia for me because the war had cut me off from my previous favorite subjects in mathematics and there were many gaps for me to fill in my knowledge, but I was involved in work which was not entirely connected with these areas. On the other hand, I was learning and working in very modern and interesting subjects.

We arrived by train in Los Angeles by about the end of April after giving lectures in Lawrence, Kansas, at the invitation of G. B. Price, whom we had met in London during the war and with whom I had common matrix interests. We also stopped in Berkeley where the Lehmers organized an excursion to San Francisco for us. Finally we reached our destination, the Institute for Numerical Analysis, housed in a temporary building on the campus of UCLA. After a rather cold winter in Princeton, I delighted in the California climate, though Jack unfortunately suffered from a severe allergy which spoiled a good bit of our stay. Working with us was Szász, a very powerful analyst from whom one could learn a good deal, and later came Rademacher, who became quite interested in my work on the theorem of Latimer and MacDuffee and used it in one of his papers. I became very active. I wrote about six papers on various subjects in matrix theory and other items. I lectured at Caltech, I saw Erdélyi installed there. I also lectured at two AMS meetings, one at Vancouver and another one at Madison, Wisconsin. It was a very active and yet very relaxed time, on the whole. Before our time at INA was over, Dr. J. Curtiss, the man who had originally brought us over to the U.S. and who had founded this new institute, frequently mentioned that we should stay on. But this was not a decision to be made in a hurry, and in any case we were committed to return to London. This we did in early September. Life in Great Britain was still very harsh and the shortages were almost worse than during the war itself. With all the moving about, I was not even completely settled in a position, but I was very busy, lecturing in Southhampton and visiting Cambridge. At that time we met Zassenhaus for the first time. I had returned to integral matrices and had some conversations with him. I did a great deal of difficult refereeing. With my growing knowledge of numerical analysis and of completely new combinatorial algorithms like linear programming, I was able to rescue some pioneering papers from being turned down. A less educated referee would not have realized their value to rather modern research. This was particularly the case with some matrix iteration work in a paper of Stein and Rosenberg which has become a classic. I corresponded with P. Stein at that time directly about changes in the paper. That started a very fine mathematical friendship with this very creative mathematician, who without my assistance and stimulation might not have been able to complete his work. Later I realized the connection of another theorem of his with the Lyapunov stability criterion for matrices and called that theorem the Stein theorem. At that time I was rather out of touch with

Stein, but I suddenly realized that it was not fair not to let him know that he was the originator of another much-used theorem and I sent him a reprint of my relevant paper. He was overwhelmed with delight and asked me to formulate a related problem for him to work on. I had at that moment formulated a problem for myself to work on, but when his letter came I handed it over to Stein who wrote two very fine papers on it which may have been his last creative work. A very gifted student of mine at Caltech, R. Loewy, later continued on this work, starting with his thesis. I had not much contact with Stein after that, but soon after his death his son wrote a very appreciative letter to me. I had never met that son, but his letter pleased me.

A lot of things happened during that year, which was to be our last year in London. Dr. Curtiss wrote to us urging us to return to work with his group, on a permanent basis, not at Los Angeles—partially because of Jack's allergy problem, which never bothered him again on later visits there—but at the headquarters of the National Bureau of Standards in Washington. After a good bit of consideration and even more tempting offers from Dr. Curtiss, we accepted. In the meantime, I continued to work very hard on three problems: bounds for eigenvalues of finite matrices, eigenvalues of sums and products of finite matrices, and integral matrices. While I later withdrew from the first one, the preparatory work on the two other ones compiled during that year was quite considerable, and they are my main problems to this day.

Washington, D.C.

In the fall of that year we moved to Washington. There we ran into Ostrowski, whose year with the Bureau was about to end. It was actually through our recommendation that he had been invited there. He was at all times and still is a very powerful mathematician who can carry out an enormous amount of work. This he did and so contributed greatly to the work of our group, particularly since he is genuinely interested in numerical analysis. I told him about my work on the Gersgorin circles, hoping to receive some further ideas on how to proceed there, but the opposite happened: he immediately set out to wipe out a large part of the relevant theory, showing me his results only after having returned his galleys for publication. (I must admit that the same thing has happened to me a number of times with other colleagues. But I find it hard to stay silent when people ask me what I am working on and I do not really mind when they do this to me, particularly since so far I have been able to find new problems for myself at all times.)

Life for me became very busy. My title was consultant in mathematics and this I truly was, because everybody dumped on me all sorts of impossible jobs, from refereeing every paper that was written by a member or visitor to the group, to answering letters from people who claimed to have "squared the circle," to helping people on their research. In my own research I was given much freedom because our chief, Dr. Curtiss, who also worked me very hard at times, appreciated the fact I was able to keep the important visitors occupied with meaningful research—that at times kept them busy for a long time after returning home—but I was also able to look after a number of talented young postdoctoral employees. For the latter I was at times the only bridge between their university mathematics and the new type of work they had to get used to. These were people like Alan Hoffman, M. Newman, and several others. I had extensive research contacts with both Hoffman and Newman, leading to joint publications. Other visitors of high standard were Wielandt, Stiefel, Fan, Kato, H. Cohn, P. Stein, and Ostrowski. Ostrowski was probably more interested in numerical analysis aspects than the others. He is an immensely powerful mathematician who had worked in many subjects, though perhaps his most famous work is in valuation theory. He spent much time at the Bureau, both in Washington and at the field station in UCLA. He wrote an enormous lot of papers there, just as elsewhere. His mathematical power is enormous. Some of the papers written at the Bureau were connected with work of mine at that time. The same was true about Fan, Kato, and Stein, primarily in matrix theory. Stiefel had not originally worked on problems of interest to the Bureau. He was a pupil of H. Hopf and had done famous work in topological algebra. Since this had at one time also been my subject, I felt very attached to him. In fact, during his first visit to the Bureau he produced a proof of a former result of mine and published his proof in the Bureau's own journal. Stiefel was amazed that on two occasions the work he brought with him from Zürich had also been carried out at the Bureau at that time. But I was also collaborating with Kato on additive commutators and on the infinite Hilbert matrix.

H. Cohn was in number theory, and at that time, from the computational point of view, he was a pioneer in computational algebraic number theory. Among the staff members working with Jack were mainly Henrici and P. Davis. In addition, we were sent out to California to our now well-established field station once or twice each year. There were a number of members of the regular mathematics department attached to our outfit, like Hestenes, Beckenbach, R. Arens, L. Paige, C. B. Tompkins, and E. Straus. There were crowds of interesting summer visitors—Mark Kac; Feynman, whom I met for the first time out there; Rosser; F. John. The Lehmers were in charge at various times. My particular working colleague for a few years was Motzkin. I worked with him on the so-called *L*-property, a concept introduced by Kac. It concerns a special set of matrix pencils. I feel very obligated to Kac for bringing this most important idea to my attention. Motzkin and I wrote a number of papers on this which are very much appreciated. I very much enjoyed working with Motzkin. He had a very clear mind and enormous skill. Then a number of other authors joined in, like Kaplansky, Wielandt, Wiegmann, R. C. Thompson, Kato, Zassenhaus, Wales, Gerstenhaber, quite recently R. Guralnick, and my present thesis student Helene Shapiro. As long as I worked in Washington, Motzkin and I made progress on our work whenever one of us visited with the other. But after we moved out to California, with the difficult drive between UCLA and Caltech we did not seem to be able to meet much. I then continued on my own and am likely to continue. My growing interest in integral matrices brought me in contact with M. Newman, a number theoretician trained by Rademacher. He continued with me on some problems connected with the integral group ring I had started on my own. This work too seems to be continued by each of us separately now. Another young postgraduate working with me in Washington was K. Goldberg. He had a good bit of creative talent in number theory and was an expert programmer. He wrote an interesting thesis on the Hausdorff formula under my guidance while still at the Bureau.

In 1955 we both took leave for a semester to teach at the Courant Institute of Mathematical Sciences at NYU Jack taught a course in numerical analysis and I taught a course on matrix theory. This was a good experience for me, because not much later we went back into academic life where we have been ever since. In our last year at the Bureau, Jack ran a training program in numerical analysis in which many experts participated as teachers while many of the people attending were teachers elsewhere. I gave a brief course on bounds for eigenvalues of matrices and wrote my notes up immediately afterwards. I then suggested to Jack to invite the other teachers to do the same and be the editor of a book. This was finally accomplished leading to the *Survey of Numerical Analysis*.

Another thing I started was a fellowship program for postdoctorals. Our first fellow was the great matrix expert M. Marcus. It was a flourishing program up to quite recently, but I do not know whether it is still continued. Writing all this makes me realize what an interesting life I had there and I feel great gratitude toward the place.

There are still other items that I was associated with at the Bureau which seem worth mentioning. When we returned there in 1949, the Director, Dr. E. Condon, had started on the now quite well known *Handbook of Physics* and was collecting authors for the various chapters. This was to be written somehow in our spare time. He expressed the hope that I would take on all of mathematics which was to be the initial section. I rather enjoyed this task, since I like writing. But with all my other duties, I proved rather too slow, so I wrote only three chapters, namely the chapters on algebra, operator theory and ordinary differential equations, but was helpful in finding suitable authors for the other chapters. This enormous undertaking was actually completed. It was later reprinted in a second edition, and is quite a well-known opus, the Condon–Odishaw *Handbook of Physics*. Next, there was the Symposium on Simultaneous Linear Equations and the Determination of Eigenvalues. This symposium was part of the Bureau's semicentennial celebration in 1951 and was run mainly by myself, with some help from L. J. Paige at UCLA (now a top administrator of the University of California) and under the official leadership of J. B. Rosser. The reason I am mentioning this is that the symposium was the first on numerical aspects of matrix theory and became the first in a chain of other symposia, of which the next was held at Wayne State University under the leadership of W. Givens and the next at Gatlinburg under the leadership of A. Householder, then at Oak Ridge. From then on, the meetings have gone by the name Gatlinburg meetings, even if they take place thousands of miles away. The proceedings of my 1951 symposium were published by the Bureau.

As soon as the SEAC, the Bureau's high-speed computing machine in Washington, was operating I tried to find suitable problems in number theory which were not easily approachable by hand computing. Although I myself helped very little in the actual programming, my guidance was not unsuccessful. In particular, I had some influence on a problem connected with Fermat's conjecture, suggested to me by H. Hasse, on consecutive power residues. The help on this came from a famous British computer J. C. P. Miller, who visited at that time. He and another expert computer, Ida Rhodes, spent many hours on the program and achieved a considerable output. This work was then taken over by the Lehmers who worked on the UCLA Bureau machine, the SWAC, and in addition wrote a number of valuable papers connected with the Fermat problem.

During the summer months we were allowed to hire student trainees, high-school kids or young students who did some programming for us or helped otherwise. This is how we met the brilliant E. C. Dade whose programs always ran without bugs. He wrote his first research paper when still an undergraduate at Harvard, on a problem coming out of our set-up.

There is one more item concerning my activities at the Bureau which I would like to discuss, mainly because I am now coming to my life at Caltech when I had to continue the same type of activity: this was administrative work. Prior to my employment at the Bureau I had never done any. I did not think I was suited for it, I was plenty busy anyway, and there were always people who were eager to do it. However, my chief at the Bureau pulled me into such work. Firstly, there was the organization of the symposium I mentioned before. But he also wanted to use my knowhow and experience for hiring and promoting personnel. He had great confidence in my loyalty and honesty, and his confidence increased these qualities. This work was more interesting than I had expected it to be, although not always pleasant.

In spite of what I have been explaining, my job at the Bureau was not exactly the right job for me. It was certainly a very interesting job. I learned a lot there, contributed a lot, and was treated with great courtesy. I never asked for a raise or a promotion; they just came to me. Our salaries were not kept secret; we expected that our bosses at all times looked after everybody's interests. As soon as it was possible, I was given tenure. There was a great team spirit in our group, nevertheless with occasional upsets, of course.

With three of her Ph.D. students, left to right, Joseph Parker, Raphael Loewy, and Fergus Gaines.

Caltech

When the invitation to Caltech came I felt very pleased and honored and I know that I had stayed at the Bureau long enough. Coming from a civil service job back to academic life meant a tremendous change, almost as much as the opposite change, which we had made years before. First of all, Caltech is a teaching institution, however high its research standards are. I myself was given a research position with permission to teach. This created a difficult situation for me. It is not entirely pleasant not to teach when everybody else is, and besides, I simply love to teach and feel that I have a good bit of natural talent for it. Furthermore, our department was greatly understaffed at that time and to some degree still is. However, I had not taught for a long time, and working with Ph.D. students on their theses is quite different from working with postdoctorate young people, as I had primarily done earlier. The students are not yet fully trained. They are frightened that they may not make it. They feel frustrated if a problem does not work out right away. There is always the possibility that they may break down. Altogether, it is a greater responsibility. However, it is with great pride that I can say it has always worked out well so far.

Of course, there is another thing that makes an academic surrounding so different from the civil service. That is the fixed hours in the civil service. In the evening or during weekends one can hardly return to one's research. At the university nobody gives you a fixed time schedule, apart from the fixed teaching schedule. So what happens is that one works practically all the time! When I wake up during the night, partially refreshed from a few hours of sleep, my mind goes back at once to my unsolved problems and I sometimes really make some progress, but feel worn out the next day. But no doubt more work is accomplished when fixed hours are removed.

By the time I came to Caltech I had no doubt that some administrative work was expected of me and indeed, our chairman at that time did expect it and treated me with confidence and appreciation.

I then took on more teaching than I had anticipated. I did not teach undergraduate classes, though undergraduates came to my graduate classes all the time and they were excellent students. I had thesis students almost from the start. They took an enormous amount of time and energy, but working with them was something wonderful, almost all the time. They also took many great problems which I had created away from me and sometimes I did not dare to return to these problems for years, not wanting to interfere with their work. But all this did not seem to matter. I always found other problems. I let the students do their own work, only giving guidance, and hence, they felt really satisfied at the conclusion of their work. It always amazes me that I found problems so readily for these students, considering the complicated research life that I had led which brought me to thesis students so late. I feel most grateful that this opportunity was finally given to me.

Outside of my contacts with the Caltech students, I have, of course, contacts with my colleagues. These contacts fall into several groups. There are the postdoctoral students some of whom became colleagues later on in junior positions. I was able to establish common research interests with all of those young people who were working in my lines. I know from my own experience that the step from a complicated thesis to independent research is a very hard one, and many people cannot make it. Some stay in the subject of their thesis forever, which can be either good or bad, depending on the case. But to find a truly new subject, to cut the umbilical cord from one's teachers is a truly exciting and great thing. I have definitely put myself out to give moral and technical support to our young postdoctorates. That I profited at the same time from the knowhow and talents of these people is understood. Clearly, these people were chosen because they were promising members of our department. Some of the ones I collaborated with were later given junior faculty positions. Some left later. One very glamorous exception was given tenure very soon. The exceptional case was the immensely brilliant E. C. Dade, mentioned previously. He was trained at Harvard and at Princeton and wrote his thesis in algebraic geometry. He is fully trained in every mathematical subject and in many other subjects as well. I always describe him as a man who could have written the thesis for any student in any subject we had. He is also a great programmer. We had met him when we were still in Washington, when he was chosen as a Westinghouse scholarship recipient. (We also met the great Paul Cohen in the same capacity there.) As I mentioned earlier, he then worked with us at the Bureau during summers as a trainee. We realized his capabilities and asked Bohnenblust, our chairman at that time, to bring him out here as a Bateman Fellow. This succeeded, and Dade rapidly went from fellow to assistant professor, to associate professor, and to full professor before he was thirty years old. During his first years here, he did research entirely with me in algebraic number theory. The new lease on life that was given to group theory a little later by Feit and Thompson lured him away from my subject, but made him more acceptable to other members of the department. But they were not able to keep him and he left. I myself continue a small bit of work by correspondence with him. The other postdoctorates with whom I was able to establish mathematical contacts were Dixon, Estes, Kisilevsky, Guralnick, and P. Morton. Fortunately, Estes and Guralnick, both highly gifted and scholarly, are now at the nearby University of Southern California.

Now I come to my older colleagues at Caltech. Several members of the department would have welcomed it greatly had I joined their mathematical pursuits; but since the latter were much removed from my own, I could not yield to this. I was too much settled in my work by then, and in the past, even at times of unemployment, I had always had my own ideas of what I preferred to do. But I did have working contacts with Morgan Ward, whose death was a great blow to me, and I do have frequent contact with Apostol who is an excellent number theoretician. I also have frequent contacts with Ryser, a man in combinatorics with a great interest in matrix theory, as well as with M. Hall and D. Wales. Then there is DePrima, an analyst, but with a feeling for other subjects as well. We got interested in several problems concerning matrices with complex entries and had some enjoyable work interactions, partially in connection with theses worked on by students of both of us. There was also my chairman at the start of my stay at Caltech, Bohnenblust (Boni), who became very interested in my problems on pairs of symmetric matrices and provided very fine theorems on them. He also discussed work of some of my thesis students with them and myself.

The famous California climate lures visitors out here easily. But some of them also come to work with us. Among them is Zassenhaus, a man of great mathematical power and knowledge, a student of E. Artin. Then there is A. Fröhlich, a student of Heilbronn; Wielandt, a student of Schur, more in group theory, but easily lured into matrix theory, Varga in analysis and complex matrix theory, D. W. Robinson, working on commutators and integral matrices, S. Pierce, in matrix theory and in number theory, and J. H. Smith, a pupil of Iwasawa, in many subjects.

There is so much mathematics going on nowadays that it is not possible to keep up with much of it. One must restrict one's creative work to small areas if this is at all possible. However, I myself do not favor the idea of working in a small area. One gets more famous if one does, just as a medical specialist can do more important work and earn more as well. Yet the general practitioner has occasionally the greater, though more arduous, life. In any case, I like to nibble at all subjects, although this is now getting harder every day. Partially, circumstances have forced me into doing this. At the start of my life all I wanted to work in was number theory. But this was frustrated through many circumstances. In fact, it took a long time before I could return to my dream subject. But apart from the complications in my career, I developed rather early a great desire to see the links between the various branches of mathematics. This struck me with great force when I drifted, on my own, into topological algebra, a subject where one studies mathematical structures from an algebraic and from a geometric point of view simultaneously. From this subject I developed a liking for sums of squares, a subject where one observes strange links between number theory, geometry, topology, partial differential equations, Galois theory, and algebras.

The theses written under my guidance reflect the main areas of my own research. At present these are: commutativity and generalized commutativity of finite matrices, which includes the difficult problems concerning eigenvalues of sums and products of matrices, and on the other hand, integral matrices, i.e. matrices whose elements are whole numbers. These two subjects sound quite different, but they have important intersections, a fact on which I am working very hard, with some success, interpreting facts in number theory via facts in matrix theory, which involves noncommutativity. This is nothing new in principle, but has not been exploited sufficiently until recently. Some facts in modern number theory have been better understood by considering numbers as one-dimensional matrices, and then generalizing to matrices of higher dimension, thus giving more meaning to the original results. I became interested in these methods as soon as I heard of them. Some go back to Poincaré who had great ideas in more subjects than people realize. I have gone my own way on this kind of work.

My husband and I are not in the same line of work, since he is an analyst and I an algebraist. However, most of the time we are able to talk to each other about our respective activities, and there are a number of items of fairly large area in which we can work together. When we first met, he was very much interested in rather abstract parts of analysis, and since I had broken away, temporarily, from number theory and was working in topological algebra, I found talking to him very easy. Later, the problems which arose in his scientific war job brought our scientific ideas together again. Applied mathematics as it is nowadays needs analysis to some degree less than matrix theory. So we are never altogether out of contact mathematically. Like myself, he is a dedicated teacher, a family trait with him. At Caltech this has led him to a number of book publications which are well received. He is also well known for his work on approximation theory, special functions and the application of elliptic functions.

The theses of my students have, as I implied earlier, on the whole, been concerned with problems of my own. Two of my students have started their work alone. One student, Hobby, worked on group theory, his thesis leading to the solution, by Golod and Shafarevich, of the class field tower problem. Some of my students have not even realized that I work at present in two quite different subjects; some of them had a definite wish not to be concerned with number theory. At one time I had three students not working in number theory, while at the same time my research was entirely in number theory. With some of my students I continue research contacts, notably R. C. Thompson, Bender, Gaines (and even his colleague Laffey), Maurer, Uhlig, and Hanlon.

Something that has brought me great pleasure over the years is my contact with Japanese mathematicians. This goes back to my student days, when I heard of the beautiful results Takagi had found in class field theory. I took the liberty of writing to him for reprints and in due course I received them. Later I sent him a copy of my thesis, after it was published. In the meantime one of his best students, Iyanaga, visited Europe. He had sent me a copy of his thesis, and he visited me in Vienna. In due course other Japanese mathematicians visited me or sent reprints. One of them was Shoda. One of the latter's reprints contained results on expressing a matrix of determinant 1 as a multiplicative commutator over the field of the matrix's entries, and of expressing a matrix of trace zero as an additive one. This immediately struck me as a very beautiful and important result. Many years later I was able to use it for my own work. I then started to teach it and Shoda's work was completed by my student R. C. Thompson in his thesis and in later work and led to a great deal of work by others as well. Shoda became aware of this and appreciated this very deeply.

I met Takagi personally in Zürich at the International Congress in 1932 and he visited me and my family later in Vienna where he, of course, also visited Furtwängler. (However, these two class field giants did not discuss class field theory!)

At the conclusion of World War II, when things had normalized somewhat, I sent a post card to him inquiring about his safety. Many Japanese mathematicians heard about this message and appreciated it immensely. Takagi had done a tremendous amount for Japanese mathematics and is revered for this.

I had always hoped to visit Japan some day, and this finally became a reality a few years ago when I was invited to the number theory conference in Kyoto held in honor of Takagi. My husband went with me and was able to attach himself to a group of mathematicians in his line of work and was given a very good reception.

But it was not only the achievements of the Japanese in number theory and algebra that came to my attention. I had met some Japanese visitors in Menger's circle working in analysis, logic, and statistics and, much later, Tosio Kato whose work I admire tremendously.

Honors

In the course of my career, I have received numerous honors of more or less importance. Several of these made me feel truly rewarded for my hard work over the years. At the time of my retirement, two

At a 1976 symposium, mathematicians from distant points came to honor Olga Taussky-Todd. Pictured (left to right) are R. C. Thompson, University of California, Santa Barbara; R. Varga, Kent State University; H. Schneider, University of Wisconsin; and D. Carlson, Oregon State University.

At the symposium a volume of papers was dedicated to her. Here we see her being congratulated by R. Varga.

journals on linear algebra—the *Journal of Linear Algebra and Applications* and the *Journal of Linear and Multilinear Algebra*—published issues dedicated to me, and a number of isolated papers appeared in various journals which were dedicated to me. The *Journal of Number Theory*, went even further: they published a book entitled *Algebra and Number Theory*, published by the Academic Press and edited by H. Zassenhaus. This book contains an autobiographical sketch, which is entirely a technical survey of some of my work.

Something I consider a real honor is the fact that some papers of mine were read to the last details by truly great people, like Carl Siegel, who informed me of some tiny slips. Another honor was when I was awarded a Ford Prize for my paper on sums of squares. Then there was a symposium arranged here at Caltech by Professor Varga. In this symposium, a number of papers connected with my work were given, but in particular one speaker, H. Schneider, reported on my influence via three particular areas on which I had worked. He talked about just three of them because they particularly concerned him and he is not likely to know about several others, but I was very grateful and elated about it all. I myself had known about my influence on matrix theory in this country, but I did not expect that anyone would spell it out as he did. Actually, Schneider went even further and published an article entitled: "On Olga Taussky's influence on mathematics and mathematicians." The eminent J. P. Serre wrote a paper called "Sur un problème d'Olga Taussky." And the Mathematical Association of America recently published a book containing a selection of papers in algebra which had been published throughout the history of the *Monthly* and the *Mathematics Magazine*, and all my major papers from the *Monthly* were included.

In 1963, I was given the "Woman of the Year" award by the *Los Angeles Times*. Apart from the strain that the ceremonies and interviews inflicted on me, it gave me great pleasure. I knew that none of my colleagues could be jealous about it (since they were all men), and that it would strengthen my position at Caltech. My husband was delighted about it and enjoyed the ceremonies. Otherwise it did nothing to me. Recognition that has pleased me far more were those instances where a specific piece of my research or a lecture I had given were involved, or where something I had done for a student was involved.

While I slowly became an established member of the mathematical community, I was awarded fellowships and grants, was a member of the Council of the London Mathematical Society, and was three times elected a member of the Council of the American Mathematical Society. A particularly appreciated event was the award of Fulbright Professorships to both of us at my own Alma Mater in Vienna. Caltech gave us leave for this. There I met Professor E. Hlawka who was still a student when I left. He was later to spend some time at Caltech and we have enjoyed his tremendous mathematical strength and warm friendship ever since.

So, in 1965, my husband and I were appointed Visiting Professors at the University of Vienna. This brought me back to Austria, and the Austrians have supplemented this invitiation by a number of rather impressive honors since then: membership in the Austrian Academy of Sciences, and later the Gold Cross of Honor, First Class, awarded by the Austrian government and handed over to me by the Consul-General of Austria in Los Angeles in a ceremony at Caltech. In addition, the University of Vienna has renewed my doctorate by awarding me a Golden Diploma.

This account was written in installments quite some time before 1980, the date on which it was delivered to the Caltech Archive. A few alterations and insertions have now been made under the advice of John Todd and Professor of History J. Grabiner. Several years have passed, but I do not plan to report much on what happened since then, although a lot did happen. For the sake of continuity, there have been some repetitions. In 1977 I became Professor Emeritus, a rather honourable title, but I was 'retired,' a phrase I absolutely abhor. Nobody, absolutely nobody, ought to be burdened with it, unless by fate or by oneself. Apart from not lecturing I am carrying on as before, maybe with rather more duties, some administrative, some highly technical. I have always helped others and I continue doing this. But I have not coauthored a paper for a number of years now. I have an enormous mathematical correspondence, people from many countries flood me with enquiries, I am asked to write a great many letters of recommendation and evaluation for universities and grant-giving agencies, and I am an editor for several journals. Some of these activities bring me pleasure and even additional knowledge. One of the most pleasing duties of this kind is my correspondence with J. Ochoa from Madrid. His interest is in integral matrices and he has very original ideas and seems entirely self trained. I continue to be invited to lecture at prestigious conferences like the Noether 100th birthday symposium at Bryn Mawr College in 1982 and at the Gödel symposium in Salzburg to happen in July 1983. I worked with two brilliant thesis students, but above everything else I completed a piece of research with which I am exceedingly pleased and proud. It concerns an application of integral matrices to two of Gauss's most famous projects in number theory, the composition of binary integral quadratic forms and the principal genus. Gauss, with his enormous talents, worked through these problems, but the application of non-commutative methods gives additional insight.

When I was sufficiently mature to think about my career, and this came to me rather early, I knew that I was dedicated to an intellectual life, with science, in particular mathematics, my main interest. However, from early childhood on, poetry and writing came to me in a natural way. But it seems to me that both in the work of others and in my own I look for beauty, and not only for achievement. Only an expert will understand what I mean by this.

In conclusion, I want to say: a person who started with the enthusiasm that came to me and did not diminish through hardships, difficulties, and disappointments 'is not given a choice' and 'shadows of the future one does not see.'

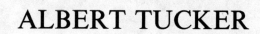

ALBERT TUCKER

ALBERT TUCKER

Interviewed by Stephen B. Maurer

The mathematical career of Albert W. Tucker, Professor Emeritus at Princeton University, spans more than fifty years. Best known today for his work in mathematical programming and the theory of games (e.g., the Kuhn–Tucker theorem, Tucker tableaux, and the Prisoner's Dilemma), he was also in his earlier years prominent in topology. Outstanding teacher, administrator and leader, he has been President of the MAA, Chairman of the Princeton Mathematics Department, and course instructor, thesis advisor or general mentor to scores of active mathematicians. He is also known for his views on mathematics education and the proper interplay between teaching and research. Tucker took an active interest in this interview, helping with both the planning and the editing. The interviewer received his Ph.D. under Tucker in 1972 and teaches at Swarthmore College.

A Career as an Actuary?

MP: *Al, let's start at the beginning: tell us where you grew up and how you got interested in mathematics.*

Tucker: I grew up in Ontario, Canada, where I was born in 1905. My father was a Methodist minister. We lived in several small towns on the north shore of Lake Ontario. It was while I was going to high school that it first seemed I had talents in mathematics. It was a three-teacher school, with the principal teaching science and mathematics. A few weeks into my Euclidean geometry course, the principal decided to give us a test. For his own convenience, he used part of a provincial examination. This contained both questions of knowledge and "originals." He had not previously given us any originals and didn't expect us to answer them. Well, I didn't know this, so I answered the originals. That night the principal came to see my father and wanted to know if my father had been coaching me, because he knew my father had taught mathematics for a year or two. My father said no, he had not. Then the principal said, "I think your son must be a mathematical genius. I think he can have a very promising career as an actuary!"

From that time on my parents thought of me, their only child, as a budding mathematician. For myself, what I realized was that although I did well in all my subjects, I did well in mathematics without trying.

Mathematics or Physics?

MP: *You attended the University of Toronto. Tell us about that.*

Tucker: I entered the University of Toronto in 1924. In those days, there was a Pass Course, which took three years, and Honors Courses, which took four. I enrolled in the Honors Course in Mathematics and Physics. There were about seventy-five enrolled in this Honors Course in my year. Almost all our courses were in mathematics and physics. Other than that I had courses in chemistry and astronomy and four or five elective courses, including one in so-called "Religious Knowledge."

My own idea at that time, as far as I had a goal, was to become a high-school teacher of mathematics and physics. I knew very little about what an actuary was, but on the other hand I had had high-school teachers I thought very highly of.

At the end of the first year I was first in my class. (I had also been first on the provincial scholarships examinations before entering, in mathematics and in Latin.) I didn't know it at the time, but a professor who was leaving to go back to his native Ireland, J. L. Synge, who had taught me a course in conic sections, left a note to the chairman of the department that there was a young man in the First Year by the name of Tucker who bore watching.

The chairman of physics was also watching. In my Second Year he taught me History of Physics. During the summer he had attended a conference in Italy where for the first time he heard about quantum mechanics. He had the fashion, a very good one which I followed later on, of having students report on various topics. He assigned me to report on quantum mechanics. At that time nothing was available except a few published papers. I read these. I don't think I understood any of them, but I put together some sort of report which greatly impressed him. He called me into his office and urged me to switch from the straddle between mathematics and physics to pure physics.

He and the chairman of the mathematics department communicated about me, though they were not on good speaking terms, and agreed that I should go on in one or the other but not in both.

I took their advice, but it was a very hard decision for me. I was more attracted to physics. It seemed to have much more glamor. These were the early days of relativity and quantum mechanics. But I found I was able to talk more satisfactorily with the professors of mathematics. I felt when I talked to them that I knew what we were talking about, whereas the physicists were always talking in terms of analogies. This was before physics had become mathematicized in the modern sense. If there had been real mathematical physics at Toronto in the modern sense, I probably would have opted for that.

So I chose mathematics. When I did this, I realized I was going somewhere other than into high-school teaching. There were jobs in the schools only for those with a joint specialty in mathematics and physics.

"Princeton Was the Place I Wanted to Go"

MP: *How did you decide on Princeton for graduate school?*

Young mathematicians of Graduate College, Princeton University, 1932. Front row: J. L. Barnes, N. Jacobson, J. B. Rosser, G. Bol, C. B. Morrey, B. Hoffman. Second row: T. Graham, R. J. Walker, W. C. Randels, A. W. Tucker. Third row: E. W. Titt, G. Garrison, E. F. Beckenbach, M. M. Flood. Top row: D. Marfield. Statue: A. West.

Tucker: Early in my Fourth Year, the mathematics chairman, Dean DeLury, called me in and told me I should be thinking about graduate study and that I ought to go abroad. He felt Oxford, which was preferred by my father, was not a good place for mathematics. Cambridge was good, but best of all was Paris, he thought. When I didn't take to the idea of Paris, he suggested Göttingen or Bologna—he knew I was very interested in geometry. But I was really frightened of studying in a foreign language. So I wrote to Cambridge and got information about courses there. These were mainly nineteenth-century style courses. Of course, I didn't know anything about the quality of these courses, but somehow they didn't impress me.

So in order to postpone making a decision, I stayed on at Toronto for a fifth year as a Teaching Fellow to get a Master's degree. I had a very good year of teaching. Originally I was given one course, but soon the other two teaching fellows dropped out and I was teaching three! I taught Advanced Calculus to the first small group of aeronautical engineers and Interest and Bond Values in a laboratory session for Pass Course students. Also, I taught Mathematics for Economists. This was to Third Year students, mainly students who had started in mathematics and physics. The trouble was there was no adequate textbook at that time. Now mark you, I had not ever studied economics. Of course, the students hadn't very much knowledge of economics either! So we learned together. But this was a year course, and after two or three months I had done all the mathematical economics I could lay hands on. So I finished the year teaching pretty much straight statistics, which I felt these students ought to know. There again I had really not had much statistics myself.

As a Teaching Fellow I got a tiny, bare office. The first time I entered that office, the one piece of reading material in the room was a Princeton graduate catalog! I looked at that catalog and saw the courses that were listed—a course by Veblen on projective geometry, a course by Lefschetz on algebraic geometry, a course by Alexander on combinatorial analysis situs, a course by Eisenhart on differential geometry and another course by Eisenhart on Riemannian geometry. Instantly, I decided that Princeton was the place I wanted to go.

Well, I went to Dean DeLury and told him this. He said, "Oh, I don't think that's a very good idea," and started in again on going abroad. "But if you insist on going to the United States," he concluded, "there are only two places, Harvard and Chicago."

I wrote to Harvard and Chicago for catalogs and compared offerings, and I decided the geometry courses there were not nearly as attractive as at Princeton. So I went back to Dean DeLury and told him what I had done and that I felt Princeton was the place for me. He said—this was the only time I really saw him angry—"Mr. Tucker, somehow you don't seem able to take advice!"

I left his office thinking there goes any further mathematics study. It never occurred to me that I could just apply. I was so naive that I thought I could only go to one of these places if in some sense Toronto sent me there.

It happened before I left for the day that I saw one of my favorite teachers, Professor Chapelon. He sensed something was wrong and got me to explain. He said, "Well, let me see these catalogs." It turned out he had gone to the same lycée in Paris as Lefschetz and thought Lefschetz was a terrific mathematician. He also thought highly of others at Princeton. So he went to Dean DeLury, and said, "I think that you should not discourage Mr. Tucker from going to Princeton," and explained.

To Dean DeLury's great credit, he immediately reversed himself and told me he would write at once to his good friend H. B. Fine. He did, but unknown to us, the letter arrived shortly after Dean Fine's death. Weeks went by without an answer, and again I became very discouraged. Finally, a letter came to DeLury from Eisenhart, explaining that Mrs. Fine had recently turned over the letter to him. It was now too late for me to apply in the regular way. But, they needed a part-time instructor in mathematics, and if DeLury could recommend me in that capacity, I could come and also start graduate study.

Thanks to all the teaching I was doing, Dean DeLury had no problem recommending me. So I was appointed a part-time instructor at Princeton for the 1929–30 school year, with a salary of $1000 and free tuition.

MP: *I take it you enjoyed your first teaching at Toronto. Did you have any other early indications of your interest in teaching?*

Tucker: Pretty early on at University, I spent hours writing up my course notes. In many of the courses there were no textbooks. I wrote up my lecture notes as though next year, if necessary, I could teach the course! That was somehow my aim in learning: to be able to explain the material or teach it. I suppose this was early evidence of my very strong pedagogical impulses.

How to Write a Thesis

MP: *At Princeton you got your Ph.D. under Lefschetz. How did this come about?*

Tucker: Both my first two pieces of research, the first, a paper written under Eisenhart, and the second, my thesis with Lefschetz, came about through trying to improve their books! With Eisenhart, during my first year at Princeton, I took his course in Riemannian geometry. Along about the middle of the year, in the chapter on Riemannian subspaces, I saw what I regarded as a flaw in the presentation of covariant differentiation. I went to see him after class and made this criticism to him. He said very courteously, why don't you write this out. The next week I gave him three or four pages. He made certain criticisms and suggested I rewrite it. This was repeated several times. Finally, one day when I had given him about the fourth rewriting, he said, "Mr. Tucker, I would like to submit this for publication in the *Annals of Mathematics*." Well, you could have knocked me over with a feather! I had had no thought at all that I was writing a paper. I was just trying to make my point with him.

Lefschetz, at the end of my first year, gave me some chapters of the manuscript of his first topology book to look over for errors during the summer in Toronto. Well, when I came back in the fall, I gave him comments not just on typographic errors and other small things which were not right, but I proposed that he *rewrite* these chapters along different lines! Oh, he was very scornful of all this, but I persisted about it. He went off for the second term (he and Alexandroff in Moscow changed places), but while he was gone I kept on working up my ideas of how his book should be written. At the end of the second summer I presented him with a lengthy screed on it. By then his book had been published, but again I was just trying to win my point. So Lefschetz said, "You better write this up and get done with it, because until you do that I see you're not going to go on and do anything else!" He actually set up a weekly seminar for me to present this material. As this went on, he got more and more enthusiastic about it. Finally he said I ought to make it my thesis, which I did. But before he said that, I had not been thinking of it as research.

Often graduate students have asked me "How do you get started writing a thesis?" I would say, there are lots of ways, but here is one way I have had good experience with myself. Take something you are interested in, mull it over, and make it your own. There's a good chance that in doing this you will find new ways of looking at the material, and this will turn into something that's publishable.

Incidentally, Lefschetz was the one who introduced the word topology, for the title of this first book of his, published in 1930 in the Colloquium Series of the AMS. There was an earlier volume in that series, written by Veblen, called *Analysis Situs*. Lefschetz wanted a distinctive title and also, as he would say, a snappy title, so he decided to borrow the word *Topologie* from German. This was odd for Lefschetz since he was French-trained and *analysis situs* was Poincaré's term; but once he decided on it, he conducted a campaign to get everyone to use it. His campaign succeeded very quickly, mainly I think because of the derivative words: topologist, topologize, topological. That doesn't go so well with analysis situs!

Also, Lefschetz was the one who invented the term algebraic topology, for his second Colloquium volume. The subject had been called combinatorial analysis situs or, later, combinatorial topology.

MP: *Let's talk now about the changes in your own work. Your work in mathematical programming and games does not appear to be closely related to your work in topology. Is it? If not, how did you get involved in these new areas?*

Tucker: Looking back now, I feel I have always been interested in combinatorial mathematics. When I was called a topologist, it was the combinatorial cell structure that interested me. Someone who studied combinatorial topology was called a topologist; I should have been called a combinatorialist, but the term just didn't exist then. So in 1948, when I had the opportunity to move into other parts of mathematics, I probably didn't consciously recognize them as combinatorial, but I think intuitively I was attracted to them for this reason.

The story of how I became involved in games and programs has been told by Harold Kuhn in that very fine survey article he has done on nonlinear programming [3]. Briefly, the Pentagon was very impressed with George Dantzig's 1947 invention of the simplex method and wanted to set up a university-based project to study linear programming further. In May 1948, Dantzig came up to Princeton from Washington to consult with von Neumann about such a project. At the end of the day, George needed a ride to the train station at Princeton Junction. I just happened to be introduced to him then and offered him a ride, during which he gave me a five-minute introduction to linear programming, using as an example the transportation problem. What caught my attention was the network nature of the example, and to be encouraging, I remarked that there might be some connection with Kirchhoff's Laws for electrical networks, which I had been interested in from the point of view of combinatorial topology. Because of this five-minute conversation, several days later I was asked if I would undertake a trial project that summer, and I agreed. The two graduate students I got to work with me were Harold Kuhn and David Gale. Thus began an Office of Naval Research project that continued over two decades.

Many people think there was a sudden change in direction for me in 1948, but it was really things I had been interested in before that led into the things we did in this seemingly new direction.

Combinatorial Mathematics

MP: *Let me pursue the nature of the "change" a little further. You've pointed out that you were always combinatorial in your interests. Nowadays you are also very algorithmic in your approach. There is a quote you very much like by Hermann Weyl, and which you taught me: "Whenever you can settle a question by explicit construction, be not satisfied with purely existential arguments." Did you have this constructive, algorithmic attitude early on as well?*

Tucker: This attitude has been a gradual thing with me. From 1948 to about 1957 I was really interested in existential results, so my approach to programming was in terms of "convex geometry," and I had never bothered to examine the simplex algorithm carefully! There was somehow in my mind —something very common with mathematicians—a compartmentalization between numerical results and theoretical results.

In 1957 I was a consultant to a project at Dartmouth of the MAA Committee on the Undergraduate Program to write an experimental text for a second year course for students in the biological and social sciences. My job was to help write a chapter on linear programming. Well, up to that point, in any talks I gave on linear programming (I had not yet given any linear programming courses), if I wanted to start off gently, I began with an example that could be solved graphically in 2 dimensions. But when it came to describing the subject in a book, I felt that one had to present a solution method that would generalize to higher dimensions. The first thing to look at, naturally, was the simplex method. And the more I looked at the simplex method, the more I became fascinated by it. I began to see that it had very interesting structure. (One can prove from this structure, by purely combinatorial and algebraic means, that the algorithm must terminate. But by definition, the algorithm terminates only if it reaches a tableau of one or another specified form, forms from which the existence or nonexistence of certain feasible or optimal solutions is obvious. Thus the proof of termination proves the fundamental "alternative theorems" of the subject, while simultaneously showing how to compute the correct alternative in any given case—something the original proofs did not do.)

"Unify and Simplify"

For me it was a revelation to see an algorithm which, if you let it, would develop the theory for you. Ever since that time, it has been my aim to make theory and the numerical methods of solving problems as unified as possible. I guess you can say in some sense it has always been my aim to unify and simplify. I believe that the simplification very often occurs through obtaining meaningful examples, examples which, if you understand them, don't need a lot of theory—the examples carry the story.

I believe I began teaching a linear programming course in the late 1950's. It was by dint of teaching this to undergraduates that the algorithmic side continued to develop for me. But it was only after teaching the course several years that I got it organized into what I think now is a very nice form [4]. I feel that all along things which I have done that might be called research have been intertwined with my teaching, and I don't know where to draw the line between one and the other.

"Develop Courses for Students"

MP: *In addition to a linear programming course, you have developed several other courses during your career. Tell us something about them.*

Tucker: When I started as an instructor at Princeton in 1933, I had the opportunity to develop two new courses. One of these was a junior course in elementary combinatorial topology. I taught that almost every year until World War II, and several times since then. That particular course has been turned into a textbook by Donald W. Blackett [2]. Various students took the course and later became topologists.

Princeton University Department of Mathematics, 1951. Front row: A. W. Tucker, E. Artin, S. Lefschetz, A. Church, W. Feller. Back row: J. T. Tate, J. W. Tukey, D. C. Spencer, R. C. Lyndon, V. Bargmann. Absent: S. Bochner, R. H. Fox, N. E. Steenrod, E. Wigner, S. S. Wilks.

The other course was one in "college geometry"—now rather out of fashion. I taught various geometric transformations in the plane, for instance, the seventeen infinite Euclidean patterns. It was a low-brow survey of geometries in the sense of Felix Klein: projective transformations, affine transformations, inversions, that sort of thing.

After I became involved in games and programs, I did most of my course development in that direction. In addition to the linear programming course, I developed an undergraduate course in the theory of games and one on combinatorial mathematics, mainly graph theory. In the more recent years before I retired, I taught the linear programming course every fall semester and alternated in the spring semester between the other two.

The most unusual course I developed was a course in geometric concepts. Soon after World War II, Princeton, like many universities, introduced general-education or distribution requirements. The natural sciences requirement could only be met by laboratory courses, so the one place where a mathematics course could fit was the catch-all Area IV called "History, Philosophy, Religion." Well, I had been a member of the faculty committee which drew up the plan, and which urged every department to develop distribution courses, so I took it on myself to try to design a mathematics course to fit in Area IV. The course I worked out was called "Evolution of Geometric Concepts." In this course I tried to trace geometric ideas from conics before Euclid down to present-day topology. The course had no prerequisites, and I concentrated on material that could be treated verbally and pictorially. I talked about things such as the Pascal configuration and the Lorentz transformation, but would depend upon plausibility arguments rather than proofs. For instance, the Pascal result clearly holds for a hexagon inscribed in a circle so that opposite sides are parallel. If you make an oblique projection of this, you get a general configuration. I don't think there is any other course that I have taught as often and for which I have the same fondness.

I feel that the chance to develop a course is a tremendous opportunity. There is a lot of work involved, but it's very rewarding. Students feel that a course that is being developed for *them* is much more meaningful than a course that is just being taught from some textbook. Also, if the instructor handles the responsibility of developing a course in an intelligent and sincere fashion, he will learn a great deal and it will make him much more interested in the job of teaching.

The Purpose of a Ph.D.

MP: *What about your philosophy of teaching on the graduate level? What do you see as the purpose of a Ph.D.?*

Tucker (laughing):   We could spend all night talking about that! I was one of the people who took an interest a number of years ago, when there was a great shortage of college teachers, in the idea of having a Doctor of Arts degree. This would not require an original contribution to knowledge but could be attained by satisfactory work over a reasonable period, like a Master's degree. I felt very strongly that if someone did a publishable piece of research, the *publication* was the acknowledgment, the credit, the reward, and that a degree to bless that was not necessary. So I was quite happy to have a doctorate degree given to anyone who reached a certain level of mathematical maturity, and I really didn't care what sort of doctorate it was called. If it was done by a research thesis, fine, but if it was done by a so-called scholarly thesis, fine also.

As a thesis adviser, I felt that my principal role was one of encouragement. Almost all my Ph.D. students seemed quite self-reliant. Very often I really did nothing for them mathematically; I simply was the straight-man against whom they could bounce their ideas.

I sometimes would suggest a general area for a student, and I had some fortunate successes. I had one student, E. F. Whittlesey, who had had to drop out because of family financial problems and came back many years later to retake his General Examination, which he had failed the first time. He passed nicely the second time and then came to me and said "I want to do a thesis with you in topology." Well, I was no longer working in topology. I felt that I was obliged, however, to meet his request. He had been very much interested in the undergraduate topology course I mentioned earlier, and he said he would like to do something with the sort of cutting and pasting I had used to classify two-dimensional manifolds. So I said, "Why don't you try to do a classification of two-dimensional *complexes.*" Well, it had been proved by Reidemeister that you could find a finite two-dimensional complex that would have any given finitely generated group as its fundamental group, so this topological problem looked as if it might be as difficult as the problem of classifying finitely generated groups.

Of course, I didn't really think Whittlesey would solve this problem. I regarded it as unsolvable. I thought he might find a subclass that he could classify, and that then this would become an important subclass, just as earlier lens spaces had become an important subclass of three-dimensional manifolds. But he went off to where he was teaching, and I neglected to tell *him* this! I don't know if this was

Father and Sons: Alan, Albert, and Tom Tucker—All Mathematicians.

carelessness or what, but it was apparently very fortunate that I didn't tell him. About three months later he called me and said he had solved it. Well, of course I didn't believe him, but I asked him to come to Princeton at the first opportunity. The next weekend he came, we worked on it until noon Sunday, and by then I was convinced that he *had* solved the problem. I told my colleague Ralph Fox about this, and he didn't believe it either. But when the thesis was turned in, he was the second reader and approved the thesis. His remark to me then was, "What devil got into you, to set him that problem?" and I had to explain. Anyway, that, of all my experiences with thesis supervision, was the most fortuitous.

Another unusual case was Marvin Minsky. He was a graduate student at Princeton in the early 50's. He was given support through a research assistantship with the Office of Naval Research project I directed. I very quickly discovered that he was very talented and had all sorts of original ideas.

One day towards the end of his first year at Princeton I asked him if he had any plans for his thesis. He said no, but he supposed he would do a thesis in topology because that's what so many of the other students were doing. "But what would you really like to do?" I asked. He indicated he would like to develop some of the ideas he had been interested in as an undergraduate concerning the relation between computers and the brain. "Well," I said, "why don't you!" He replied, "The Department would never accept a far-out thesis like that!" "No, as far as I'm concerned," I said, "the only requirement for a Ph.D. thesis is that it should be an original contribution to human knowledge; there's no limitation about far-out!" "But who would supervise it?" Minsky asked. "I'm willing to supervise it. I can't help you with the material," I explained, "but I will serve the formal purposes." And so we agreed that he would try to develop his ideas and put them in thesis form.

I did have some qualms about this, because there would have to be a second reader of the thesis, and the report of the readers would have to be accepted by the Department. But I really felt strongly that Minsky should develop his ideas on "artificial intelligence," as it is now called.

Well, he did. In the end his thesis was about 300 pages. As I recall, the title was "Neural Networks and the Brain Problem". He did it all on his own. When I saw him I would ask how things were going, but this was really just general encouragement. I was the first reader and John Tukey was the second. We also had an independent reading done by the chairman of the biology department. The point of this was to have him assess whether or not the physiological assumptions that Minsky made were reasonable ones. He said they were, and put it in writing. So the report was made to the Department and there were no objections.

There is a profile of Minsky published a year ago in the *New Yorker* [1], in which mention is made of his exceptional thesis and of the informal club-like atmosphere of Fine Hall that he shared with creative contemporaries, such as John McCarthy and Lloyd Shapley.

MP: *Al, let me ask one more question about teaching, of a more personal nature. Both your sons, Alan and Tom, are active mathematicians. In bringing them up, did you stack the deck?* [*N.B.: Alan is also called Al by some, but must be called Alan here, for obvious reasons.*]

Tucker: I'm very happy that my sons seem to share my mathematical tastes, but this has not come about through any pressure from me. It's quite clear that through some osmosis they acquired values that inclined them towards mathematics and its teaching. The thing I can't really understand is why they both have combinatorial interests akin to my own. They did not attend Princeton and I at no time tutored them. Of course, I would chat with them about the mathematics they were studying. At one time Alan was having difficulty, he thought, in finding a thesis adviser. But I knew that he was enjoying the work that he did summers at the Rand Corporation, where he was associated with Ray Fulkerson. So I suggested that perhaps he could start a thesis with Fulkerson and complete it at Stanford, where he was a student. This indeed happened, thanks especially to George Dantzig.

Founding of the Annals of Mathematics Studies

MP: *I understand you've been involved in a number of editorial activities over the years which are not well known but which you feel have been important. Tell us something about them.*

Tucker: Well, in 1933, when I was appointed to the mathematics faculty at Princeton, I was assigned the job of handling the manuscripts which came in to the *Annals of Mathematics*, until they were either accepted or rejected. Lefschetz and von Neumann were the editors, but Lefschetz made most of the decisions. At that time the Annals had no paid secretary. So when a manuscript came in, Lefschetz would take a quick look at it. In cases where he knew the author or the subject matter, he would make a snap decision whether to accept it or reject it. The other papers were turned over to me, and I had to find referees for them. Then I rode herd over the referees. I have no good substitute to suggest for the refereeing system, but it's a pain in the neck for just about everybody concerned. Anyway, if there was correspondence, it was always signed by Lefschetz; I never formally appeared as having anything to do with it. But it was all dictated and handled by me.

Also in 1933 I was put in charge of the department mimeograph machine, merely because there had been some problem about people using the machine carelessly without supervision, and Dean

Eisenhart, the chairman of the department, believed in running a tight ship. The mimeograph was used to run off course notes, and not much else. But there were a large number of course notes. The Institute for Advanced Study had just recently been established and shared old Fine Hall with the Princeton mathematics department. Institute professors such as Oswald Veblen, Hermann Weyl, John von Neumann and Marston Morse had been accustomed to lecturing, and even though they were not required to lecture at the Institute, they just did it anyway out of habit. Also, the Institute professors had distinguished assistants, and these assistants would take notes of the lectures. Some of these sets of lecture notes were on research; for example, the first publication of von Neumann on the work he was doing on linear operators came in such notes.

People would hear about these course notes, even in Europe, and they would write for copies. So this became a business. We priced the notes originally to cover just the cost of ink and paper; the work of mimeographing and collating was done by students. It was a very amateur enterprise. But when it became clear that one of the two mathematics secretaries was spending a sizeable fraction of her time on correspondence about these notes, something had to be done. Several remedies were tried starting in 1937, but the final successful change was made with the creation in 1940 of the Annals of Mathematics Studies, which were published by the Princeton University Press as photo-offsets. The course notes became volumes in the Studies. At the same time the Annals of Mathematics had been having difficulty with what to do with long papers. Such papers were transferred to the Studies as monographs, and this is why the series was called the Annals of Mathematics Studies.

So, just because of jobs I was assigned as low man on the totem pole, I had a major role in the creation of this distinguished series.

My other experience at that time in editing—it's another long story with chance elements—was in helping to establish and run the Princeton Mathematical Series (an equally distinguished series of hardback full-length advanced books, also published by the Princeton University Press). Because it was felt that I was too young, inexperienced and unknown to be sole editor of this series (this was 1938 and I had just been given tenure), two other editors were also designated, H. B. Robertson and Marston Morse.

There are several interesting stories about how books got into the Princeton Series, especially books by European authors published during World War II. On the other hand, I feel that the Annals Studies, which have now passed 100 volumes, is a greater contribution than the Series. The Series was not too different from what other publishers might do, but at the time the Annals Studies were started there was *nothing* in the United States in the way of low-cost paperback editions of serious mathematics. Gödel's consistency of the continuum hypothesis might never have been published if there had not been the Annals Studies in which it could be done at a low cost.

MP: *We have before us a picture of a combinatorial problems seminar at the IBM Research Center. I understand this is an unusual picture.*

Left to right: Merrill M. Flood, John L. Selfridge, Herbert J. Ryser, Ralph E. Gomory, Albert W. Tucker, Edward F. Moore, George B. Dantzig, and James H. Griesmer.

"Mathematics Must Become More Algorithmic"

Tucker: Well, the picture is not unusual, but the workshop, which was held in the summer of 1959, was the first as far as I know in the area of combinatorial mathematics. The participants—there were about fifteen of us—spent a considerable amount of our social time in trying to define what we meant by combinatorial mathematics. We were all agreed that networks or graphs fell in combinatorial mathematics, and also that many aspects of matrices were combinatorial. And of course we agreed that the traditional combinatorics arising out of permutations and combinations was part of the subject. But outside that we all had our own conceptions of combinatorial mathematics. It seems strange to me now that it was such a short time ago that this area of mathematics was being defined and recognized.

MP: *Al, what do you see as the greatest challenge facing mathematics in the coming years?*

Tucker: The computer revolution. Mathematicians have set great stock in abstract mathematics in which concepts and rigor have been the dominant things. But now algorithms are really important. There have been algorithms around from Euclid's algorithm on, but they have been regarded as rather unusual. I think that mathematics will have to become more and more algorithmic if it is going to be active and vital in the creative life. This means it is necessary to rethink what we teach, in school, in college, and in graduate school. In our emphasis on deductive reasoning and rigor we have been following the Greek tradition, but there are other traditions—Babylonian, Hindu, Chinese, Mayan—and these have all followed a more algorithmic, more numerical procedure. After all, the word algorithm, like the word algebra, comes from Arabic. And the numerals we use come from Hindu mathematics via the Arabs. We can't regard Greek mathematics as the only source of great mathematics, and yet somehow in the last half century there has been such emphasis on the greatness of "pure" mathematics that the other possible forms of mathematics have been put down. I don't mean that it is necessary to put down the rigorous Greek style mathematics, but it *is* necessary to raise up the status of the numerical, the algorithmic, the discrete mathematics.

MP: *We've talked very little about you personally. What are some of your hobbies?*

Tucker: I certainly do like to travel, and fortunately I've had many professional opportunities for this. Even during vacation trips I like to visit with mathematicians and give talks. My favorite place for travel has been Australia. I've been there four times as a visiting lecturer. The city of Perth in Western Australia is my favorite city. That's where I would live if it weren't so far away from everything else that I'm tied to.

The other hobby I might mention is that I like detective stories. It isn't that I read them so much to try to guess the end; I really read them for just relaxation. I have quite a collection of paperbacks. I like best the classical British detective stories, which I started reading when I was a student in Toronto.

Early on I had liked chess, but I swore off chess when I discovered that after playing a keen chess game I had difficulty sleeping at night. I was continuing to concentrate on the game. So I switched to reading, and found that somehow detective stories provided me with the sort of relaxation I liked.

REFERENCES

1. Jeremy Bernstein, Profiles (Marvin Minsky), The New Yorker, 80 (Dec. 14, 1981).
2. Donald W. Blackett, Elementary Topology: A Combinatorial and Algebraic Approach, Academic Press, New York, 1967, rev. ed., Academic Press, London, 1982.
3. Harold W Kuhn, Nonlinear programming: a historical view, SIAM-AMS Proceedings, vol. IX, AMS, Providence, R. I., 1976.
4. Evar D. Nering and Albert W. Tucker, Linear Programs and Related Problems (in preparation).

SOLOMON LEFSCHETZ: A REMINISCENCE

by Albert W. Tucker

Solomon Lefschetz is the mathematician I have known best, and perhaps because I have known him best, admired most. But it is very difficult for me to talk about Lefschetz in a way that I feel will do justice to the tremendous respect, admiration and affection that I have for him.

I don't know of any other mathematician whose career has been so intertwined with another's as mine has been with Lefschetz'. I did my Ph.D. thesis with him, and it was understood when I joined the Princeton faculty that I was going to work with him in my research. Then in 1945, when he succeeded Eisenhart as chairman of the department, I was his lieutenant for all undergraduate departmental administration. Lefschetz retired in 1953 and I succeeded him as chairman, which was clearly his wish. But when he retired, this was only a formal thing; he kept coming to Fine just as much as before.

We were very close friends. I was one of the few persons who was frequently in the Lefschetz home. By and large they didn't have parties or invite people in a formal way to their house. But often I would give him a ride home, or walk with him, and then he would invite me in. In a very real sense I regarded him as a parent, someone I had the same regard for as for my father, but of course someone I was able to talk to much more readily than my father.

Lefschetz' research accomplishments are very wide-ranging. He started in algebraic geometry, earning his Ph.D. from Clark University in 1911. Within a few years he perceived that the things he was trying to deal with in algebraic geometry were really problems of topology. He wrote a monograph on the relationship between the two which was given the Bordin Prize in Paris in 1919. From that he moved into topology. He published in 1926 his famous paper on the Lefschetz fixed point formula. I already talked about how he introduced the terms topology and algebraic topology. Also, he had some remarkable students, Paul Smith, Steenrod, and Fox among them. The so-called Princeton School of topology was his doing. Alexander was a help; it was he who brought Lefschetz to Princeton from Kansas. But Alexander was reserved and didn't find it congenial to work with students, whereas Lefschetz was very gregarious and was always the center of a group of people.

Later on he made another switch. This came about during World War II, when he was asked to be a consultant to the Navy on problems of nonlinear differential equations. He went on from that to make another career in that area, especially studying global properties. Much of this work was done after he retired.

He worked very hard at his mathematics. He did most of his work early in the morning. He would get up at 5 A.M., work to 10 or 11 A.M at home, and then come in to the office.

He was very outspoken; indeed, many people found him quite offensive, and sometimes this had unfortunate results: several graduate students at Princeton left because of the harsh talkings-to he gave them. He meant well, but he usually spoke really without thinking. As he said once, people first have to make up their minds and then find their reasons. That was so typical of Lefschetz himself: he made up his mind very impulsively and then he gathered various arguments.

He was very quick and very imaginative. But he had great difficulty making a rigorous argument. I've heard it said that any proof Lefschetz would give would be wrong, but any result he would announce would be right. He had a tremendously sound intuition, but he was just so restless and impatient that he wouldn't take the time to make rigorous arguments.

Another thing about him: even when he knew he was wrong, he would never admit it, at least not then and there. We had some very fierce arguments and I would go away thinking I hadn't gotten anywhere at all with him. But a day or so later I would find that he had accepted the argument I had made and was going ahead with that. But he never *said* that he agreed with me! The one way I realized, years later long after Lefschetz retired, that he was getting old was that I could win an argument from him in one day!

I guess one of the reasons I got along with him so well, when many others did not, is because we were both direct and stubborn. My first brush with Lefschetz occurred quite early on in my first year as a graduate student. I was taking his course in topology from which he was writing his first book. He had given a proof of something, and I had the temerity to speak up and challenge his proof. So he invited me to go to the board and give my proof, which I did. Then he proceeded to heap all sorts of scorn on what I had said. This made me wonder; after all, I hadn't had very much experience with people such as Lefschetz. But at the next class he began by saying that he wanted to go over something from the previous time. He proceeded to give exactly the proof I had given, and then went on. He made no reference at all to me, but all my fellow graduate students knew that it was my proof. This experience somehow gave me the courage to stand up to him when I thought he was wrong. That was something Lefschetz certainly liked with people. People who would give him back as good as he gave impressed him; the people who just curled up under the rather harsh things that he often said, he had no respect for.

So, Lefschetz was emotional and quick on the trigger, but I felt that in all the dealings I had with him that he was extremely fair, and on important issues he took an objective, balanced view.

I also want to say that Lefschetz had a great disability in that he had artificial hands, and yet he never complained. He would never refer to his artificial hands. He would never say, "You'll have to do that for me, I can't do it." Instead he made a simple polite request for anything he needed to have done: "Please cut my meat," or "Please open that door." The courage that he had—he would go into New York by himself and ask strangers in the subway to take a token out of his pocket and put it in the turnstile. He wouldn't say he had artificial hands, but would simply repeat the request until, just by the power of his persistence, he would get this done.

Let me end with the following story about Lefschetz. I tell it as a mathematical joke, but it also illustrates very nicely some of the things I have said.

On one occasion during World War II, there was a one-day meeting of the AMS held in New York City, and Lefschetz and I and Oskar Zariski, who happened to be at the Institute for Advanced Study at the time, traveled into New York together on the train. Lefschetz and Zariski were talking about a certain paper, which had recently appeared in algebraic geometry, which they thought was a very good paper. Lefschetz remarked that he wasn't sure if he would classify the paper as algebra or topology. You see, Lefschetz took the view that algebraic geometry was part of topology, and Zariski that it was part of algebra. So Zariski, to tease Lefschetz a bit, asked, "How do you draw the line between algebra and topology?" Quick as a flash, Lefschetz came back with, "Well, if it's just turning the crank, it's algebra, but if it's got an idea in it, it's topology!"

Now please keep in mind, I'm telling you this as a joke. I'm not expressing any opinions. Whether Lefschetz was expressing an opinion is another matter. I think he was partly telling it in fun, but also to express in some way his own ideas.

REFERENCE

1. Solomon Lefschetz, Reminiscences of a mathematical immigrant in the United States, *Amer. Math. Monthly*, 77 (1970).

STANISLAW M. ULAM

STANISLAW M. ULAM

Interviewed by Anthony Barcellos

Stanislaw M. Ulam spent the winter quarter of the 1978–79 academic year at the Davis campus of the University of California. As Distinguished Visiting Professor of the College of Letters and Science, Ulam presented weekly seminars that touched on such topics as probability, finite state automata, evolutionary genetics, and coding theory. Professor Ulam's audiences reflected the broad span of his interests: mathematicians, physicists, engineers, and biologists, among others.

Reminiscing about his career in an interview with the *Two-Year College Mathematics Journal* during his Davis sojourn, Stan Ulam displayed a disarming modesty and personal charm. It was difficult to perceive in him the intimidatingly self-confident man described in his autobiography *Adventures of a Mathematician* [1]. The discussion touched on Ulam's origins in Poland, his work on the Manhattan Project during World War II, and the course of his subsequent research and interests. Drawing on his experiences in several scientific disciplines, Ulam spoke on the current status of mathematics and sciences and speculated on their possible development.

Decision to Become a Mathematician

MP: *Was your decision to become a mathematician a conscious one, or something that developed over a period of time?*

Ulam: At the age of ten I was interested in astronomy, then in physics, and finally in mathematics. By the time I was fifteen I was reading number theory; there was a fascinating book by Sierpinski—in Polish, of course. And then I read about set theory. At that time I thought that if it's at all possible, or practical, to become a mathematician, I would want to be one. Of course, from the practical point of view, it was very difficult to decide on studying mathematics—only mathematics—at the university because of the exigencies of a career: there were very few positions. To make a living in mathematics was very, very difficult.

So I entered an engineering school, the Polytechnic Institute, and ordered a so-called "general faculty" which actually contained a lot of mathematics courses. Then Professor Kuratowski, a very famous topologist, certainly influenced some of my early choices in topics in mathematics. I met other mathematicians more my contemporaries—although a few years older—like Mazur, and Banach. Banach was a professor at the University, but he gave courses at the Polytechnic Institute.

Very soon, just because—perhaps by luck—I managed to solve a few problems which were open, I became more sure of myself and decided to study—instead of electrical engineering—mathematics itself, come what may. I continued my work, continued writing my papers, and by the time I received my doctorate I had nine or ten papers published.

MP: *What were the fields of study that interested you most at first and how have those changed over the years?*

Ulam: Well, set theory—and topology. That slowly, of course, changed with the years—one should say decades, almost. But I have been interested in probability theory and always, so to say, platonically interested in theoretical physics.

Pure vs. Applied

MP: *During your career the work you've done has been both in very abstract mathematics and in various applied fields. Do you yourself perceive a fundamental difference between pure and applied math?*

Ulam: I really don't. I think it's a question of language, and perhaps habits. Even between pure mathematics and theoretical physics the thinking process bears many similarities. As I try to say in my seminar here in Davis: Mathematicians start with certain facts—which we call axioms—and deduce consequences, theorems. In physics, in a sense, it's the other way around: The physicists have a lot of facts, lots of relations, formal expressions, which are the results of experiments; and they search for a small number of simple laws—we could call them axioms in this case—from which these results can be deduced. So in some ways it's an inverse process, but the course of thinking about it and the intuitions have great resemblance in both cases. And the question of habits, so-called rigor, which mathematicians require is often absent in physics. If one is tolerant, however, you could say that what physicists do is quite rigorous, but with different primitive notions than the ones too naively pursued.

Now you actually didn't ask me about physics so much as about applied and pure mathematics. Even in applied mathematics the really good work is not merely a service type activity, but invention of new tools, new methods, new applications. For somebody like Gauss, you know, distinctions are really very hard to perceive; he was perhaps the greatest number theorist who ever lived, and then he did some marvelous applied work—the method of least squares, for one thing.

MP: *Currently you're a professor at the University of Florida at Gainesville, professor emeritus of the University of Colorado at Boulder, and you're still a consultant for the Los Alamos Scientific Laboratory of the University of California. How do you divide your time these days?*

Ulam: I'm afraid most of the time I'm just staying at home, because in Florida I spend three or four months at most. I don't stand humid heat very well, so come April I usually leave. Now Los Alamos is very near where I live in Santa Fe (New Mexico). In Colorado I still have an office. Lots of written material is there because there's no room in my house for tons of written material.

Los Alamos and the Bomb

MP: *What do you work on at the Los Alamos Scientific Laboratory?*

Stan Ulam wonders: "Will the shape of mathematics change? Will there be "large" theorems such that individual theorems will be left out as exercises, or corollaries?"

Ulam: Mainly problems that are not concerned with weapons. I did some work on nuclear propulsion of space vehicles, and general mathematical studies—also, what you might call mathematical biology. It's a little presumptuous term, perhaps at the present time, but people are beginning to use mathematics even conceptually for biological schemata observation. Not many solvable equations appear in practical applications; there is very little done on that.

There's little odds and ends. I'm actually now trying to write a book on unsolved problems. It's a sequel to the book on this subject which I wrote twenty years ago [2].

MP: *To stay with the subject of Los Alamos, how did it feel when you came from a background in theoretical mathematics to join a large group of engineers and physicists working on the very practical problems of the Manhattan Project and the atomic bomb?*

Ulam: It was fascinating. I must say it felt very good in the sense that it was interesting. It wasn't really—in the group in which I found myself—very "engineering" in the ordinary sense, because the thing was so new and so unknown in many aspects that it was almost like a purely theoretical discussion. The problems really had mathematical interest even though the crux was always the physics of it. That was one of the most interesting periods of my life, intellectually. There was a realization of the possible enormous changes which could be brought about through use of nuclear energy.

MP: *How do you feel about the consequences of the Los Alamos work? Are you satisfied with what has happened?*

Ulam: It's hard to say—satisfied from what point of view? It's hard to say, certainly, "satisfied" or "dissatisfied" with facts of nature. These things exist. I believe—some people say—that the advent of nuclear bombs prevented a third world war and will hopefully prevent such an unimaginable catastrophe—actually surpassing by orders of magnitude all the horrors of the past war. Some people say that, and perhaps it's true. Let's hope. Now other uses of nuclear energy are beneficial: use of radioisotopes and even use as energy sources. I myself believe strongly in the use of fission reactors for producing energy. It seems to me that there could be safeguards for disposing of the waste. After all, there are several hundred reactors running even now, mainly for energy of some sort; there hasn't been a single major accident. Some people say it's by far the safest way to produce energy—safer than coal, in terms of production of coal and the use of coal.

* * *

After an unsatisfying postwar stint at the University of Southern California, Ulam was invited to return to Los Alamos. Russian acquisition of the atomic bomb spurred efforts at Los Alamos to perfect the "super," as the hydrogen bomb was called in its development stage. Ulam agreed that the matter was urgent and was an important part of the research work. He provided the key that finally made possible the ignition of a thermonuclear device.

Ulam lays no claim to the title "Father of the H-Bomb" which Edward Teller has willingly worn. Teller exerted himself strenuously on behalf of the bomb's development and was its self-appointed champion on all fronts. He was continuing the struggle to salvage his plans in the face of increasingly negative theoretical results when Ulam produced his vital contribution. As Ulam tells it, in his *Adventures of a Mathematician*:

... Teller continued to be very active both politically and organizationally at the moment when things looked at their worst for his original "super" design, even with the modifications and improvements he and his collaborators had outlined in the intervening period.

Perhaps the change came with a proposal I contributed. I thought of a way to modify the whole approach by injecting a repetition of certain arrangements . . . .

The next morning I spoke to Teller. At once Edward took up my suggestions . . . I wrote a first sketch of the proposal. Teller made some changes and additions, and we wrote a joint report quickly . . . . The report became the fundamental basis for the design of the first successful thermonuclear reactions and the test in the Pacific called "Mike." [1, pp. 219–20]

[A]s a result of my work on the hydrogen bomb, I became drawn into a maze of involvements. [I]n some circles I became regarded as Teller's opponent, and I suspect I was consulted as sort of a counterweight. Some of these political activities included my stand on the Test Ban Treaty and testimony in Washington on that subject. The cartoonist Herblock drew in the *Washington Post* a picture of the respective positions of Teller and me in which I fortunately appeared as the "good guy." [1, p. 251]

Stanislaw M. Ulam

It's A Wise Father That Knows His Own Bomb

—from Straight Herblock (Simon & Schuster, 1964)

However, the role he played in the establishment of the nation's nuclear policy appears to loom less large in his mind now than at the time he penned his autobiography:

MP: *You figured very strongly in the debate on the management of nuclear resources.*

Ulam: No, not really. I was once involved in some testimonies about testing in the atmosphere—the Test Ban Treaty. My friends' and my own opinion was that atmospheric testing was not necessary. And finally the U.S. Senate ratified by an overwhelming margin an agreement not to have nuclear explosions in the atmosphere.

MP: *You were viewed at that time as something of a counterweight to Edward Teller.*

Ulam: No, there were many people who argued for the test ban, and I think Teller was a minority.

MP: *In the scientific community?*

Ulam: In the scientific community acquainted with the appropriate technology. On the whole, people thought that atmospheric testing was not necessary. I don't know how Teller feels about it now.

MP: *You mentioned your belief in the safety of nuclear reactors. Do you think that the general public has any real understanding about these issues, that they're properly educated on them and have sufficient information to make reasoned judgments?*

Ulam: I think that they do not have any information. Too much is the result of emotion. In this case, unfortunately, I think that it slows down the attempts of the United States to become independent in production of energy. I do not know why one does not build many more reactors.

MP: *This leads me to the broader question of the current status of mathematics, science, and technology. So many people have become very anti-technology and feel that it's responsible for most of our problems. Do you see any likelihood of this trend reversing in the near future?*

Ulam: I don't know. I'm not a prophet. Certainly what you say is true, but many of the phenomena that have been going on are due, it seems to me, to feelings of inadequacy—individuals who are baffled by the facts of science. I think some of this is one of the reasons for the unrest in the world—feelings of inadequacy. I don't know how to counteract this or how to proceed in education to make people feel better about the fact—which is now, I think, unavoidable—that one does need special technological and scientific frameworks to organize the world with its enormous population, and so many demands on an ultimate shortage of the old type of fuels.

MP: *This is a very popular question to ask mathematics professors: Do you have any opinions on the "disastrous failure" or the "qualified success" of the new math?*

Ulam: Yes, I had some feelings about the new math right away. I thought that in principle, ideally, it was an interesting thing to attempt to instill or inculcate in children a sort of more abstract way of reasoning. Unfortunately, in practice that requires very special teachers. More than that: Many people —including, for example, myself—need examples, practical cases, and not purely formal abstractions and rules, even though mathematics consists of that. They need contact with intuition. Variety almost by itself confuses the student. I think a great problem is teaching mathematics as a question of grammar rather than the structure. Sometimes, especially with teachers who themselves are not too good at it, it was a negative change and discouraged, I think, many bright children from going into more mathematical things. [This formalism] was a big problem. That's how I felt in the beginning and I think that by now it uses much less of this.

Large Theorems

MP: *To what mathematical questions would you most like to know the answers?*

Ulam: Well, I'll tell you. It's very strange; it's a question I ask myself—exactly what you ask me now. And it's very hard to give just one or two. But I certainly would like to know the outlines of a future basis of set theory—so to say, the sequel of the discovery of Gödel of undecidability. Then there are problems in number theory; some mathematicians, of course, mention Riemann's hypothesis, or Goldbach's conjecture. Are there "theorems" of this type which will be proved undecidable on the basis of present systems of number theory?

What interests me more now is not any special theorem, but rather whether the shape of mathematics will change: Will there be "large" theorems such that individual theorems will be left out as exercises, or corollaries? Well, I have to make it more precise.

* * *

In his autobiography Ulam addresses at greater length the topic of the future "shape"of mathematics. Remarking on the overthrow of the set-theoretical assumptions on which modern mathematics was founded, Ulam strives to express a sense of the new and broader concepts of "true" and "false" which may be formulated to replace the deficient current notions:

> Gödel, the mathematical logician at the Institute for Advanced Studies in Princeton, found that any finite system of axioms or even countably infinite systems of axioms in mathematics, allows one to formulate meaningful statements within the system which are undecidable—that is to say, within the system one will not be able to prove or disprove the truth of these statements. Cohen opened the door to a whole class of new axioms of infinities. There is now a plethora of results showing that our intuition of infinity is not complete. They open up mysterious areas in our intuitions to different concepts of infinity. This will, in turn, contribute indirectly to a change in the philosophy of foundations of mathematics, indicating that mathematics is not a finished object as was believed, based on fixed, uniquely given laws, but that it is genetically evolving. This point of view has not yet been accepted consciously, but it points a way to a different outlook. Mathematics really thrives on the infinite, and who can tell what will happen to our attitudes toward this notion during the next fifty years? Certainly, there will be something—if not axioms in the present sense of the word, at least new rules or agreements among mathematicians about the assumption of new postulates or rather let us call them formalized desiderata, expressing an absolute freedom of thought, freedom of construction, given an undecidable proposition, in preference to true or false assumption. Indeed some statements may be undecidably undecidable. This should have great philosophical interest. [1, pp. 283–4]

* * *

Ulam: I think that computers will bring about great changes in the aspect of both stating and proving theorems. Finally, the most interesting thing is the schema of the human brain itself. What kind of mathematics will our gradually acquired knowledge of the workings of the brain suggest? That's what I think is the most fascinating of all. Some Greek poet said, "There are many wonders, but the greatest of all is human thought." Was it Aeschylus?

MP: *That reminds me of Johnny von Neumann—his classical learning.*

Ulam: A most remarkable man. Since his death the application of his work and his influence is growing steadily. He was recognized very much during his life, but his very great fame started developing really, I think, after 1957 when he died. Too bad he didn't live to see the enormously increasing role of computers. He was an early prophet of this.

MP: *Recently the four color theorem was proved with computer assistance. How do you feel about this?*

Ulam: In some cases it might be that this sort of thing will become more frequent. Certainly I believe in the heuristic or experimental value of computers where one by working examples will get intuitions about the more general fact. Ultimately the computers will be able to make formal proofs and operate symbolically the way we do now in thinking about mathematics. There's no question at all. Now there are computers playing fair games of chess. They have a sort of 2000 rating. [This ranking would correspond to a very good amateur.]

MP: *Do you think it would be fair to call such computers increasingly intelligent? Or does the word have any meaning in this context?*

Ulam: Well, I think that actual intelligence is very difficult to define even for people. Don't you agree? There are so many different types of what you might call intelligence in individuals. Some people have intelligence in certain directions and are very dumb in some other directions. Isn't that true? Usually if you call a person intelligent it's sort of faint praise. One wants more. "He's intelligent." That's not such a great compliment.

Monte Carlo Methods

MP: *Your mention of computers and heuristic reasoning, working out special cases and examples to give one a feeling for things, naturally brings to mind your Monte Carlo method.*

Ulam: Yes, about the time when I left Los Alamos just when the war ended, I had the first thoughts about it. When I came back to Los Alamos I developed it some more and then, mainly in collaboration with von Neumann, I established several regions of application.

* * *

One may choose a computation of a volume of a region of space defined by a number of equations or inequalities in spaces of a high number of dimensions. Instead of the classical method of approximating everything by a network of points or "cells," which would involve billions of individual elements, one may merely select a few thousand points at random and obtain by sampling an idea of the value one seeks. [1, p. 200]

* * *

MP: *What do you want to do now?*

Ulam: What does anyone want to do? Enjoy a few more years of normal life. I want to write this book (of problems), finish it. I am still—*still*—thinking about problems. It's also interesting to see what's happening in foundations of physics, in particle physics, and the very strange phenomena in astronomy. Also in biology, tremendous things will happen, maybe more rapidly now than in any other science.

MP: *I have one more fairly prosaic question.*

Ulam: It's the answers which are prosaic. Oscar Wilde said, "No question is indiscreet; the answer might be indiscreet."

MP: *I would like you to cite, if you can, the things you like best among what you've done. You've mentioned the Monte Carlo method and everyone seems to recognize that as important.*

Ulam: I know, but intellectually it wasn't a great deal.

MP: *What do you like most that you've done?*

Ulam: You mean, sort of narcissism?

MP: *Yes, something of that sort.*

Ulam: It's hard. I don't compare things, but a few I thought were—by luck—not unimportant, not totally unimportant. I believe in the role of luck in scientific research. I like some works I did in collaboration with other people. I wrote many joint papers with Mazur, Schreier, Banach, Borsuk, Hyers, Everett, Oxtoby, etc. In general I somehow like to talk to people and work together.

MP: *Is it that you have the "habit of luck" yourself or that you associate with people who do?*

Ulam: That's a strange thing. Some people say, "Ah, it cannot be luck because why does it happen several times in a row?" and so on. I don't know; that's a good question. But, clearly, it's not a question of the power of the brain alone. The times must be right, and by chance you come upon something. Even somebody like Einstein, or, as people say, Newton. Who was it that said that Newton was so lucky because only once can you discover the fundamental laws of the universe? Actually there are infinitely many fundamental laws, perhaps. Certainly luck plays a role, even at the highest level, not to mention the level of a working mathematician like myself.

S. M. Ulam

[1] Adventures of a Mathematician, Scribner's, New York, 1976.
[2] Collection of Mathematical Problems, Interscience, New York, 1960.
[3] John von Neumann, 1903–1957, Bull. Amer. Math. Soc., vol. 64, no. 3, pt. 2 (1958) 1–49.

BIOGRAPHICAL DATA

Garrett Birkhoff

Born in Princeton, New Jersey, January 10, 1911. A.B., Harvard University, 1932. Positions held: Harvard University, 1936–. Member of the American Academy of Arts & Sciences and the National Academy of Sciences.

David Harold Blackwell

Born in Centralia, Illinois, April 24, 1919. A.B., University of Illinois, 1938; A.M., University of Illinois, 1939; Ph.D., University of Illinois, 1941. Positions held: Institute for Advanced Study, 1941–42; Southern Agricultural and Mechanical College, 1942–43; Clark University, 1943–44; Howard University, 1944–54; University of California, Berkeley, 1954–. Member of the American Academy of Arts and Sciences and the National Academy of Sciences.

Shiing-Shen Chern

Born in Kashing, China, October 26, 1911. B.S., Nankai University, 1930; M.S., Tsing Hua University, 1934; D.Sc., University of Hamburg, 1936. Positions held: Tsing Hua University, 1937–43; Academia Sinica, 1946–48; University of Chicago, 1949–59; University of California, Berkeley, 1960–. Member of the National Academy of Sciences, the American Academy of Arts and Sciences and Academia Sinica.

John Horton Conway

Born in Liverpool, England, 1937. Fellow of Gonville and Caius Colleges, former Fellow of Sidney Sussex College, Cambridge University; Reader in Pure Mathematics at the University of Cambridge. Fellow of the Royal Society.

Harold Scott MacDonald Coxeter

Born in London, February 9, 1907. B.A., Cambridge University, 1929; Ph.D., Cambridge University, 1931. Positions held: University of Toronto, 1936–. Member of the Royal Netherlands Academy of Arts and Sciences and Fellow of the Royal Society and of the Royal Society of Canada.

Persi Diaconis

Born in New York City, January 31, 1945. B.S., City College of New York, 1971; M.A., Harvard University, 1973; Ph.D., Harvard University, 1974. Positions held: Stanford University, 1974–.

Paul Erdös

Born in Budapest, March 26, 1913. Ph.D., University of Budapest, 1934. Positions held: Technion, Haifa; University of Colorado; University of Budapest. Member of the Hungarian Academy of Sciences.

Martin Gardner

Born in Tulsa, Oklahoma, October 21, 1914. B.A., University of Chicago, 1936. Positions held: Scientific American, 1957–80.

Ronald Lewis Graham

Born in Taft, California, October 31, 1935. B.S., University of Alaska, 1958; M.A., Ph.D., University of California, Berkeley, 1962. Positions held: Bell Laboratories, 1962–.

Paul Richard Halmos

Born in Budapest, March 3, 1916. B.S., University of Illinois, 1934; M.S., University of Illinois, 1935; Ph.D., University of Illinois, 1938. Positions held: University of Illinois, 1938–39, 1942–43; Institute for Advanced Study, 1939–40, 1940–42; Syracuse University, 1943–46; University of Chicago, 1946–61; University of Michigan, 1961–68; University of Hawaii, 1968–70; University of Indiana, 1970–84; University of Santa Clara, 1984–. Member of the Hungarian Academy of Sciences and Fellow of the Royal Society of Edinburgh.

Peter John Hilton

Born in London, April 7, 1923. M.A., Oxford University, 1948; D.Phil., Oxford University, 1950; Ph.D., Cambridge University, 1952. Positions held: Birmingham University (England), 1958–62; Cornell University, 1962–71; University of Washington, 1971–73; Case Western Reserve University, 1972–82; State University of New York at Binghamton, 1982–. Member of Brazilian Academy of Sciences.

John George Kemeny

Born in Budapest, May 31, 1926. B.A., Princeton University, 1947; Ph.D., Princeton University, 1949. Positions held: Princeton University, 1946–53; Dartmouth College, 1953–; President, Dartmouth College, 1970–82. Member of the American Academy of Arts and Sciences.

Morris Kline

Born in New York City, May 1, 1908. B.Sc., New York University, 1930; M.Sc., New York University, 1932; Ph.D., New York University, 1936. Positions held: New York University, 1930–36, 1938–42, 1945–; Institute for Advanced Study, 1936–38.

Donald Ervin Knuth

Born in Milwaukee, Wisconsin, January 10, 1938. B.Sc., Case Institute of Technology, 1960; M.S., Case Institute of Technology, 1960; Ph.D., California Institute of Technology, 1963. Positions held: California Institute of Technology, 1963–68; Stanford University, 1968–. Member of the American Academy of Arts and Sciences and the National Academy of Sciences.

Benoit Mandelbrot

Born in Warsaw, November 20, 1924. Engr., Polytechnic School, Paris, 1947; M.S., California Institute of Technology, 1948; Prof. Engr., California Institute of Technology, 1949; Ph.D., University of Paris, 1957. Positions held: Institute for Advanced Study, 1953–54; Institut Henri Poincaré, 1954–55; University of Geneva, 1955–57; Lille University and Polytechnic School, Paris, 1957–58; International Business Machines, 1958–.

Henry Otto Pollak

Born in Vienna, December 13, 1927. B.A., Yale University, 1947; M.A., Harvard University, 1948; Ph.D., Harvard University, 1951. Positions held: Bell Laboratories, 1951–.

George Pólya

Born in Budapest, December 13, 1887. Ph.D., University of Budapest, 1912. Positions held: Swiss Federal Institute of Technology, 1914–40; Stanford University, 1942–. Member of the Académie des Sciences, Paris; Académie Internationale de Philosophie des Sciences, Bruxelles; Hungarian Academy of Sciences; American Academy of Arts and Sciences; and the National Academy of Sciences.

Mina Spiegel Rees

Born in Cleveland, Ohio, August 2, 1902. A.B., Hunter College, 1923; A.M., Columbia University, 1925; Ph.D., University of Chicago, 1931. Positions held: Hunter College, 1926–48, 1953–61; Office of Naval Research U.S. Navy, 1946–53; City University of New York, 1961–. National Welfare Medalist, NAS.

Constance Bowman Reid

Born in St. Louis, Missouri, January 3, 1918. A.B., San Diego State College; M.A., University of California, Berkeley.

Herbert Ellis Robbins

Born in New Castle, Pennsylvania, January 12, 1915. A.B., Harvard University, 1935; A.M., Harvard University, 1936; Ph.D., Harvard University, 1938. Positions held: Harvard University, 1936–38; Institute for Advanced Study, 1938–39; New York University, 1939–53; Columbia University, 1953–. Member of the American Academy of Arts and Sciences and the National Academy of Sciences.

Raymond Smullyan

Born in Far Rockaway, New York, 1919. B.S., University of Chicago, 1955; Ph.D., Princeton University, 1959. Positions held: Dartmouth College, 1954–56; Princeton University, 1958–1961; Yeshiva University, 1961–68; City University of New York, Lehman College, 1968–84; Indiana University, 1982–.

Olga Taussky-Todd

Born in Olomouc, Czechoslovakia, August 30, 1906. Ph.D., University of Vienna, 1930. Positions held: University of London, 1937–44; National Bureau of Standards, 1948–57; California Institute of Technology, 1957–. Member of the Austrian Academy of Sciences.

Albert William Tucker

Born in Oshawa, Ontario, Canada, November 28, 1905. B.A., University of Toronto, 1928; M.A., University of Toronto, 1929; Ph.D., Princeton University, 1932. Positions held: Princeton University, 1933–.

Stanislaw Marcin Ulam

Born in Lwow, Poland, April 3, 1909. M.A., Polytechnic Institute, Poland, 1932; Dr. Math. Sc., Polytechnic Institute, Poland, 1933. Positions held: Institute for Advanced Study, 1936; Harvard University, 1939–40; University of Wisconsin, 1941–43; Los Alamos Scientific Laboratory, 1943–65; University of Colorado, 1965–76; University of Florida, 1974–. Member of the American Academy of Arts and Sciences and the National Academy of Sciences. Died May 13, 1984.

INDEX OF NAMES

Petrie, J. E., 63
Piaget, Jean, 212
Picard, Émile, 210, 272
Picasso, Pablo, 294
Pierce, George W., 4, 7, 15
Pierce, S., 333
Plancherel, M., 319
Poe, Edgar Allen, 301
Poincaré, Henri, 8, 143, 207, 218, 224, 272
Pollak, Henry, 227–244, 364
Pólya, George, 58, 88, 113, 122, 177, 179,
    245–53, 276, 320, 364
Pontrjagin, L. S., 144, 322
Pósa, Louis, 89
Pound, Ezra, 105
Price, G. Baley, 328
Pritchett, Gordon, 50
Pushkin, Alexander Sergeivitch, 137
Putnam, William Lowell, 286

Rademacher, Hans, 328
Radin, Charles, 68
Radon, J. R., 74
Ramanujan, S., 55
Ramsey, Frank, 113
Randels, W. C., 340
Rees, Mina Spiegel, 255–67, 365
Reichardt, H., 322
Reid, Constance Bowman, 8, 28, 177, 269–80,
    284, 365
Reidemeister, Kurt, 276, 345
Reiner, I., 328
Rényi, A., 90
Renz, Peter, 98–99, 103
Richardson, Lewis F., 215
Richardson, R. G. D., 8
Riemann, Bernhard, 87, 113, 179, 248–49, 278,
    323, 341–42, 359
Riesz, Friedrich, 10
Ringel, Gerhard, 58, 63
Robbins, Herbert E., 281–97, 365
Robertson, H. B., 347
Robinson, Abraham, 130, 319
Robinson, D. W., 333
Robinson, Julia, 258, 271–72, 274, 276, 278
Robinson, Raphael M., 271–272
Rodrigues, A., 40
Roitberg, Joseph, 135, 141, 144, 149
Rosenfeld, A., 60
Rosser, Barkley, 263–64, 340
Rosser, C. B., 330
Rossini, Gioacchino, 89
Rota, Gian-Carlo, 3
Roth, K. F., 88
Rothschild, Bruce, 113
Roy, S. N., 288
Royden, Halsey, 247
Rubinstein, Artur, 128

Rüdenberg, Lilly, 276
Russell, Bertrand, 55, 163, 318

Sabidussi, Gert, 86
Sagan, Carl, 102
Samelson, Hans, 323
Sarton, George, 262
Savage, James, 25
Schattschneider, Doris, 59
Schecter, Bruce, 111
Schiffer, Max, 169
Schläfli, L., 57
Schlick, M., 318
Scholz, A., 316, 320, 322
Schneider, H., 334–35
Schreier, Otto, 316, 322, 360
Schrödinger, Erwin, 4
Schur, Issai, 84, 317, 320, 333
Selberg, Atle, 86
Serre, Jean-Pierre, 144, 322
Shafarevich, I. R., 322, 333
Shakespeare, William, 137, 158
Shannon, Claude, 24, 44
Shapiro, Helene, 330
Shelah, Saharon, 85, 88–89
Shepp, Larry, 203
Shoda, K., 334
Siegel, Carl Ludwig, 277, 335
Sierpinski, W., 355
Simoes, P., 40
Simon, Bill, 96
Slepian, David, 243
Smith, J. H., 333
Smith, Paul, 349
Smullyan, Raymond, 96, 299–307, 365
Snow, C. P., 164
Solomon, Louis, 10
Sommerfeld, Arnold, 277
Speiser, A., 4, 319
Spencer, D. C., 344
Stammbach, Urs, 135, 141
Stanford, C. V., 54
Steen, Lynn Arthur, 135, 153, 183
Steenrod, Norman, 344, 349
Stein, P., 328–29
Stein, Sherman, 101, 107
Steiner, Jakob, 57, 243
Stevenson, Adlai, 30
Stewart, B. M., 88
Stewart, Robert, 218
Stiefel, E., 323, 329
Stiny, George, 15
Stockton, Frank, 301
Stone, A. H., 96, 99
Stone, Marshall, 7, 83, 262
Straus, Ernst, 91, 330
Sturm, J. C. F., 220
Sun, Dan, 35, 39